Frequency References, Power Management for SoC, and Smart Wireless Interfaces

Andrea Baschirotto • Kofi A.A. Makinwa
Pieter Harpe
Editors

Frequency References, Power Management for SoC, and Smart Wireless Interfaces

Advances in Analog Circuit Design 2013

Editors
Andrea Baschirotto
Department of Physics "G. Occhialini"
University of Milan
Milano, Italy

Kofi A.A. Makinwa
Delft University of Technology
Delft, The Netherlands

Pieter Harpe
Department of Electrical Engineering
Eindhoven University of Technology
Eindhoven, The Netherlands

ISBN 978-3-319-01079-3   ISBN 978-3-319-01080-9 (eBook)
DOI 10.1007/978-3-319-01080-9
Springer Cham Heidelberg New York Dordrecht London

Library of Congress Control Number: 2013946413

© Springer International Publishing Switzerland 2014
This work is subject to copyright. All rights are reserved by the Publisher, whether the whole or part of the material is concerned, specifically the rights of translation, reprinting, reuse of illustrations, recitation, broadcasting, reproduction on microfilms or in any other physical way, and transmission or information storage and retrieval, electronic adaptation, computer software, or by similar or dissimilar methodology now known or hereafter developed. Exempted from this legal reservation are brief excerpts in connection with reviews or scholarly analysis or material supplied specifically for the purpose of being entered and executed on a computer system, for exclusive use by the purchaser of the work. Duplication of this publication or parts thereof is permitted only under the provisions of the Copyright Law of the Publisher's location, in its current version, and permission for use must always be obtained from Springer. Permissions for use may be obtained through RightsLink at the Copyright Clearance Center. Violations are liable to prosecution under the respective Copyright Law.
The use of general descriptive names, registered names, trademarks, service marks, etc. in this publication does not imply, even in the absence of a specific statement, that such names are exempt from the relevant protective laws and regulations and therefore free for general use.
While the advice and information in this book are believed to be true and accurate at the date of publication, neither the authors nor the editors nor the publisher can accept any legal responsibility for any errors or omissions that may be made. The publisher makes no warranty, express or implied, with respect to the material contained herein.

Printed on acid-free paper

Springer is part of Springer Science+Business Media (www.springer.com)

Preface

This book is part of the Analog Circuit Design series and contains contributions of all 18 speakers of the 22nd workshop on Advances in Analog Circuit Design (AACD). The local chairs were Dominique Morche (from CEA-Leti) and Angelo Nagari (from ST-Ericsson). The sponsors of the workshop this year have been CEA-Leti, Minatec, STMicroelectronics, and ST-Ericsson. The workshop was held at Minatec in Grenoble, France, in April 16–18, 2013.

The book comprises three Parts, covering advanced analog and mixed-signal circuit design fields that are considered highly important by the circuit design community:

- Frequency References
- Power Management for SoC
- Smart Wireless Interfaces

Each Part is set up with six papers from experts in the field.

The aim of the AACD workshop is to bring together a group of expert designers to discuss new developments and future options. Each workshop is followed by the publication of a book by Springer in their successful series of Analog Circuit Design. This book is the 22nd in this series. The book series can be seen as a reference for all people involved in analog and mixed-signal design. The full list of the previous books and topics in the series is given next.

We are confident that this book, like its predecessors, proves to be a valuable contribution to our analog and mixed-signal circuit design community.

Milano, Italy	Andrea Baschirotto
Delft, The Netherlands	Kofi A.A. Makinwa
Eindhoven, The Netherlands	Pieter Harpe

Contents

Part I Frequency References

1 A Monolithic CMOS Self-compensated LC Oscillator Across Temperature 3
A. Helmy, N. Sinoussi, A. Elkholy, M. Essam,
A. Hassanein, and A. Ahmed

2 A Piezo-resistive, Temperature Compensated, MEMS-Based Frequency Synthesizer 23
J.T.M. van Beek, C. van der Avoort, A. Falepin, M.J. Goossens,
R.J.P. Lander, S. Menten, T. Naass, K.L. Phan,
E. Stikvoort, and K. Wortel

3 A MEMS TCXO with Sub-PPM Stability 41
Aaron Partridge, Hae-Chang Lee, Paul Hagelin,
and Vinod Menon

4 Dual Core Frequency Reference for Mobile Applications in 65-nm CMOS 55
Emmanuel Chataigner and Sébastien Dedieu

5 UHF Clocks Based on Ovenized AlN MEMS Resonators 71
Augusto Tazzoli and Gianluca Piazza

6 Towards Portable Miniature Atomic Clocks 83
David Ruffieux, Jacques Haesler, Laurent Balet,
Thomas Overstolz, Jörg Pierer, Rony Jose James,
and Steve Lecomte

viii Contents

Part II Power Management for System-on-Chip

7 From AC to DC and Reverse, the Next Fully Integrated Power Management Challenge 103
Michiel Steyaert, Hans Meyvaert, and Piet Callemeyn

8 Fully Integrated Switched-Capacitor DC-DC Conversion 129
Elad Alon, Hanh-Phuc Le, John Crossley, and Seth R. Sanders

9 Battery Management in Mobile Devices 147
Francesco Rezzi, Luca Collamati, Maurizio Costagliola, and Massimo Cutrupi

10 Is Digital SMPS Ready to Eliminate Analog Regulators for Portable Applications Power Management? 169
S. Cliquennois and A. Nagari

11 A 2.2A, 4 MHz Switch-Mode Battery Charger for a Cellular Power Management Unit 189
Jay Ackerman, Mike Baker, Ryan Desrosiers, Vipul Katyal, Marc Keppler, John McNitt, Russ Radke, Mark Rutherford, Scott Savage, and Kerry Thompson

12 Power Gating and State Retention Applied to SOC Standby Power Management 209
David Flynn

Part III Smart Wireless Interfaces

13 Unconventional Receiver Architectures 229
Rinaldo Castello and Antonio Liscidini

14 Smart Self-interference Suppression by Exploiting a Nonlinearity 249
Erwin Janssen, Hooman Habibi, Dusan Milosevic, Peter Baltus, and Arthur van Roermund

15 The Design of Ultralow-Power MEMS-Based Radio for WSN and WBAN 265
Aravind Heragu, David Ruffieux, and Christian Enz

16 mm-Wave Silicon: Smarter, Faster, and Cheaper Communication and Imaging 281
Ali M. Niknejad, Amin Arbabian, Steven Callender, JiaShu Chen, Jun-Chau Chien, Shinwon Kang, Jungdong Park, and Siva Thyagarajan

17	**An IEEE 802.15.4A Ultra-Wideband Transceiver for Real Time Localisation and Wireless Sensor Networks**	297

Dries Neirynck

18 Architectures for Digital Intensive Transmitters in Nanoscale CMOS 311

Mark Ingels

The topics covered before in this series

2012	Valkenburg (The Netherlands)	Nyquist A/D Converters Capacitive Sensor Interfaces Beyond Analog Circuit Design
2011	Leuven (Belgium)	Low-Voltage Low-Power Data Converters Short-Range Wireless Front-Ends Power Management and DC-DC
2010	Graz (Austria)	Robust Design Sigma Delta Converters RFID
2009	Lund (Sweden)	Smart Data Converters Filters on Chip Multimode Transmitters
2008	Pavia (Italy)k	High-Speed Clock and Data Recovery High-Performance Amplifiers Power Management
2007	Oostende (Belgium)	Sensors, Actuators and Power Drivers for the Automotive and Industrial Environment Integrated PAs from Wireline to RF Very High Frequency Front Ends
2006	Maastricht (The Netherlands)	High-Speed AD Converters Automotive Electronics: EMC issues Ultra Low Power Wireless
2005	Limerick (Ireland)	RF Circuits: Wide Band, Front-Ends, DACs Design Methodology and Verification of RF and Mixed-Signal Systems Low Power and Low Voltage
2004	Montreux (Swiss)	Sensor and Actuator Interface Electronics Integrated High-Voltage Electronics and Power Management Low-Power and High-Resolution ADCs

(continued)

(continued)

2003	Graz (Austria)	Fractional-N Synthesizers
		Design for Robustness
		Line and Bus Drivers
2002	Spa (Belgium)	Structured Mixed-Mode Design
		Multi-bit Sigma-Delta Converters
		Short-Range RF Circuits
2001	Noordwijk (The Netherlands)	Scalable Analog Circuits
		High-Speed D/A Converters
		RF Power Amplifiers
2000	Munich (Germany)	High-Speed A/D Converters
		Mixed-Signal Design
		PLLs and Synthesizers
1999	Nice (France)	XDSL and Other Communication Systems
		RF-MOST Models and Behavioural Modelling
		Integrated Filters and Oscillators
1998	Copenhagen (Denmark)	1-Volt Electronics
		Mixed-Mode Systems
		LNAs and RF Power Amps for Telecom
1997	Como (Italy)	RF A/D Converters
		Sensor and Actuator Interfaces
		Low-Noise Oscillators, PLLs and Synthesizers
1996	Lausanne (Swiss)	RF CMOS Circuit Design
		Bandpass Sigma Delta and Other Data Converters
		Translinear Circuits
1995	Villach (Austria)	Low-Noise/Power/Voltage
		Mixed-Mode with CAD Tools
		Voltage, Current and Time References
1994	Eindhoven (The Netherlands)	Low-Power Low-Voltage
		Integrated Filters
		Smart Power
1993	Leuven (Belgium)	Mixed-Mode A/D Design
		Sensor Interfaces
		Communication Circuits
1992	Scheveningen (The Netherlands)	OpAmps
		ADC
		Analog CAD

Part I
Frequency References

Kofi Makinwa

The first part of the book discusses recent developments in the design and implementation of frequency references. Traditionally, frequency references have been based on quartz crystal resonators, but since these cannot be readily co-integrated on chip, frequency references based on LC tanks and MEMS resonators are becoming more and more popular, especially as their performance continues to improve. The papers in this section describe frequency references based on quartz crystals, MEMS resonators, LC tanks and atomic clocks.

The first paper describes a frequency reference based on an LC resonant tank. Instead of driving the tank at the usual 180 ° phase shift, a so-called self-compensated oscillator drives it at a fixed temperature-null phase. The result is a measured stability of ± 50 ppm from -20 °C to $+70$ °C after a low-cost room-temperature trim.

The second paper, by Joost van Beek et al., describes a frequency reference based on a MEMS resonator and a programmable PLL. Unusually, the MEMS resonator is read out piezo-resistively instead of capacitively. As a result, its output amplitude is insensitive to resonator scaling, which, in turn, facilitates the use of small high frequency resonators. A frequency reference based on a 55 MHz resonator achieves an inaccuracy of ± 20 ppm over temperatures ranging from -20 °C to $+85$ °C.

The third paper, by Aaron Partridge et al., also describes a frequency reference based on a MEMS resonator and a programmable PLL. A MEMS thermistor co-integrated with the resonator provides the information necessary to compensate for the resonator's temperature dependency. A frequency reference based on a 48 MHz resonator achieves an inaccuracy of less than 1ppm over temperatures ranging from -40 °C to $+85$ °C.

The fourth paper, by Augusto Tazzoli and Gianluca Piazza, describes a frequency reference based on an AlN MEMS resonator with co-integrated heaters. By driving the heaters appropriately, the resonator can then be operated at a (near) constant temperature. Using this approach, a 586 MHz oscillator was shown to exhibit a temperature stability of 1.7 ppm from -45 °C to 85 °C.

The fifth paper, by Emmanuel Chataigner and Sebastian Dedieu, describes a dual-core frequency reference intended for use in mobile devices, which often require two clocks, a low-noise high-frequency one and a low-power low-frequency one. Instead of using two crystals, a re-configurable circuit incorporates a single crystal into two different oscillators, which can then be separately optimized for low-noise and high frequencies (26–52 MHz), and for low-power and low frequency (32 kHz), respectively.

The last paper discusses recent progress toward the goal of a miniature atomic clock. An ASIC was realized that generates an accurate 10 MHz output by locking a VCXO to the atomic transitions of ^{87}Rb. Using an external miniature atomic vapor cell ($100\ \text{mm}^3$), and while dissipating 30 mW (excluding the power required to heat the vapor cell) an Allan deviation of $\sigma_y = 6 \times 10^{-11}$ over a 1s stride has been demonstrated.

Chapter 1
A Monolithic CMOS Self-compensated LC Oscillator Across Temperature

A. Helmy, N. Sinoussi, A. Elkholy, M. Essam, A. Hassanein, and A. Ahmed

Abstract This paper describes a monolithic CMOS reference clock based on an LC oscillator. To achieve a low temperature coefficient, its LC tank is operated at a temperature-null phase. The result is a self-compensated oscillator (SCO) whose output can be programmed from 1 to 133 MHz and which draws 7 mA (no load) from a 3.3 V supply at 25 MHz. After a low cost room temperature trim, the SCO in both ceramic and plastic packages achieves a measured stability of ±50 ppm from −20°C to +70°C. At 133 MHz, its integrated jitter is 0.4 ps from 1.875 to 20 MHz, while at 25 MHz its period jitter is 2.7 ps.

1.1 Introduction

A highly stable and accurate reference clock will always be required for any electronic system irrespective of size and complexity as long as it needs to communicate and/or process data. Since their introduction in 1919 [1], quartz crystal oscillators (XOs) have been an industry de-facto standard and dominated the frequency control market for many decades [2]. XOs may be viewed as self-compensated oscillators since they may be manufactured to exhibit very low temperature sensitivity. By selecting a specific crystal cut, defined by two rotation angles phi and theta around the crystallographic axes, the temperature dependence and aging properties may be optimized. An AT-cut quartz crystal with its characteristic cubic temperature dependence and an inflection point at room temperature, can achieve frequency stabilities typically better than ±50 part per million (ppm) across the industrial temperature range. However, such performance is only achievable with precise crystal manufacturing and assembly in a hermetic, most probably ceramic, package. Miniaturization efforts have yielded commercially available XOs as

---

A. Helmy (✉) • N. Sinoussi • A. Elkholy • M. Essam • A. Hassanein • A. Ahmed
Timing Products Division, Si-Ware Systems, Cairo, Egypt
e-mail: ayman.ahmed@si-ware.com

A. Baschirotto et al. (eds.), *Frequency References, Power Management for SoC, and Smart Wireless Interfaces: Advances in Analog Circuit Design 2013,* DOI 10.1007/978-3-319-01080-9_1, © Springer International Publishing Switzerland 2014

small as $1.6 \times 1.2 \times 0.5$ mm at the expense of assembly complexity and cost. However, integrating an XO with a silicon chip is not commercially available to date. Long lead times, may be more than 10 weeks, are required to develop quartz crystals for new reference frequencies leading to long development cycles and slower market deployment. Thus, the physical and package limitations of quartz crystals, the ever increasing demand for higher integration levels and lower cost have motivated many research efforts to explore new replacement technologies.

The first silicon based Micro Electro-Mechanical System (MEMS) resonator has been introduced in 1967 [3]. Silicon MEMS resonators usually have temperature coefficients of approximately -20 ppm/°C. Several temperature compensation techniques have been utilized successfully to neutralize this frequency deviation. An integrated temperature sensor and a Σ-Δ fractional-N PLL multiplier that is digitally controlled across temperature has been used as a common technique in several commercial MEMS Oscillator (MO) programmable reference clocks [4, 5]. Accuracy levels of ±25 ppm across the industrial temperature range have been achieved. Accuracy levels of ±0.5 ppm have been recently reported using a highly accurate thermistor-based temperature-to-digital converter [6]. In all cases, the MEMS resonators have been hermetically sealed under vacuum on the wafer level to improve quality factor and control aging. The large operating temperature coefficient of MEMS resonators imposes challenges in maintaining frequency stability in response to fluctuations in resonator temperature not tracked by the system temperature sensor. This triggered researchers to develop temperature self-compensated MEMS resonators including the use of alternative materials [7], stresses [8] and variable gaps [9]. However, such techniques rely on multiple physical effects with opposite temperature dependence resulting in reduced temperature sensitivity. However, difficulty in controlling the manufacturing process precision resulted in highly nonlinear temperature behavior without reaching adequate frequency stability that would justify dropping a temperature compensation system. Furthermore, these techniques tend to significantly complicate the resonator fabrication process impacting production repeatability, compensation circuitry, production testing and ultimately cost. An MO is a two die solution, commonly the MEMS resonator die stacked on top of the active silicon die, packaged in plastic to reduce cost. Careful trimming of each device is done at production to tune the temperature compensation circuitry to achieve the target frequency stability and initial accuracy with a clear tradeoff between test cost (determined by testing temperature(s), number of temperature insertions and test time) and frequency stability. Overall, an MO even though has managed to achieve impressive stability performance, offers short lead times through programmability and resistance to shock and vibration, remains not clearly differentiated from an XO in terms of manufacturing cost to compete in the highly price sensitive consumer market. Moreover, the integration challenge has not been solved using MEMS resonators, not only due to lack of readily integrating a MEMS device with active circuitry but also the difficulty in designing and operating an MO and compensating for it across temperature.

Integrated CMOS RC-based oscillators suffer from poor frequency stability across temperature owing to high resistor and capacitor temperature sensitivities. Since the reporting of a temperature compensated Wien-type RC oscillator in 1968 [10], temperature sensitivities of RC-based compensated oscillators have always been within 1–5 % [11–13] which is far from meeting the ± 100 ppm frequency stability required by consumer applications. However, a very recent effort [14] is reporting ± 100 ppm for an RC-based 32 kHz reference clock. A different approach has been recently reported in [15]. The approach relies on the well-defined rate at which heat diffuses through silicon i.e., on the thermal diffusivity of silicon [16]. A 16 MHz oscillator implemented based on this approach achieves ± 0.1 % absolute inaccuracy from $-50°C$ to $125°C$.

LC tanks exhibit a non-linear frequency temperature dependence where the first order temperature coefficient, f_{TC}, may be in the range of $(-50--100)$ ppm/°C. Thus, the main challenge in designing an LC-based reference oscillator is to compensate for this deviation across temperature, supply voltage, load and maintain performance across all manufacturing process corners. To successfully achieve the required compensation accuracy, it is imperative not just to have an accurate temperature measurement system but to also have precise knowledge of the oscillator frequency performance across temperature and its different frequency tuning knobs. An LC-based programmable reference clock [17] has been commercialized that utilizes a highly accurate digital temperature compensation system. Temperature readings are used to compute a multi-segment polynomial factor to neutralize the frequency deviation of the oscillator by continuously tuning an RF Digitally Controlled Oscillator (DCO). Alternatively, an analog compensation approach has been utilized effectively in [18–20] to achieve active, through tuning varactors, and passive, through introducing temperature dependent resistors in series to the tank capacitive components. It is obvious that an LC-based solution has given larger flexibility in frequency programmability yet new challenges in compensating the higher frequency sensitivity to temperature, humidity and stress. However, the cost structure has been obviously improved since the solution has been reduced to a single die that may be smaller, thus lower cost, than an MO die. It is then obvious that a monolithic CMOS solution will offer the lowest cost, be inherently integrated and allow for highest flexibility in programmability.

This work has been motivated by the same efforts to replace XOs with an integrated and cost effective solution that satisfies the ± 100 ppm requirements for a wide range of consumer applications. The solution is an LC-based reference oscillator designed to operate at a very special electrical phase operating point that has a very low f_{TC} and in turn exhibits an SCO [21] and [22]. Section 1.2 describes the theory behind the SCO, design and implementations challenges as well as means of controlling the tank temperature performance. Section 1.3 proposes a new single point trimming (SPT) algorithm that is utilized at room temperature to achieve a cost effective trimming solution for the SCO. Challenges and basic concept details are highlighted through this section. Measurement results are discussed in Sect. 1.4 and finally conclusions are drawn in Sect. 1.5.

1.2 LC-Based Self Compensated Oscillator

1.2.1 Background and Theory

The natural resonant frequency of an ideal LC tank with no losses is defined as $\omega_o = 1/\sqrt{LC}$ where L is the tank inductance and C is the tank capacitance. The implementation of an ideal tank with zero losses is practically impossible due to the physical limitations of having an infinite quality factor (Q). Integrated inductors usually have low Q values owing to inductor metal resistive losses (r_L) and substrate losses (r_{SUB}). Similarly integrated capacitors have finite Q values. Accordingly, an integrated LC tank exhibits an overall Q and impedance that are temperature dependent. Thus, building an oscillator with an integrated LC tank would have very poor frequency stability across temperature. A resonant LC tank requires an amplifier to overcome the tank losses and produce a sustainable oscillation. Classical oscillator designs satisfy the required Barkhausen criterion by operating the LC tank at a real impedance value where the phase of the tank impedance is zero. The phase of the impedance of the simple LC tank illustrated in Fig. 1.1a is expressed as:

$$\varphi_{Tank} = \angle Z_{Tank} = tan^{-1}\left(\frac{\omega L r_C - \frac{r_L}{\omega C}}{r_L r_C + \frac{L}{C}}\right) - tan^{-1}\left(\frac{\omega L - \frac{1}{\omega C}}{r_L + r_C}\right) \tag{1.1}$$

where r_C represents the capacitor resistive losses. Under a zero phase oscillation condition, the frequency of oscillation of the LC tank is given by:

$$\omega = \omega_o \sqrt{\frac{1 - \frac{r_L^2 C}{L}}{1 - \frac{r_C^2 C}{L}}} \tag{1.2}$$

The resistive losses of the inductor exhibit a temperature coefficient (TC) that is dependent on the characteristics of the inductor metal traces material and r_L can be expressed as $r_L = r_{L_o}(1 + \alpha(T - T_o))$ where r_{L_o} is the value of r_L at T_o and α is the temperature coefficient of r_L. Inductor losses are much higher than capacitor resistive losses. Thus, for small r_C, the f_{TC} of the tank is expressed as:

$$f_{TC} = \frac{\partial \omega}{\partial T}\frac{1}{\omega} \approx -\frac{C r_L}{L}\frac{\omega_o^2}{\omega^2}\frac{\partial r_L}{\partial T} \tag{1.3}$$

With a positive linear dependence of r_L on temperature, the LC tank exhibits a negative TC with a large quadratic frequency variation across temperature. The phase of the impedance of an LC tank expressed in Eq. 1.1 is illustrated in Fig. 1.1b where phase is plotted across frequency for different temperature values. The temperature dependence of the quality factor of the inductor (Q_L) produces variable

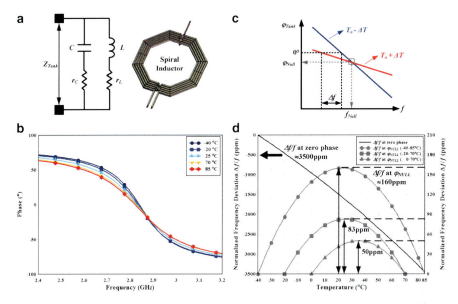

Fig. 1.1 (a) Simple LC Tank model with inductor r_L and capacitor r_C resistive losses, (b) LC tank impedance phase vs. frequency for different temperatures, (c) illustration of TNULL concept for a specific temperature range, (d) comparison of $\Delta f/f$ for an LC tank operating at zero phase and at φ_{NULL} across a -40–$85\,°C$ temperature range

phase plots with different slopes across frequency. The lower the temperature, the higher Q_L and the higher the phase slope magnitude. Phase plots across temperature intersect at a phase (φ_{NULL}) generating a very desirable oscillation phase operating point where the oscillation frequency deviation across temperature is minimized and is almost null. A classical oscillator exhibits thousands of ppms of frequency deviation across temperature at an LC tank zero phase operating point. However, an oscillator with an LC tank operating at φ_{NULL} shows a few tens of ppms of frequency deviation. Thus, an oscillator with an LC tank designed to operate at φ_{NULL} is an SCO that is intrinsically stable across temperature. A Temperature NULL (TNULL) can be defined across a specific temperature range of interest as the intersection of the tank phase curves at the temperature extremes as illustrated in Fig. 1.1c. The operating phase φ_{NULL} is the point of minimum temperature sensitivity across this temperature range with $d\omega/dt = 0$ at T_o; the center of the temperature range. The TNULL oscillation frequency can be derived by finding the intersection of two phase curves at temperature extremes $T_o + \Delta T$ and $T_o - \Delta T$:

$$\omega_{NULL} = \omega_o \sqrt{1 + \frac{Cr_{L_o}^2 (1 - \alpha^2 \Delta T^2)}{L}} \qquad (1.4)$$

and φ_{NULL} can be expressed as:

$$\varphi_{NULL} \approx -\tan^{-1}(2r_{L_o} C \omega_{NULL}) \qquad (1.5)$$

The frequency deviation across a specific temperature range relative to the frequency of oscillation at T_o defines a figure of merit (FOM) that determines the quality of the TNULL.

$$FOM = \frac{\Delta\omega}{\omega_{T_o}\Delta T}ppm/^{\circ}C \qquad (1.6)$$

where $\Delta\omega$ represents the frequency deviation across ΔT and ω_{T_o} represents frequency at T_o. Equation 1.5 illustrates that an SCO designed using the TNULL concept will have an LC tank oscillating at a negative phase compared to the zero phase conventional phase operating point. The FOM improves as Q_L increases and as the temperature losses of the inductor decrease. The normalized frequency deviation $(\Delta f/f)$ across a specific temperature range is defined as the frequency deviation across this range divided by the frequency at one of the extremes of this temperature range. The TNULL quality is determined by the magnitude and shape of $\Delta f/f$ at φ_{NULL} across a specific temperature range. Figure 1.1d compares the analytical results for $\Delta f/f$ of an LC tank operating at a zero phase and an LC tank operating at φ_{NULL} across a -40–85°C temperature range. The LC tank is designed with $L = 7.13$nH, $C = 440$ fF, $Q = 10.4$ and $\alpha = 0.003$ K^{-1}. At zero phase, the LC tank exhibits a negative frequency deviation with temperature as predicted by Eq. 1.3 with an approximate total deviation of 3,500 ppm from -40°C to 85°C. However, operating at the TNULL shows superior frequency stability. Plots in Fig. 1.1d show a positive parabolic behavior across temperature with 50, 83 and 160 ppm of total frequency deviation for 3 different temperature ranges; $(0$–$70)^{\circ}$C, $(-20$–$70)^{\circ}$C and $(-40$–$85)^{\circ}$C respectively.

A very interesting fact originates from Eq. 1.2 where a FOM of zero can be achieved by designing an LC tank where the capacitor losses across temperature track exactly the inductor losses or equivalently $Q_L = Q_C$ across a specific temperature range. Under this condition, the resonant frequency of the LC tank will always be equal to the natural resonant frequency independent of temperature. Additionally, the position of φ_{NULL} defined in Eq. 1.5 will move to the classical zero phase oscillation condition [23].

The analysis so far has focused on the resistive losses of the inductor as the main contributor for frequency deviations across temperature at φ_{NULL}. However, the tank capacitance has a major role in defining the overall frequency stability across temperature. An integrated tank capacitance is largely determined by a designed Metal Insulator Metal (MIM) or Metal Finger (MF) capacitor. However, the amplifier Metal Oxide Silicon (MOS) capacitance, the oxide and fringing capacitances of the physical layout metal traces may contribute a significant part of the net tank capacitance. Thus, an LC tank is composed of the sum of several capacitor types that are physically different in nature, properties and most importantly stability across temperature. A simple and practical representation of the temperature dependence of the tank capacitance is illustrated in Eq. 1.7.

1 A Monolithic CMOS Self-compensated LC Oscillator Across Temperature

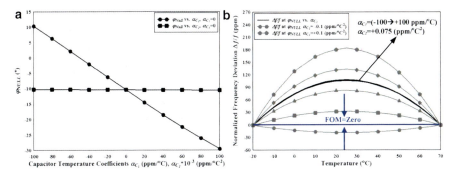

Fig. 1.2 (a) Plot of analytically calculated values of φ_{NULL} vs. α_{C_1} and α_{C_2}, (b) $\Delta f/f$ across temperature at φ_{NULL} for variations of α_{C_1} and α_{C_2}

$$C = C_{T_o}\left(1 + \alpha_{C_1}(T - T_o) + \alpha_{C_2}(T - T_o)^2\right) \tag{1.7}$$

where C_{T_o} represents the net tank capacitance at T_o, α_{C_1} and α_{C_2} are the first and second order temperature coefficients respectively. Substituting Eq. 1.7 in Eq. 1.1 the trends in the phase and frequency deviation across temperature at φ_{NULL} as a function of α_{C_1} and α_{C_2} is studied. Figure 1.2a shows plots of φ_{NULL} versus each of α_{C_1} and α_{C_2} separately. The analysis has been applied to the same tank values used previously. The family of curves in black in Fig. 1.2b show $\Delta f/f$ across temperature at φ_{NULL} for values of α_{C_1} ranging from -100 to $+100$ ppm/°C with a step of 50 and $\alpha_{C_2} = 0.075$ ppm/°C^2. Curves are almost identical signifying no impact of α_{C_1} on performance. In contrast, the family of curves in grey in Fig. 1.2b is for values of α_{C_2} varying from -0.1 to $+0.1$ ppm/°C^2 with a step of 0.05 and $\alpha_{C_1} = 0$, showing clearly that performance is modulated by α_{C_2}. These findings suggest the possibility of designing an LC tank with zero FOM.

1.2.2 Design Challenges

The major challenge in the design of the SCO is adjusting the LC tank to oscillate at its TNULL. In order to force the tank to oscillate at the non-zero phase φ_{NULL}, the oscillator circuit has to introduce an opposite phase, $-\varphi_{NULL}$, such that the oscillator loop satisfies the Barkhausen criterion. A conceptual oscillator that can accomplish the mentioned requirements has been illustrated previously in [22].

One of the main design aspects is how accurate the oscillating phase of the LC tank should be to operate at the TNULL. While operating at the TNULL, the resulting $\Delta f/f$ plot versus a given temperature range is denoted as the "TNULL characteristic" of this range. In Figs. 1.1d and 1.2b, the TNULL characteristics ($\Delta f/f$ plots) show exactly equal frequencies at the extremes of a

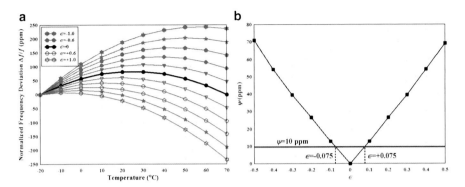

Fig. 1.3 (a) Plot of $\Delta f/f$ vs. temperature at different phase angles ($\varphi_{NULL} + \epsilon$) where ϵ varies from $-1°$ to $1°$, (b) plot of confinement degradation Ψ vs. ϵ

given temperature range. Hence, every TNULL characteristic implicitly assumes that the LC tank phase at oscillation is exactly equal to φ_{NULL}. Practically, there is a finite phase error (ϵ) and the final tank oscillating phase becomes $\varphi_{NULL} + \epsilon$. This imposes a frequency error between the two temperature extremes and in turn, the whole $\Delta f/f$ plot deviates from the ideal TNULL characteristic. In order to quantify this effect, the phase error ϵ is added to the mathematical model and changed from $-1°$ to $1°$ in steps of $0.1°$. The resulting $\Delta f/f$ plots are shown in Fig. 1.3a. As the absolute value of ϵ increases, the absolute value of the frequency error between the two temperature extremes increases. Thus, the overall confinement (ξ) of $\Delta f/f$ degrades where ξ is defined as the peak to peak frequency deviation of the $\Delta f/f$ plot across temperature. A new parameter Ψ is defined as the degradation of ξ due to ϵ referred to the value of ξ at $\epsilon = 0$ i.e. $\Psi = \xi(\epsilon) - \xi(\epsilon = 0)$. Figure 1.3b shows the variation of Ψ vs. ϵ. In order to achieve a target spec of a few tens of ppms across a -20–$70°C$ temperature range, Ψ must not exceed ± 10 ppm. Projecting this target on the plot of Fig. 1.3b, it is required to obtain φ_{NULL} with an accuracy better than $\pm 0.075°$. Moreover, this tight accuracy budget has to accommodate the stability of the operating phase due to variations in temperature, supply voltage, process and other environmental conditions. Such a stringent accuracy requirement forms one of the major design challenges in the SCO.

The second major design challenge in the SCO is the impact of the oscillator active circuitry on the TNULL characteristic which is based on the small signal analysis of the LC tank. At steady state, the circuit nonlinearities limit the oscillation amplitude and a current rich in harmonic content is injected into the tank. In [24], the relation between the oscillation frequency (ω_{os}) and the harmonic content is given by:

$$\omega_{os} = \omega_o \sqrt{\frac{1 - \frac{r_L^2 C}{L}}{1 - \frac{r_C^2 C}{L}} \left(1 - \frac{1}{2Q^2} \sum_{n=2}^{\infty} \frac{n^2}{n^2 - 1} \left(\frac{I_n}{I_1}\right)^2\right)} \qquad (1.8)$$

where I_1 and I_n are the fundamental and the *nth* harmonic of the current of the tank. This implies that the variation of the harmonic content across temperature induces frequency deviation. Hence, the active circuitry modulates the position of φ_{NULL} and the TNULL characteristic through the current harmonic content injected in the tank. This imposes the use of an amplitude control mechanism in order to reduce such an effect.

Finally, the accuracy of the inductor model across temperature can prohibit the accurate prediction of the final performance within the design phase. The macro-models provided by silicon fabrication facilities are concerned with modeling the absolute values of L and Q with relatively good accuracy. However, these models are not accurate enough to predict the frequency deviation across temperature with a relative accuracy in the order of 1e-6. A high accuracy Electro-Magnetic (EM) simulator is utilized to model the used inductor. However, the EM simulator needs accurate information on the technology cross section including the temperature dependence of all materials used.

1.2.3 Architecture and Implementation

From a circuit design perspective, it is required to realize the conceptual oscillator illustrated in [22] such that φ is programmable in steps of $0.1°$ which is quite challenging in the GHz frequency range. Figure 1.4a shows the first proposed architecture which is based on a quadrature LC oscillator. The two tanks oscillate at a non-zero phase φ such that $tan(\varphi)$ is equal to the ratio of the coupling transconductance (Gmc) to the main oscillator transconductance (Gmo) [25]. Hence, the phase φ is given by:

$$\varphi = tan^{-1}\left(\frac{Gmc}{Gmo}\right) \tag{1.9}$$

This can be further explained by the phasor diagram of Fig. 1.4b which illustrates the different phasors annotated in Fig. 1.4a. The value of φ is controlled by digitally programming Gmc. Each of the four transconductors in Fig. 1.4a are designed to be an integer multiple of a transconductor unit cell (gm). This is done in order to conserve the transconductance ratio across temperature, supply voltage, process and different environmental conditions. Gmo consists of N parallel gm cells; hence, $Gmo = Ngm$, whereas Gmc consists of a programmable array of gm cells also connected in parallel. The digital control word m programs the value of Gmc by switching gm cells in and out from the array such that $Gmc = mgm$. Thus, the tank phase is given by:

$$\varphi = tan^{-1}\left(\frac{m}{N}\right) \tag{1.10}$$

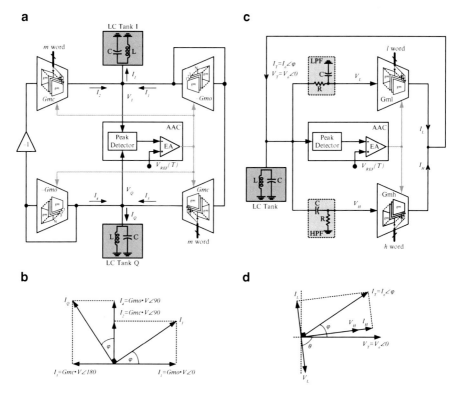

Fig. 1.4 (a) The quadrature LC oscillator architecture utilized to operate at φ_{NULL}, (b) a phasor diagram of currents in the quadrature LC oscillator, (c) the single tank oscillator architecture utilized to operate at φ_{NULL} and (d) a phasor diagram of currents in the single tank oscillator architecture

Thus, the stability requirement of φ across temperature, supply voltage, process and other environmental conditions is achieved provided that the four arrays of gm cells are appropriately matched. N has to be sufficiently large in order to achieve the required 0.1° resolution in φ. Finally, the Automatic Amplitude Control (AAC) loop defines the oscillation amplitude at a specific level according to the reference voltage V_{REF}. The aim of the AAC is to keep the four transconductors operating close to their linear regime; hence, decrease the current harmonic content injected into the tank and in turn reduce the impact of the active circuitry on the final TNULL characteristic.

Although it generates a stable and precise phase φ, the quadrature oscillator architecture imposes the use of two LC tanks. This comes at the cost of die size. Figure 1.4c shows a proposed single tank architecture. This architecture relies on an RC phase shifting network to produce the required phase. The oscillator forces the tank to operate at a non-zero phase by utilizing a mix of Low Pass Filter (LPF) and

1 A Monolithic CMOS Self-compensated LC Oscillator Across Temperature

High Pass Filter (HPF) sections inserted inside the oscillator loop. The sections drive two transconductors, Gml and Gmh, to inject two current components with different phases, I_L and I_H into the tank. Accordingly, the final tank current, I_T has a phase shift φ that is controlled by the ratio between Gml and Gmh. Fig. 1.4d shows the phasor diagram illustrating the signals annotated in Fig. 1.4c. In [22], the phasor diagram is analyzed in detail and the phase φ is derived to be:

$$\varphi = 180 - \theta - tan^{-1}\left(\frac{k}{\chi}\right) \tag{1.11}$$

where $\theta = tan^{-1}(k)$ and $k = \omega RC$. Furthermore, χ is defined as the ratio of Gml to Gmh i.e. $\chi = Gml/Gmh$. In this architecture, φ is controlled through programming the ratio χ. Adhering to the same concept of the quadrature architecture represented earlier, each of the two transconductors, Gml and Gmh is divided into a programmable array of small gm unit cells. The digital control words, l and h define the length of each array such that $Gml = lgm$ and $Gmh = hgm$. Substituting in Eq. 1.11, φ can be expressed as:

$$\varphi = 180 - tan^{-1}k - tan^{-1}\left(k\frac{h}{l}\right) \tag{1.12}$$

In addition to the ratio between the two integers h and l, φ of the single LC tank architecture depends on the parameter k which represents the RC time constant. This imposes two drawbacks of this technique. First of all, k is a process dependent parameter because it follows the process variations of the RC time constant. Hence, the implemented trimming infrastructure has to accommodate the process variations of k. Furthermore, the RC mixture has to be chosen carefully from the different CMOS process modules to produce a stable RC time constant vs. temperature. Otherwise, the TNULL characteristic is degraded.

1.2.4 Controlling the TNULL Characteristic

Although the first order mathematical model of the tank presented in Sect. 1.2.1 is quite useful in analyzing the TNULL phenomenon, it does not model a lot of effects that have a significant impact on the final TNULL characteristic. Higher order effects in the inductor model, such as the skin depth effect and proximity effect. These effects and other EM effects are captured with limited accuracy through the EM simulations of the inductor. Another missing effect in the mathematical model is the impact of the active circuitry which was referred to earlier by Eq. 1.8. Utilizing the AAC can only reduce this effect but cannot eliminate it. This effect can be captured to a great extent in the large signal circuit simulations.

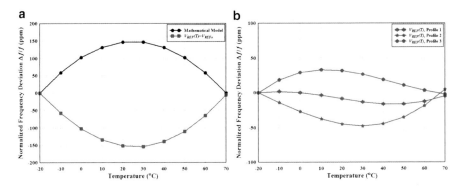

Fig. 1.5 (a) Comparison of 1st order mathematical model results for TNULL characteristic of an LC tank vs. simulation results for a quadrature oscillator operating at φ_{NULL} with $V_{REF}(T) = V_{REF_o}$, (b) simulation results for TNULL characteristics of a quadrature oscillator operating at φ_{NULL} with three different profiles for $V_{REF}(T)$

Finally, the mathematical model does not include the mechanical stimuli applied to the chip due to packaging and how they modulate the inductance value. All the additional listed effects are prone to the variation of temperature, process, humidity and mechanical stress. In some cases, the resulting TNULL characteristic can be unacceptable regarding the frequency stability specification. Based on Eq. 1.8, the AAC reference voltage can be used as a fine tuning knob to control the TNULL characteristic.

In order to illustrate how the TNULL characteristic can be controlled through the AAC reference voltage, an LC tank is designed around 2 GHz using a standard 0.18 μm CMOS technology. The inductor is simulated using a 2.5D EM simulator and a circuit macro-model of the inductor is optimized to fit the results of the EM simulations. The inductor and capacitor data are fed into the first order mathematical model and the output TNULL characteristic is shown in Fig. 1.5a. Furthermore, the tank is used to build an SCO based on the quadrature oscillator architecture shown previously in Fig. 1.4a. The SCO is simulated using a SPICE based simulator. The oscillator is programmed to satisfy the condition $\varphi = -\varphi_{NULL}$ for the temperature range -20–$70°C$. Figure 1.5a shows the resulting TNULL characteristic in the case of using a reference voltage V_{REF} that is constant across temperature i.e. $V_{REF}(T) = V_{REF_o}$. The TNULL characteristic deviates from the mathematical model due to the presence of the active circuitry and the higher order macro-model of the inductor. The tank must be designed such that the overall impact of the circuitry results in a good performance. The TNULL characteristic may be further controlled by using a temperature dependent reference voltage to the AAC. Three different temperature profiles of V_{REF} are simulated and the resulting TNULL characteristics are plotted in Fig. 1.5b. Results illustrate clearly that by applying the proper programmability to V_{REF}, the TNULL characteristic can be accurately controlled across a specific temperature range of interest.

1.3 Single Point Trimming

1.3.1 SCO Trimming Challenges

The SCO relies on operating at φ_{NULL} to achieve high frequency stability across a specific temperature range. However, the value of φ_{NULL} varies with process, oscillation frequency and the required operating temperature range. Thus, trimming is required to compensate for these variations. Trimming is one of the main challenges in having a highly accurate and fully integrated LC-based reference oscillator as it can limit the overall cost and accuracy of the oscillator. The main objective of the trimming of the SCO is to set the oscillator phase to φ_{NULL} while adjusting the oscillator frequency to the required output frequency.

There are many challenges to develop an accurate, robust and cost effective trimming solution for the SCO presented in this work. The main challenge is that a direct method for measuring the tank phase across temperature to determine directly the value of φ_{NULL} does not exist. Consequently, the brute force solution to find φ_{NULL} is to measure the oscillator frequency while varying the tank phase setting (PS) at the two extreme temperature points of the required operating range. The tank (PS) that minimizes the frequency difference between the two extreme temperature points is considered φ_{NULL}. However, this two temperature point trimming solution is not cost effective due to the high cost of the two required temperature insertions. Additionally, a very long testing time is usually required for a large number of accurate frequency measurements. This work, proposes an SPT algorithm that overcomes this challenge and enables achieving a highly accurate and cost effective SCO reference oscillator.

1.3.2 Single Point Trimming Basic Concepts

Conceptually, the temperature dependence of any oscillator can be estimated at T_o by applying a square wave temperature modulating signal as shown in Fig. 1.6a. In response to the temperature modulating signal, the oscillator output becomes a frequency modulated (FM) signal that depends on the frequency temperature sensitivity or slope (K_T) at this temperature. The oscillator frequency (F_{osc}) can then be converted into a digital word (D_{osc}) using an accurate frequency-to-digital converter (FDC) that performs FM demodulation to the oscillator output. The value of K_T at T_o can be estimated from the frequency difference $F_{osc}(T_o + \Delta T/2) - F_{osc}(T_o - \Delta T/2)$ where ΔT represents a variation in temperature around T_o. Equivalently, the difference between $D_{osc}(T_o + \Delta T/2)$ and $D_{osc}(T_o - \Delta T/2)$ can be utilized to generate the same estimate. The difference between these two digital words gives an accurate digital representation of K_T that is utilized in the proposed SPT trimming algorithm. The concept is illustrated in Fig. 1.6a where K_T of the SCO is estimated using a temperature modulating signal and an FDC. At φ_{NULL}, the SCO temperature dependence is a parabolic curve and T_o is the center of the

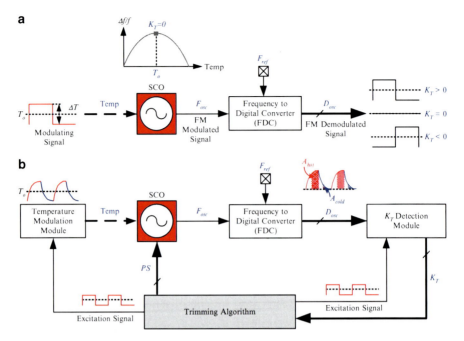

Fig. 1.6 (**a**) Conceptual illustration of measuring the temperature sensitivity of the SCO by applying a temperature modulating signal, (**b**) block diagram of the trimming infrastructure utilized to measure K_T of an SCO

required temperature range as illustrated previously in Fig. 1.1d. Thus, at T_o, $K_T = 0$ represents the minimum SCO temperature dependence.

Practically, the thermal modulation frequency is limited by the thermal time constant of the packaged part under test. Thus, the temperature modulating square wave is low pass filtered as illustrated in Fig. 1.6b. The thermal time constant depends mainly on die and package sizes. As the thermal modulation frequency decreases, the trimming routine becomes slower. Thus, testing cost increases impacting the overall cost of the SCO. Consequently, it becomes very important to reach the optimum tank phase setting (PS_{Opt}) that is equivalent to φ_{NULL} in a minimum number of thermal modulation cycles. Oscillator thermal and flicker phase noise affect the accuracy of the SPT especially when K_T approaches zero as the magnitude of the demodulated signal becomes very close to zero too. The impact of phase noise at very low frequency offsets on K_T is insignificant as sensing K_T depends on the difference between two frequencies. Thus, most of the noise is cancelled. However, phase noise at high frequency offsets is effective and can be suppressed by incorporating an integrate and dump filter. Only a fraction of the heating and cooling periods is utilized by the integrate and dump filter due to the low pass filter effect induced by the slow thermal time constant of the package under test. A fraction of the heating cycle period is integrated into A_{hot} and similarly a fraction of the cooling cycle period is integrated into A_{cold} as shown in Fig. 1.6b. The difference between A_{hot} and A_{cold} is thus a digital word that represents K_T more accurately in the presence of oscillator phase noise and large thermal time constant. In this manner, only phase noise at frequency offsets close to the modulation frequency can slightly affect the accuracy of K_T sensing and consequently the results of the SPT.

The objective of the phase trimming algorithm is to search for PS_{Opt} that adjusts the SCO K_T to a user's slope control word (KCW). The selection of KCW depends on the operating temperature T_o, the predetermined temperature range and the desired SCO temperature dependence curve. Usually a value of KCW that is equal to zero or very close to zero is used to optimize the SCO temperature dependence using room temperature only (RTO) trimming ($T_o = 25°C$) for the different temperature ranges illustrated in Fig. 1.1d.

The proposed SPT utilizes a single insertion temperature point and employs integrated on-chip heaters for temperature modulation to detect K_T as highlighted previously. Moreover, it utilizes smart algorithms rather than extensive sweeps to set PS_{Opt}. A digital frequency locked loop (DFLL) is used to adjust the oscillator frequency by changing the oscillator frequency setting (FS) based on a user's frequency control word (FCW). At the end of the trimming routine, the oscillator trimmed parameters PS and FS are programmed in a one-time programmable (OTP) read only memory (ROM) module. The OTP ROM holds the trimmed parameters in normal-operation after trimming is complete and is automatically loaded at power up.

1.4 Measurement Results

The proposed SCO was fabricated in a 0.18 μm CMOS process with a single poly and Six aluminum metal layers. A thick top metal option was used to have higher quality factor spiral inductors. The chip architecture is illustrated in Fig. 1.7a where the SCO operates at 2 GHz and a number of integrated heaters modulate the frequency of the SCO during SPT. A highly programmable bank of capacitors is used to adjust the SCO absolute frequency with an accuracy of ± 2.5 ppm. The SCO is followed by a chain of programmable dividers that adjusts the output frequency such that the output clock frequency may be adjusted from (1 to 133) MHz. The chip has an output buffer that can drive a 15 pF load to the supply rails. Any supply voltage from 1.71 to 3.6 V can operate the chip since an internal band-gap referenced low drop-out (LDO) regulator produces a 1.6 V supply to all blocks. The chip includes a serial data interface that is used to communicate with the chip during testing to trim each part. The proposed SPT digital infrastructure is integrated on chip. Trimming and frequency programmability settings are stored on an OTP ROM module. Figure 1.7b shows the photo of the die (1.55×1.05 μm). The die was packaged in a four-pin 5.0×3.2 mm ceramic package and a custom DFN $5.0 \times 3.2 \times 0.8$ mm plastic package.

To test the frequency stability of the implemented SCO across temperature, 50 ceramic and plastic packaged parts operating from a 3.3 V supply were trimmed to 25 MHz. The proposed SPT algorithm was used at room temperature to operate

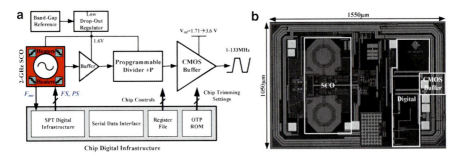

Fig. 1.7 (**a**) Chip architecture of the implemented SCO with a 1–133 MHz CMOS output clock, (**b**) SCO die photomicrograph implemented in 0.18 μm 1P6M CMOS

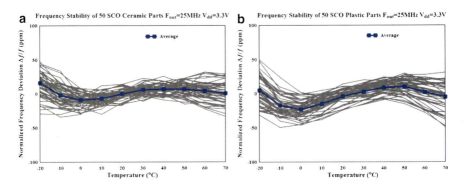

Fig. 1.8 Normalized frequency stability across temperature of parts trimmed to 25 MHz (**a**) 50 ceramic packaged parts and (**b**) 50 plastic packaged parts

the parts at the TNULL. The integrated on-chip heaters were used to modulate the SCO temperature while searching for the optimum settings of the oscillator PS_{Opt} and FS_{Opt} to achieve the target output frequency and temperature stability across the target temperature range. An accurate 1 kHz reference clock was used during trimming as the reference to the FDC thus minimal external equipment is required. After successfully completing the trimming routine, oscillator and divider settings are programmed to the OTP ROM. Parts frequency stability performance was characterized across temperature from (−20–70)°C with a 10°C step. The frequency was measured using an accurate 53131A Agilent frequency meter with a gate time of 100 ms. The results of the 50 parts in a non-hermetic ceramic package and the 50 plastic packaged parts are illustrated in Fig. 1.8a, b respectively. All 100 parts are showing excellent frequency stability performance within ±50 ppm across a (−20–70)°C temperature range. Additionally, there is no obvious difference between the performance in ceramic and plastic packages. This is a strong indication that the plastic package materials' electrical and mechanical properties are not impacting the characteristics of the TNULL much.

It is worth noting that results show strong correlation with the analytically predicted performance discussed earlier. The supply sensitivity of one of the

1 A Monolithic CMOS Self-compensated LC Oscillator Across Temperature

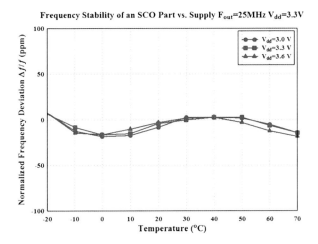

Fig. 1.9 Frequency stability of a 25 MHz SCO across supply and temperature

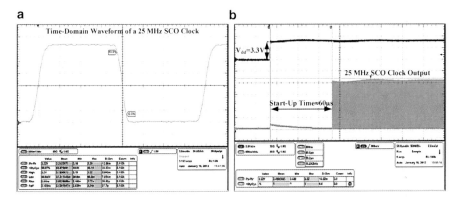

Fig. 1.10 (a) Time-domain waveform of a 25 MHz SCO driving a 15 pF output load and (b) measured start-up time from power-up for a 25 MHz SCO

trimmed parts was measured over a ±10 % variation around the nominal 3.3 V and the normalized frequency was plotted for each supply voltage across a (−20–70)°C temperature range. Measurement results are illustrated in Fig. 1.9 where the SCO shows a ±5 ppm frequency variation from 3.0 to 3.6 V.

The output waveform and start-up time of the chip were measured using a 20 GS/s oscilloscope. The output waveform of the 25 MHz SCO clock driving a 15 pF load is illustrated in Fig. 1.10a. Results in Fig. 1.10b show an approximate start-up time of 60 μs. This latency is dominated mainly by the start-up behavior of the chip including the start-up time of the band-gap reference and LDO. The start-up time of the SCO is two orders of magnitude less than the start-up time of MEMS based oscillators and XOs that are dominated by high Q resonator's start-up latency and temperature compensation system response time.

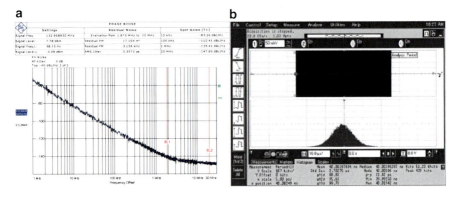

Fig. 1.11 (a) Measured SSB phase noise of a 133 MHz SCO reference clock and (b) measured period jitter for a 25 MHz SCO reference clock for 50 k cycles

Table 1.1 Summary of measured SCO performance

Parameter	Value
Output clock frequency (MHz)	1–133
Power supply (V)	1.71–3.6
Current consumption at 25 MHz (mA)	7.1
Normalized frequency deviation $\Delta f/f$ across (−20–70) °C temperature range (ppm)	±50
Normalized frequency deviation $\Delta f/f$ across a ±10 % supply variation @ 3.3 V (ppm)	±5
SSB phase noise PSD for a 133 MHz output clock at offset	
1 kHz (dBc/Hz)	−53
10 kHz (dBc/Hz)	−83
100 kHz (dBc/Hz)	−110
1 MHz (dBc/Hz)	−135
20 MHz (dBc/Hz)	−148
Integrated RMS phase jitter:	
1.875–20 MHz (ps)	0.36
12 kHz–20 MHz (ps)	7.3
Period Jitter for a 25 MHz output	
1-σ (ps)	2.7
Peak-Peak (ps)	21.9
Start-up latency (μs)	60

A Rhode & Schwarz FSU3 spectrum analyzer was used to measure single-side band (SSB) phase noise of an SCO trimmed to produce an accurate 133 MHz output clock. Results are illustrated in Fig. 1.11a and the spectrum clearly shows no spurious tones and a noise floor of −148 dBc/Hz. The achieved phase noise may not be suitable for telecom applications yet satisfies consumer and many networking applications including Ethernet. The time domain jitter analysis is done on a 25 MHz output using a 20 GS/s oscilloscope. Figure 1.11b shows a screenshot of the results indicating excellent period jitter of 2.7 ps and a Gaussian histogram signifying very low deterministic jitter. A summary of the SCO measured performance is listed in Table 1.1.

1.5 Conclusion

XOs have enjoyed for decades the merits of self-compensation through precise manufacturing. The MO has not been able to match XO performance through a significantly lower cost structure and requires a smart temperature compensation IC to achieve competitive performance and programmability. Both XOs and MOs cannot offer true monolithic integration with ICs and cannot match the superior cost structure of an LC-based reference clock. The main challenge for LC reference clocks is to achieve the required frequency stability. An SCO operating at a specific TNULL phase has the potential of becoming the best solution to address consumer applications. Implementation of an SCO has many design challenges that may be surmounted through careful tank and oscillator design. Measurements of the SCO show excellent period jitter performance and a frequency stability of ± 50 ppm over $(-20–70)°C$ in both ceramic and plastic packages. Such performance was achieved using a low cost on-chip SPT algorithm at room temperature. Finally, frequency stability of all reference clocks relies on the dimensional stability of the frequency determining element. The sensitivity of these elements to any perturbations will evidently impose packaging requirements to achieve good long term stability and aging. Thus, the reliability of a reference clock remains an important metric in comparing different technologies that the authors will cover in future work and publications.

Acknowledgements The authors would like to thank and appreciate the design engineers of the Timing Products division, Digital, Layout and Product Engineering teams at Si-Ware Systems for their continuous efforts in the realization, characterization and testing of the SCO.

References

1. V. Bottom, A history of the quartz crystal industry in the USA, in *Thirty Fifth Annual Frequency Control Symposium*, 1981, pp. 3–12
2. C.S. Lam, A review of the recent development of MEMS and crystal oscillators and their impacts on the frequency control products industry, in *IEEE Ultrasonics Symposium*, (Beijing, 2008), pp. 694–704
3. H. Nathanson, W. Newell, R. Wickstrom, J. Davis, J.R., The resonant gate transistor. Electron Devices, IEEE Trans. **14**(3), 117–133 (1967)
4. *DSC8001 Series PureSilcon^{TM} Programmable CMOS Oscillator Datasheet, MK – Q – B – P – D – 090110–03–2* ed. by (Discera, San Jose, 2010)
5. *MEMS Replacing Quartz Oscillators*, Application Note SiT-AN10010 Review 1.1, (SiTime Corporation, Sunnyvale, 2009)
6. M.H. Perrott et al., A temperature-to-digital converter for a MEMS-based programmable oscillator with $< \pm 0.5$-ppm frequency stability and $<$ 1ps integrated jitter. Solid-State Circuit IEEE J. **48**(1), 276–291 (2013)
7. M. Renata et al., Temperature-compensated high-stability silicon resonators. Appl. Phys. Lett. **90**(24), 244107–3 (2007)

8. Wan-Thai Hsu, C.T.-C. Nguyen, Geometric stress compensation for enhanced thermal stability in micromechanical resonators, in *Proceedings, 1998 I.E. International Ultrasonics Symposium,* (Sendain, 1998), pp. 945–948, 5–8 Oct 1998
9. Wan-Thai Hsu, C.T.-C. Nguyen, Stiffness-Compensated temperature-insensitive micromechanical resonators, in *Technical Digest,* 2002 I.E. International Micro Electro Mechanical Systems Conference, (Las Vegas, 2002), pp. 731–734, 20–24 Jan 2002
10. R.F. Adams, D.O. Pederson, Temperature sensitivity of frequency of integrated oscillators. IEEE J. Solid-State Circuit **SC-3**(4), 391–396 (1968)
11. A.V. Boas, A. Olmos, A temperature compensated digitally trimmable on-chip IC oscillator with low voltage inhibit capability, in *Proceeding IEEE International Symposium Circuits and Systems (ISCAS),* vol. 1 (2004), pp. 501–504
12. K. Sandaresan, P.E. Allen, F. Ayazi, Process and temperature compensation in a 7-MHz CMOS clock oscillator. IEEE J. Solid-State Circuit **41**(2), 433–441 (2006)
13. Y. Tokunaga et al., An on-chip CMOS relaxation oscillator with voltage averaging feedback. IEEE J. Solid-State Circuit **45**(6), 1150–1158 (2010)
14. *eoSemi Unveils Silicon Oscillator Technology, Announces Shipment of First Silicon to Select Customers* (Congleton, 2012), 13 Mar 2012
15. S. Mahdi Kashmiri, Kamran Souri, Kofi A. A. Makinwa, A scaled thermal-diffusivity-based 16 MHz frequency reference in 0.16 μm CMOS. *IEEE J. Solid-State Circuit* **47**(7), 1535–1545 (2012)
16. J. Ebrahimi, Thermal diffusivity measurement of small silicon chips. J. Phys. D. Appl. **3**, 236–239 (1970)
17. *Si500S Single-Ended Output Silicon Oscillator* Review 1.0 5/11, (Silicon Laboratories, Austin)
18. M. McCorquodale et al., A 25-MHz self-referenced solid-state frequency source suitable for XO-replacement. Circ. Syst. I: Regul. Pap. IEEE Trans. **56**(5), 943–956 (2009)
19. M.S. McCorquodale et al., A silicon Die as a frequency source, in *Proceeding of IEEE International Frequency Control Symposium,* (2010), pp. 103–108, 1–4 June 2010
20. *3DN Series CrystalFree^{TM} Oscillator Preliminary Data Sheet* (Integrated Device Technology, San Jose, 2012)
21. A. Ahmed, B. Hanafi, S. Hosny, N. Sinoussi, A. Hamed, M. Samir, M. Essam, A. El-Kholy, M. Weheiba, A. Helmy, A highly stable CMOS Self-Compensated Oscillator (SCO) based on an LC tank temperature null concept, in *Proceeding IEEE International Frequency Control Symposium,* 2011, pp. 1–5
22. N. Sinoussi, A. Hamed, M. Essam, A. El-Kholy, A. Hassanein, M. Saeed, A. Helmy, A. Ahmed, A single LC tank self-compensated CMOS oscillator with frequency stability of ±100 ppm from −40°C to 85°C, in *Proceeding IEEE International Frequency Control Symposium,* 2012, pp. 1–5
23. B. Hanafi, S. Hosny, A. Ahmed, Method, system and apparatus for accurate and stable LC-based reference oscillators, U.S. Patent 8, 072, 281B2, (2011) 6 Dec 2011
24. J. Groszkowski, *Frequency of Self-Oscillations,* (Oxford/Pergamon, 1964)
25. A. Mirzaei, M.E. Heidari, R. Bagheri, S. Chehrazi, A.A. Abidi, The quadrature LC oscillator: a complete portrait based on injection locking. Solid-State Circuit IEEE J. **42**(9), 1916–1932 (2007)

Chapter 2
A Piezo-resistive, Temperature Compensated, MEMS-Based Frequency Synthesizer

J.T.M. van Beek, C. van der Avoort, A. Falepin, M.J. Goossens, R.J.P. Lander, S. Menten, T. Naass, K.L. Phan, E. Stikvoort, and K. Wortel

Abstract This paper describes a frequency synthesizer based on a MEMS resonator. Uniquely, the piezo-resistive properties of silicon are exploited to read out the resonator, resulting in low impedance levels at resonance frequencies up to several 100 MHz. A 55 MHz MEMS oscillator with a phase noise of -128 dBc/Hz @ 1 kHz offset and a -140 dBc/Hz noise floor has been realized. The oscillator is combined with a programmable PLL to realize a complete frequency synthesizer that can generate output frequencies ranging from 25 MHz to 200 MHz. It achieves ± 20 ppm frequency accuracy over temperatures ranging from $-20°C$ to $+85°C$, and draws 15 mA from a 2.5 V supply at an output frequency of 25 MHz.

2.1 Introduction

An emerging class of high performance Phase Locked Loop (PLL) based frequency synthesizers uses MEMS resonator technology replacing the bulky quartz resonator as frequency referencing element [1]. The extraordinary small size, high level of integration, low cost and high volume manufacturing capability that is possible with MEMS appear to open exceptional possibilities for creating miniature-scale precision reference oscillators at low cost. The MEMS resonator can be combined with a PLL in a single module using a standard low cost plastic package. It can be expected that a MEMS-based PLL has a superior noise performance and frequency stability compared to self-referenced CMOS synthesizers, since the MEMS-based oscillator is based on mechanical resonance exhibiting a much higher Q-factor than

---

J.T.M. van Beek (✉) • C. van der Avoort • M.J. Goossens • S. Menten
T. Naass • K.L. Phan • E. Stikvoort • K. Wortel
NXP Semiconductors, Eindhoven, The Netherlands
e-mail: j.t.m.van.beek@nxp.com

A. Falepin • R.J.P. Lander
NXP Semiconductors, Leuven, Belgium

A. Baschirotto et al. (eds.), *Frequency References, Power Management for SoC, and Smart Wireless Interfaces: Advances in Analog Circuit Design 2013*, DOI 10.1007/978-3-319-01080-9_2, © Springer International Publishing Switzerland 2014

LC based electrical resonators. At the same time, it is expected that the use of MEMS reduces the size and cost and increases the level of system integration compared to quartz referenced PLLs, since the processes and materials being used are often CMOS compatible and use the CMOS manufacturing infrastructure. MEMS technology allows the realization of high performance frequency synthesizers with low noise and a high degree of frequency stability without the need for an external quartz crystal. Thereby reducing the footprint of the synthesizer, reducing the number of I/O pins on the package and associated solder connections, and ease of design-in.

As a rule of thumb, the reference frequency, f_{ref} produced by the MEMS oscillator should be chosen as high as possible in order to minimize the noise and spur contribution of the MEMS reference oscillator to the PLL's output frequency [2]. The first reason is that a PLL which employs a sequential phase-frequency detector and charge pump (PFD/CP) is in reality a sampled system, due to the nature of the PFD. As a consequence, the sampling process places an upper limit on the open-loop bandwidth of the PLL in relation to f_{ref}. The second reason to keep f_{ref} high is to minimize the amplitude of spurious components at offset frequencies N. f_{ref} with N being an integer number. The spurious signals result from leakage currents at the voltage controlled oscillator (VCO) input and at the loop filter or from CP imperfections. The third and most important reason to maximize f_{ref} is because the equivalent synthesizer phase noise floor is effectively multiplied by M^2, where M is the division ratio of the main divider, when converted to the output of the VCO. So, to minimize the noise contribution from the synthesizer blocks one needs to minimize the divider ratio M. By maximizing f_{ref} the division ratio M is minimized for a given output frequency. Higher reference frequency at the PFD can be used resulting in smaller multiplication factors. The expected improvement in the phase noise of $20.\mathrm{Log}(M)$ is somewhat reduced due to the fact that the PFD adds more noise at higher frequencies. A phase noise improvement of $10.\mathrm{Log}(M)$ is however realistic. In practice an upper limit of f_{ref} is set by the fact that driving the PFD for standard CMOS technology beyond a frequency of 130–150 MHz will cause strong degradation of the phase noise [3] and a f_{ref} in the range of 50–150 MHz seems ideal.

Conventional MEMS resonators are based on capacitive transduction: the resonator motion is detected through a capacitance measurement that measures the change of capacitance between the resonator perimeter and a sense electrode held at fixed position. Capacitive resonators have the intrinsic disadvantage of having relatively high impedance compared to piezoelectric resonators, such as quartz. These limitations find their origin in the low level of electro-mechanical coupling that can be achieved with capacitive transduction since measured change in capacitance as a result of the resonator's motion is extremely small, and tends to decrease even further at high resonance frequencies as a result of reduced resonator dimensions at higher frequencies. As such, capacitive transduction of MEMS resonators seems to be in conflict with the desire to have high oscillation frequencies of the reference oscillator interfacing the PLL.

However, an advantage of capacitive transduction is that it allows for the use of single crystal silicon (SCS) as the resonating medium without the need of having lossy metal electrodes present at its surface, as is the case in piezoelectric MEMS resonators, such as the ones based on aluminum-nitride. Silicon resonators typically show a much higher *Q-factor* and very little aging compared to their AlN counterparts. Furthermore, SCS resonators are relatively easy to process using SOI wafers and can easily be vacuum packaged on wafer level. Therefore, it is desired to investigate transduction techniques that increase the coupling factor of SCS resonators without having to resort to the lossy piezoelectric and metal thin films exhibiting low Q-factor. A promising transduction scheme that is compatible with SCS exploits the piezo-resistive properties of silicon to sense the mechanical vibration of the MEMS resonator. This concept allows for realizing miniature resonators with a high frequency fundamental tone combined with high output signal and associated low effective impedance. Piezo-resistive resonators are very well suited to realize MEMS based reference oscillators in the 50–150 MHz frequency range that is ideal for interfacing with high performance PLL based frequency synthesizers.

2.2 Oscillator Based on Piezo-resistive Resonator

Instead of detecting gap modulation, as is done in capacitive based resonators, a direct measurement of mechanical strain that is build up inside the resonator body is used to sense its motion. The mechanical strain is detected by means of the piezo-resistive effect. Although silicon is not piezo-electric, it exhibits a strong piezo-resistive behavior and therefore resonators made from Si, and SCS in specific, are well suited to adopt this transduction principle. In this way, strain rather than gap variation is the parameter that is being sensed. The output current can be tailored by the DC bias current that is sent through the resonator.

The unique property of this type of resonator is that its output signal is insensitive to geometric scaling and is therefore suitable for achieving high resonance frequencies because the transduction efficiency does not depend directly on the resonator size. Fundamental mode resonators with resonance frequency at 1.1 GHz with $Q = 550$ [4] and higher order modes up to 4.5 GHz with $Q = 11200$ [5] and more recently even up to 40 GHz with $Q = 130$ [6] have been demonstrated. It is shown that at 1.1 GHz the effective impedance is reduced by orders of magnitude as a result of the piezo-resistive instead of capacitive readout. In [7] a piezo-resistive SCS 10 MHz resonator is demonstrated with a $Q = 125.000$ underpinning the fact that very high Q-factor can be achieved using piezo-resistive transduced SCS resonators.

The resonator described in this work is a dogbone shaped resonator having a fundamental tone at 55 MHz, measuring $20 \times 40 \,\mu m^2$, and is etched and released in a 1.5 μm-thick SOI layer. The MEMS resonator is sealed in a low pressure ambient using NXP's proprietary thin-film capping technology, which is a low-temperature ($<400°C$) and low-cost CMOS-compatible process. Figure 2.1 shows a cross-section of a capped resonator. The cavity under the cap can sustain <40 mbar of

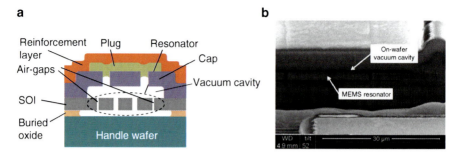

Fig. 2.1 (**a**) Schematic cross-section of on-wafer vacuum package. (**b**) FIB cross-section

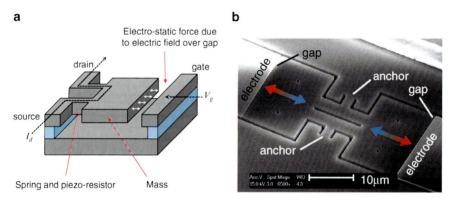

Fig. 2.2 (**a**) Schematic layout of a piezo-resistive dogbone resonator. (**b**) SEM top view of realized dogbone resonator

pressure, which is enough to enable resonance with a Q-factor of >40,000. The resonator process flow including thin film encapsulation of the resonator has been proven to be manufacturable with high process yield and to survive various accelerated lifetime tests, such as HAST, TMCL, and HTSL, as well as the steps needed for die assembly, such as wafer grinding, dicing and plastic injection molding.

The resonator vibration is sensed using the piezo-resistive effect, causing a strain induced change in the resistance of the spring piezo-resistors. The piezo-resistors can be readout by applying a current bias, I_d over the source-drain terminals that also serve as mechanical anchors of the resonator. At resonance, the modulation of the piezo-resistors results in an AC voltage across the anchors. In the realized device, depicted in Fig. 2.2b, the layout depicted in Fig. 2.2a is mirrored along the source-drain axis in order to have no nett force acting on the anchor points when the resonator heads are in resonance. This results in very little loss of vibration energy into the substrate and therefore allows for a high Q-factor [8].

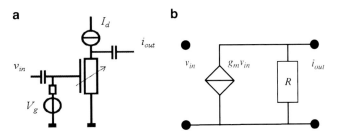

Fig. 2.3 (a) Piezo-resistive resonator equivalent electrical model including bias sources. (b) Small signal model of the resonator

The dogbone resonator is actuated by an electrostatic force, F_{el} via an electrode separated from the dogbone head by a narrow gap of 200 nm, as depicted in Fig. 2.2. The F_{el} is generated by an AC voltage, v_{gap} that is superimposed on a DC bias voltage over the gap, V_{gap}. On its turn, V_{gap} is set by the difference of the applied gate bias, V_g and the voltage on the resonator head. The voltage on the resonator head is set by I_d and the resistance, R of the dogbone and is equal to $I_d R/2$. In most cases, v_{gap} can be approximated by the externally applied AC voltage v_{in}, since the AC voltage on the resonator head is typically much smaller than v_{in}. Furthermore, the electrostatic force scales with change in gap capacitance per unit of displacement of the resonator head and is therefore dependent on gap width, g and frontal area of the dogbone head A_h,

$$F_{el} = v_{gap} V_{gap} \frac{\varepsilon_0 A_h}{g^2} = v_{gap}(V_g - I_d R/2) \frac{\varepsilon_0 A_h}{g^2} \approx v_{in}(V_g - I_d R/2) \frac{\varepsilon_0 A_h}{g^2} \quad (2.1)$$

The electrical model of the dogbone resonator is described by a time-alternating resistor R with proper bias current I_d and voltage V_g applied, as is shown in Fig. 2.3a. The F_{el} induces a mechanical strain and therefore a relative change in resistance r/R in the two piezo-resistors. The output current i_{out} is simply given by,

$$i_{out} = I_d \frac{r}{R} \quad (2.2)$$

The relative change in resistance r/R is proportional to the resonant displacement of the resonator and is described by a Lorentzian function centered around the mechanical resonance frequency, ω_0 of the resonator that is scaled by the electrostatic force $\alpha V_{gap} v_{in}$ acting on the resonator head,

$$\frac{r}{R} = \frac{\alpha V_{gap} v_{in}}{1 - \frac{\omega^2}{\omega_0^2} + j \frac{\omega}{\omega_0 Q}} \quad (2.3)$$

The pre-factor, α is a constant that sets how strong the F_{el} is concentrated in the resonator springs multiplied by the piezo-resistive gauge factor and has a negative value for the silicon crystal orientation with respect to the dogbone orientation being used. By combining Eqs. 2.1, 2.2, and 2.3 it can be seen that the relation between i_{out} and v_{in} can be described by a transconductance, g_m as is schematically depicted in Fig. 2.3b,

$$
\begin{aligned}
g_m &= \frac{\alpha}{1 - \dfrac{\omega^2}{\omega_0^2} + j\dfrac{\omega}{\omega_0 Q}} I_d V_{gap} \\
g_{m,\max} &= -j\alpha Q I_d V_{gap} \\
g_{m,norm} &= \frac{|g_{m,\max}|}{Q}
\end{aligned}
\tag{2.4}
$$

The maximum value of transconductance, $g_{m,max}$ is reached at frequency $\omega = \omega_0$. From Eq. 2.4 it can be seen that the transconductance at resonance frequency is about Q-factor times larger than off-resonance. Hence, the resonator serves as a frequency selective filter in the oscillator loop. It is noted that at resonance there is a 90° phase shift between i_{out} and v_{in}. This is fundamentally different from capacitive or piezo-electric resonators where there is no phase shift between v_{in} and i_{out} at resonance. For convenient comparison of different bias conditions $g_{m,max}$ can be normalized to Q-factor and is called $g_{m,norm}$.

The measured transconductance at resonance of the dogbone resonator is shown in Fig. 2.4 as a function of the product of V_{gap} and I_d when I_d is varied from -2.5 to $+2.5$ mA. From Fig. 2.4a it can be seen that there is indeed a linear relation between the $V_{gap} \cdot I_d$ and $g_{m,max}$ as predicted by Eq. 2.4. However, it can be seen that the slope is reduced for large values of I_d. This can be attributed to a small reduction in the Q-factor for larger currents caused by a thermal damping effect [9], as is evident from Fig. 2.4c. The bias current dependency is absent when plotting $g_{m,norm}$ as can be seen from Fig. 2.4b underpinning the validity of the simple resonator model. The resonator is typically operated using V_g values between -5 and -10 V and I_d values between $+1$ and $+2$ mA resulting in an inductive behavior of $g_{m,max}$ with values between -j80 and -j320 µA/V, which is equivalent to an impedance of 3–12 kΩ or an inductance lying between 9 and 36 µH.

The load, source, and feedthrough capacitances need to be taken into account when inserting the resonator in an oscillation loop, as is schematically depicted in Fig. 2.5. Of particular interest is the feedthrough capacitance, C_{ft}. The C_{ft} causes a feedthrough current, i_{ft} that is preceding the voltage at the resonator input v_{in} by 90°. Therefore, this current cancels in part the inductive current caused at resonance induced by $g_{m,max}$. For the oscillator to lock at the mechanical resonance frequency it is required that $|g_{m,max}| > |\omega_0 C_{ft}|$. In practice, the C_{ft} originates from gate to drain coupling between bond wires and substrate coupling on the MEMS die and is approximately 50 fF which is equivalent to an admittance of $+$ j18 µA/V at 55 MHz, which is typically well below the transconductance $g_{m,max}$ of the resonator as is required. Another important impedance is the capacitance at the resonator drain to ground, C_l which is in parallel to the resonator resistance R. Both impedances set

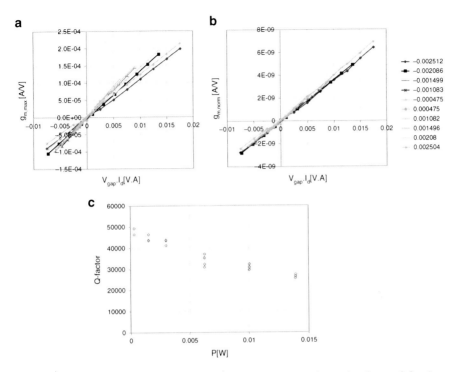

Fig. 2.4 (a) Measured $g_{m,max}$ versus $I_d \cdot V_{gap}$ for different I_d levels varying from -2.5 mA to $+2.5$ mA. (b) $g_{m,max}$ normalized to Q-factor versus $I_d \cdot V_{gap}$ for different I_d levels. (c) Measured Q-factor as a function of dissipated power inside the resonator body

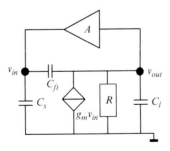

Fig. 2.5 Resonator placed in an oscillator loop, with R = 600 Ω, C_{ft} = 50 fF, C_s = C_l = 1 pF

the phase between i_{out} and the voltage, v_{out} at the input of the amplifier. In our case the resonator resistance is set at 600 Ω by choosing the appropriate doping level in the SOI layer. The C_l consists of bondpad capacitance on the MEMS die and the amplifier die and is estimated to be 1 pF resulting in a phase difference between i_{out} and v_{out} of 10° at 55 MHz. Therefore the total phase shift between v_{in} and v_{out} is

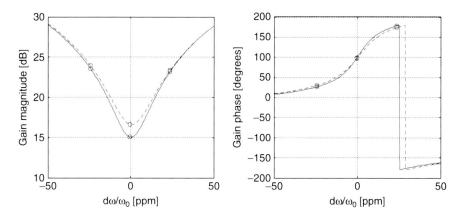

Fig. 2.6 Required amplifier gain and phase to sustain oscillation at $V_g = -8$ V, $I_d = 1.75$ mA, $Q = 40.000$, $f_0 = 55$ MHz, $R = 600\,\Omega$, $C_1 = C_s = 1$ pF, $C_{ft} = 50$ fF. Dots are for $d\omega/\omega_0 = -1/Q$, $\omega = \omega_0$, and $d\omega/\omega_0 = +1/Q$, respectively. *Blue solid line* include thermal expansion effects while the *red dashed line* does not

estimated to be $-90-10° = -100°$. Based on this simple model the voltage attenuation of the resonator at resonance frequency ω_0 can be written as,

$$\left|\frac{v_{out}}{v_{in}}\right| = \frac{|g_{m,\max} - \omega_0 C_{ft}|}{\sqrt{\left(\frac{1}{R}\right)^2 + (\omega_0 C_l)^2}} \qquad (2.5)$$

Assuming resonator bias of $V_g = -8$ V and $I_d = 1.75$ mA, corresponding to $g_{m,\max} = 270$ μA/V, results in a voltage attenuation of -17 dB according to Eq. 2.5. Therefore, the amplifier, A in Fig. 2.5 should be able to provide a gain of $+17$ dB and a phase rotation of $+100°$ in order to sustain oscillation.

Using a more sophisticated model, the complex amplifier gain is calculated as a function of oscillation frequency, as is shown by the blue solid line in Fig. 2.6. It can be seen that the phase rotation is indeed close to $100°$, as predicted by the simple model. However, the required gain at ω_0 is calculated to be 15 dB, which is 2 dB less than predicted by Eq. 2.5. The difference can be attributed to thermal forces acting on the resonator body that help to reduce the resonator attenuation. This can be concluded from a simulation where it is assumed that there is no thermal expansion of the resonator, as indicated by the red dashed curve in Fig. 2.6. This thermal expansion effect is neglected in Eq. 2.5.

From Fig. 2.6 it can be seen that the phase selectivity of the oscillator is very good: $\pm 75°$ change in loop phase results in only $\pm 1/Q$ frequency pulling. This phase margin is much more than $\pm 45°$ typical for a quartz resonator for similar pulling levels. This is a direct result of the fact that our resonator has an inductive and not a resistive behavior at resonance as is the case with piezo-electric and capacitive resonators. This complex admittance in combination with the complex

2 A Piezo-resistive, Temperature Compensated, MEMS-Based Frequency Synthesizer   31

feedthrough admittance of opposite sign causes a very large phase rotation of 360°
around resonance.

One important characteristic of the piezo-resistive transduction is that in addition
to mechanical noise, now the resonator is also a source of electrical noise originating
from the electrical energy dissipated in the resistor. Obviously, these additional
noise sources need to be taken into account when optimizing the phase noise of the
oscillator. The white noise of the resonator contributes to the phase noise floor. The
current noise originating from the resonator is expressed as,

$$\overline{i^2_{noise}} = \frac{4k_bT}{R}df \tag{2.6}$$

The ratio of this undesired noise and the desired signal generated by the
resonator sets the fundamental lower limit of the phase noise floor. From Eq. 2.1
the mean-squared signal from the resonator is expressed as,

$$\overline{i^2_{signal}} = \frac{1}{2}\left(I_d\frac{r}{R}\right)^2 \tag{2.7}$$

Therefore the noise-to-signal ratio output of the resonator is written as,

$$\frac{\overline{i^2_{noise}}}{\overline{i^2_{signal}}} = \frac{8k_bT}{I_d^2R}\left(\frac{r}{R}\right)^{-2}df \tag{2.8}$$

Or expressed in terms of $g_{m,max}$ and v_{in},

$$\frac{\overline{i^2_{noise}}}{\overline{i^2_{signal}}} = \frac{8k_bT}{\left(v_{in}g_{m,max}\right)^2 R}df \tag{2.9}$$

Half of this ratio goes into phase noise, assuming that the white noise is evenly
spread over amplitude and phase. From Eq. 2.8 is can be easily understood that lower
phase noise is achieved by increasing the DC power dissipation in the resonator and
by increasing the level of resistance modulation. In Fig. 2.7 the measured resistance
modulation is shown for a large population of more than 1,100 devices when driving
the resonator with $V_g = -5$ V and AC peak voltage $v_{in} = 1V_{pk}$. It can be seen that
under these bias conditions the average resistance modulation is $r/R = 8\%$. Assuming 2 mW of dissipation inside the resonator and a resistance modulation level of 8 %
gives a fundamental phase noise floor of -149 dBc/Hz. In practice, a few dBc/Hz
still needs to be added to account for noise contributions of the amplifier and
resonator bias sources. In Fig. 2.8 the measured phase noise of a complete oscillator
is plotted for $v_{in} = 0.5V_{pk}$, $V_g = -8$–10 V, and $I_d = 1.75$–2 mA. The measured
phase noise is approximately 7 dB above the fundamental minimum as estimated
from Eq. 2.9. The 7 dB added noise is attributed to contributions from the bias source
and amplifier. Furthermore, it is demonstrated that a low near carrier noise of only
-128 dBc/Hz at 1 kHz offset can be realized.

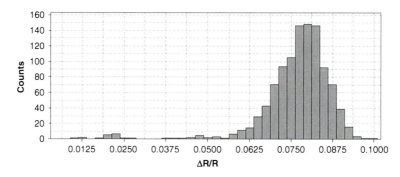

Fig. 2.7 Measured r/R modulation of a dogbone resonator for a population of 1,100 devices driven at DC bias $V_g = -5$ V and AC peak voltage $v_{in} = 1V_{pk}$. Average r/R is 8 %

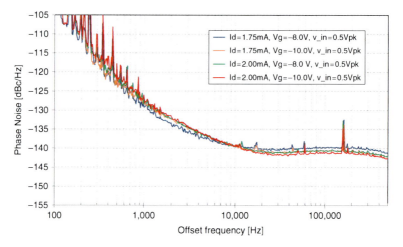

Fig. 2.8 Measured phase noise under different resonator bias conditions. Measured noise floor is below -140 dBc/Hz and is about 7 dB higher than the fundamental minimum set by the noise contribution from the resonator itself

Apart from noise requirements or non-deterministic frequency stability, an oscillator also has to fulfill requirements regarding its deterministic frequency stability and absolute accuracy. The uncompensated temperature stability of MEMS resonators is approximately -30 ppm/K and is inferior to that of quartz and can be attributed to the relatively large negative temperature coefficient, dE_{Si}/dT of the Young's modulus of SCS. In a piezo-resistive resonator the temperature is not only determined by the ambient temperature, but also by the additional heating of the resonator caused by the power dissipated through the bias current I_d. In Fig. 2.9 the resonance frequency of the resonator is shown, both as function of power dissipated in the resonator and ambient temperature. It can be seen that increasing these parameters reduces the resonance frequency of the dogbone resonator, as expected.

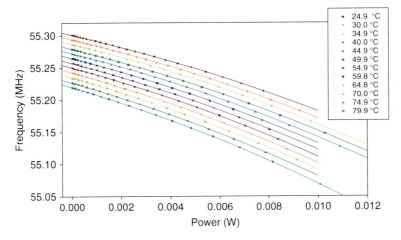

Fig. 2.9 Measured resonance frequency of a non-oxidized resonator as a function of ambient temperature and power dissipation inside the resonator body

Compensation of the temperature dependency of the Young's modulus can be achieved through material modification of the resonator. Silicon and most other materials exhibit a negative temperature dependence of the Young's modulus. When these materials are combined with a material with an opposite signed Young's modulus temperature dependence, such as SiO_2, than temperature drift compensation is achieved [10]. Considering the spring of the dogbone resonator it can be seen that the strain is evenly distributed over its cross section. For cancelling the temperature drift it is therefore necessary that a cross-section holds approximately equal amounts of Si and SiO_2, since the temperature derivative of the Young's modulus of silicon dE_{Si}/dT is about equal in magnitude to the temperature derivative dE_{SiO2}/dT. This in general means that the thickness of the oxide layer increases when resonator dimensions increase and can lead to excessive oxidation times in the case the resonator is thermally oxidized. The fact that the output current of a piezo-resistive resonator is not dependent on its dimensions, as opposed to a capacitive bulk mode resonator, makes the piezo-resistive concept well suited for this oxidation technique, since the required oxide thickness for first order temperature drift cancellation can in practice be limited to only a few 100 nm, as is shown in Fig. 2.10a. It can be seen that the temperature coefficient of the resonance frequency is increased by increasing the oxide thickness at a rate of 0.12 ppm/K/nm. Figure 2.10b shows the temperature drift of an oxidized Si resonator compared to a non-oxidized resonator, AT-cut quartz resonator, and a tuning fork quartz resonator when the oxide thickness is tuned to cancel the first order drift term. It can be seen that the oxide layer effectively compensates for the first order temperature drift even when considering accuracy to which the thickness of the SiO_2 layer can be controlled in production. Higher order drift terms still remain which are attributed to higher order drift terms of the SCS itself.

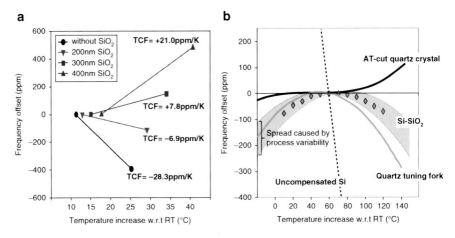

Fig. 2.10 Measured temperature drift reduction through thermal oxidation of the resonator. (**a**) Temperature drift as a function of oxide thickness. (**b**) Residual, high order temperature drift compared to quartz and non-compensated SCS

A drawback of oxidizing the resonator is that the oxidation causes a considerable shift of resonance frequency at fixed temperature, since the Young's modulus of SiO_2 $E_{SiO2} = 60$ GPa is considerably lower than the Young's modulus of Si $E_{Si} = 130$ GPa. Growing an oxide layer on the resonator reduces its stiffness and hence lowers its resonance frequency at a rate of -380 ppm/nm for our dogbone resonator [11]. This dependency is much stronger than the oxide thickness dependency of the first order temperature drift and limits the overall frequency accuracy caused by variations in oxide thickness. Uniform thermal oxidation of SCS resonators alone does not lead to the level of accuracy in resonance frequency that is typically required for reference oscillators and timing applications. However, in PLL based synthesizers the offset in resonance frequency and any residual temperature drift can easily be calibrated out during product calibration, hence elegantly solving the issue of process induced spread of the MEMS resonance frequency and temperature drift.

2.3 Frequency Synthesizer Based on Piezo-resistive MEMS Oscillator

The frequency synthesizer product comprises a MEMS die stacked on top of an ASIC die in a lead-less plastic package with a 5.0×3.2 mm^2 footprint, as is shown in Fig. 2.11. The frequency synthesizer has a LVCMOS frequency output that is programmable between 25 and 200 MHz. The current is drawn from a 2.5 to 3.3 V supply voltage and consumes less than 50 µA in standby mode.

2 A Piezo-resistive, Temperature Compensated, MEMS-Based Frequency Synthesizer

Fig. 2.11 Product assembled in a leadless 4-pin plastic package measuring 5.0×3.2 mm^2. *Left*: The MEMS resonator chip is stacked and wire bonded to the ASIC. *Middle*: Bottom view of package showing supply, ground, enable, and frequency output pins. *Right*: Top view of package

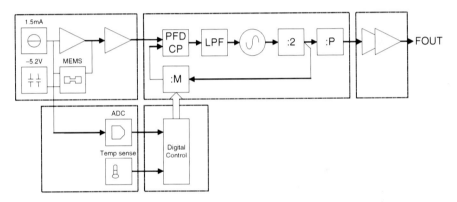

Fig. 2.12 System architecture of the frequency synthesizer

The system architecture is shown in Fig. 2.12 indicating the various functional blocks. The MEMS resonator is biased at -5 V at the gate terminal using a charge pump. On the drain port the resonator is biased at 1.5 mA coming from a current source. The resonator drain signal is amplified by a two-stage amplifier and the amplified signal is fed back to the gate terminal with proper phase shift. The output of the MEMS oscillator is fed into the PFD which drives the CP that generates a correction voltage that is fed into a VCO. The VCO is based on a 2 GHz LC tank that is continuously tunable over the full temperature range via a varactor. A varactor instead of a capacitor bank is used to stabilize the VCO frequency over temperature in order to avoid discrete frequency steps at the VCO output. The frequency range of 25–200 MHz can be programmed in 30 Hz steps by setting the proper values of the frac-M divider, the P divider, and by selecting the proper VCO band through a switched capacitor bank that is integrated in the LC tank. Several low-dropout regulators (LDOs) are in place to stabilize the supply voltage for the different circuit blocks.

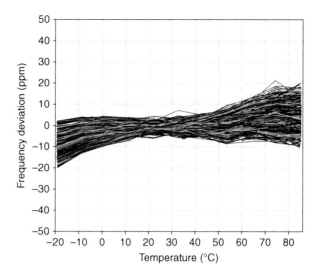

Fig. 2.13 Frequency stability against ambient temperature, and supply voltage varied from 2.4 to 3.6 V, measured on 110 samples

The frequency offset and temperature drift of the MEMS is compensated for by programming the value of the frac-M divider in the PLL. For temperature drift correction both the ambient temperature as well as the power dissipation inside the resonator needs to be known. The ambient temperature is measured with a PTAT sensor that is integrated on the ASIC. The power dissipation inside the resonator is derived from the stabilized 1.5 mA resonator current bias plus a voltage reading through an ADC connected to the resonator drain terminal. The shift in MEMS frequency is calculated by feeding the resonator drain voltage and the PTAT sensor output into a polynomial with coefficients that are pre-programmed in the MTP. The calculated MEMS frequency shift sets the value of the frac-M divider in the PLL and compensates the temperature induced frequency shift of the MEMS at the input of the PFD. Frequency calibration is done at room temperature only and is used to compensate for manufacturing spread in the MEMS resonance frequency. The frequency versus temperature relation is not calibrated for individual samples and is assumed to be the same for all MEMS resonators. This single temperature insertion calibration results in a frequency spread of less than ± 20 ppm over a temperature window of $-20°C$–$+85°C$ and supply window of 2.4–3.6 V, as is shown in Fig. 2.13 on a population of 110 samples. It should be noted that in this case a non-oxidized resonator was used.

Measured phase noise at 100 MHz output frequency is shown in Fig. 2.14. Integrated phase jitter is measured to be 0.44 ps in the 1.9–20 MHz band and 2.96 ps in the 12 kHz–20 MHz band. Current dissipation as a function of output frequency for supply voltage of 2.5 and 3.3 V under minimal loading condition on the output pin is shown in Fig. 2.15a. It can be seen that current consumption is less than 15 mA at 25 MHz and increases up to a maximum of 22 mA at 200 MHz output frequency. Zooming in on the power budget breakdown as depicted in the pie chart of Fig. 2.15b it can be seen that the VCO is with 43 % the largest contributor to

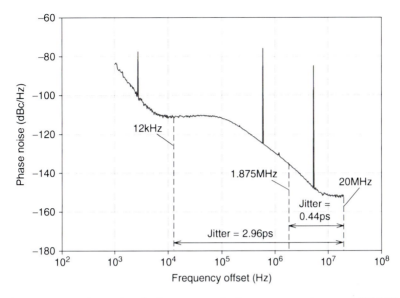

Fig. 2.14 Measured phase-noise of a frequency synthesizer programmed to output 100 MHz

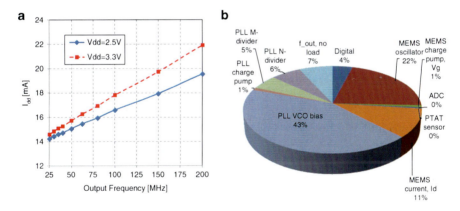

Fig. 2.15 (a) Current consumption, I_{dd} of the complete synthesizer under minimal load condition measured for $V_{dd} = 2.5$ V and $V_{dd} = 3.3$ V supply voltage. (b) Break down of power consumption over different circuit blocks when output frequency is programmed at 25 MHz

overall power dissipation followed by the MEMS core amplifier including MEMS drain bias consuming a combined 33 % of total power.

An ageing test is performed on a population of 30 products to assess frequency stability over time, as is shown in Fig. 2.16. During the ageing test the product is kept in an oven at +85°C and powered at nominal supply voltage. It can be seen that no fails occur and no significant frequency ageing is observed over a period of more than 25 days.

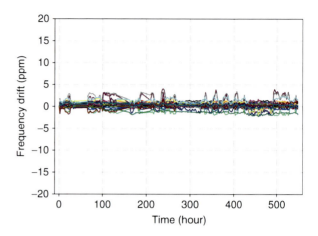

Fig. 2.16 Life-time test at 85°C on 30 samples, lasting for 25 days, showing a frequency stability of better than ±4 ppm

2.4 Conclusion and Outlook

A piezo-resistive transduced MEMS resonator has the unique property that its output signal is not reduced when scaling down its physical dimensions. This makes it suitable for realizing fundamental tone MEMS oscillators with high output frequency in the 50–150 MHz range that is ideal for interfacing with high performance PLL based frequency synthesizers. A low noise, 55 MHz fundamental tone piezo-resistive MEMS oscillator is demonstrated with a phase-noise floor below −140 dBc/Hz and near carrier noise of −128 dBc/Hz at 1 kHz offset.

The first order temperature induced frequency drift of the SCS resonator can effectively be reduced by coating the resonator with a layer of thermally grown SiO_2. The oxide thickness should be in proportion with the resonator dimensions in order to effectively remove the first order drift term. It is desired that resonator dimensions are small in order not to require excessively thick SiO_2 layers and associated complications in wafer processing. Also this aspect combines well with piezo-resistive instead of capacitive transduced resonators.

A fully functional PLL based frequency synthesizer running from a 2.5–3.3 V supply has been realized based on a 55 MHz MEMS reference resonator. The MEMS die containing the piezo-resistive resonator is stacked on top of and wirebonded to the ASIC. The stacked die is assembled in a low cost 4-pin leadless plastic package measuring $3.2 \times 5mm^2$. The frequency drift with temperature of the uncompensated MEMS resonator is effectively monitored measuring the (a) ambient temperature with a PTAT sensor and (b) power dissipation inside the resonator that is induced by the resonator bias current. Both parameters are used for programming the frac-M divider in the PLL thereby stabilizing the frequency at the VCO output. It is demonstrated that a frequency accuracy of ±20 ppm over −20°C–+85°C is realized for a population of 110 products incorporating a non-oxidized resonator using only a single insertion frequency calibration at room temperature. The timing jitter level at the output of the frequency synthesizer is sufficiently low for clocking low- to mid-end high speed digital data transmission.

Future work will focus on further reducing the jitter, especially jitter contributions within the loop bandwidth of the PLL that find their origin in the PFD/CP and MEMS core. Furthermore, it is expected that the incorporation of oxidized MEMS resonators will further improve overall frequency stability over temperature of the synthesizer to levels well below 20 ppm.

Acknowledgements We would like to thank Harry Houterman, Friso Jedema, Sumy Jose, Erik-Jan Lous, Joost Melai, Edwin Orij, Nivesh Rai, Chris Rittersma, Kirsten Rongen, Jos Sistermans, Frank Swartjes, and Peter van der Velden from NXP for their contributions on product reliability, testability, and manufacturability. Many thanks go to Tjeu van Ansem and Peter Vermeeren from NXP, and AXIOM-IC for their help in ASIC design. Micha in't Zandt from NXP, and IMEC are gratefully acknowledged for their support in processing the MEMS wafers.

References

1. J.T.M. Van Beek, R. Puers, A review of MEMS oscillators for frequency reference and timing applications. J. Micromech. Microeng. **22** 013001, 35pp (2012)
2. C.S. Vaucher, *Architectures for RF Frequency Synthesizers* (Kluwer, 2002), ISBN1-4020-7120-5, pp. 70
3. J.H. Kuypers, G. Zolfagharkhani, A. Gaidarzhy, R. Rebel, D.M. Chen, S. Stanley, D. LoCascio, K.J. Schoepf, M. Crowley, P. Mohanty, *High Performance MEMS Oscillators for Communications Applications* (Chiba, 2010)
4. J.T.M. Van Beek, G.J.A.M. Verheijden, G.E.J. Koops, K. Le Phan, C. Van der Avoort, J. Van Wingerden, D. Badaroglu Ernur, J.J.M. Bontemps, Scalable 1.1 GHz fundamental mode piezo-resistive silicon MEMS Resonator, in *IEEE International Electron Devices Meeting, IEDM, 2007*, Washington, DC, USA, pp. 411–4, 10–12 Dec 2007
5. D. Weinstein, S.A. Bhave, Internal dielectric transduction in bulk-mode resonators. J. Microelectromech. Syst. **18**, 1401–8 (2009)
6. W. Wang, L.C. Popa, R. Marathe, D. Weinstein, An unreleased mm-wave resonant body transistor, in *Proceeding MEMS, 2011*, Cancun, Mexico, pp. 1341–4
7. J.T.M. Van Beek, P.G. Steeneken, B. Giesbers, A 10 MHz piezoresistive MEMS resonator with high Q, in *IEEE International Frequency Control Symposium and Exposition, 2006*, Miami, Florida, USA, pp. 475–80, 4–7 June 2006
8. J.T.M. Van Beek, K. Le Phan, G.J.A.M. Verheijden, G.E.J. Koops, C. Van der Avoort, J. Van Wingerden, D. Ernur Badaroglu, J.J.M. Bontemps, R. Puers, A piezo-resistive resonant MEMS amplifier, in *IEEE International Electron Devices Meeting IEDM 2008* vol 1–4, pp. 667–70
9. P.G. Steeneken, K. Le Phan, M.J. Goossens, G.E.J. Koops, G.J.A.M. Brom, C. Van der Avoort, J.T.M. Van Beek, Piezoresistive heat engine and refrigerator. Nat. Phys. **7**, 354–9 (2011)
10. J.T.M. Van Beek, H. Loebl, F.W.M. Van Helmont, A MEMS resonator, a method of manufacturing thereof, and a MEMS oscillator US patent 7847649, Issue date: 7 Dec 2010
11. C. Van der Avoort, J. Van Wingerden, J.T.M. Van Beek, The effects of thermal oxidation of a MEMS resonator on temperature drift and absolute frequency, in *IEEE 22nd International Conference on Micro Electro Mechanical Systems, MEMS 2009*, Sorrento, Italy, pp. 654–6, 25–29 Jan 2009

Chapter 3
A MEMS TCXO with Sub-PPM Stability

Aaron Partridge, Hae-Chang Lee, Paul Hagelin, and Vinod Menon

Abstract This paper introduces a MEMS-based TCXO that delivers <1 ppm (parts per million) frequency stability from -40 C to $+85$ C. Its system architecture, MEMS resonator, and key circuit blocks are described. The oscillator achieves a phase noise of -134dBc/Hz at 1 kHz and -142dBc/Hz at 10 kHz from a 26 MHz carrier, with a far phase noise of -158dBc/Hz. Moreover, its integrated jitter is 0.5 ps from 12 to 20Mhz. The oscillator's frequency is programmable from 1 to 220 MHz and it draws 32 mA from 1.8 to 3.3 V supply at 26 MHz. The transition from quartz- to MEMS-based oscillators is also discussed, with a review of the oscillator architecture and accompanying benefits, e.g. programmability, improved reliability and robustness, and decreased sensitivity to vibration and EMI.

3.1 Introduction

MEMS oscillators are displacing quartz oscillators in clocking, timing, and frequency generator applications. In this paper, we discuss the first commercial MEMS oscillators that are stable to better than one part per million (ppm). The focus of this paper is on architecture of these oscillators and how this architecture is improved from that used in the legacy quartz oscillators.

In all commercial MEMS oscillators, the output is derived from a MEMS resonator where the resonator's frequency is translated with a frac-N PLL. Under control of a digital state machine, the PLL compensates for the resonator's initial frequency offset and frequency variation over temperature. Quartz oscillators usually produce an output frequency at their crystal's resonance, and do not rely on circuitry to adjust that frequency. In temperature compensated quartz oscillators, the frequency compensation is usually provided by pulling the resonator through capacitive loading, not with a fractional PLL.

A. Partridge (✉) • H.-C. Lee • P. Hagelin • V. Menon
SiTime Corp, Sunnyvale, USA
e-mail: ap@sitime.com

A. Baschirotto et al. (eds.), *Frequency References, Power Management for SoC, and Smart Wireless Interfaces: Advances in Analog Circuit Design 2013*, DOI 10.1007/978-3-319-01080-9_3, © Springer International Publishing Switzerland 2014

MEMS oscillators are thus circuit-centric, whereas simple quartz oscillators are material-centric. This choice of circuit-centric approach is driven by the needs of the MEMS resonators and by market requirements, and is made possible by the advancing capabilities of CMOS.

MEMS and quartz oscillators can be divided into three categories: XOs, that provide accuracies of ± 100 ppm (parts per million) to ± 25 ppm. Temperature compensated oscillators, TCXOs, that provide accuracies of ± 2.5 ppm to ± 500 ppb (parts per billion), and in rare cases to ± 100 ppb. And ovenized oscillators, OCXOs, that provide accuracies from ± 100 ppb down to ± 1 ppb or lower. Their prices vary inversely with their accuracy; a 50 ppm oscillator may cost $0.40 while a 1 ppb oscillator may cost $400.

3.2 Architecture

There are three application drivers that quartz addresses with mechanical processes that MEMS addresses electronically: (1) providing a wide range of application frequencies, (2) trimming the resonator frequency over production tolerances, and (3) compensating the resonator frequency over temperature.

Commercial applications specify hundreds of different frequencies. In quartz oscillators, the resonators must be manufactured precisely to these frequencies, necessitating the crystals to be cut and ground to hundreds of different thicknesses. Second order effects often require the lateral dimensions of the crystals to be optimized for each frequency.

Building MEMS resonators across widely varying frequencies would be difficult, time consuming and expensive. Commercial MEMS resonators required years of individual development, usually involving many fab-and-test cycles. Deriving designs with different frequencies is generally not straightforward. For resonators with lateral modes, the dimensions or shapes are specific for each frequency, and this implies additional tape-outs, one for each frequency. For resonators with vertical modes, multiple material thicknesses would need to be optimized for each frequency. Clearly for MEMS resonators, supplying hundreds or even dozens of frequencies could be commercially unviable.

Quartz crystals are individually trimmed to their specified frequencies, or in the case of TCXOs are trimmed to within the pull range of their specified frequencies. This mechanical trimming is usually done by ion milling or laser ablating the quartz or metallization. MEMS resonators can also be trimmed this way; however mechanical trimming can complicate the already difficult MEMS design and packaging.

Present-generation MEMS resonators show greater temperature sensitivity than AT-cut quartz, and while MEMS can be temperature compensated [1] the devices shown to-date have greater sensitivity than can be tolerated by many applications. For this reason MEMS oscillators are generally electronically temperature compensated. In this sense, they are like quartz TCXOs. But unlike quartz oscillators that are temperature compensated only for certain precision applications,

3 A MEMS TCXO with Sub-PPM Stability

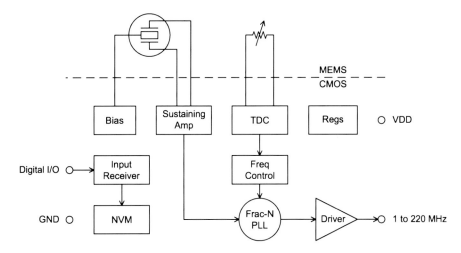

Fig. 3.1 Oscillator topology

MEMS oscillators are compensated for the majority of applications. Unlike quartz TCXOs, MEMS oscillators are generally not trimmed by resonator pulling, but instead with fraction PLL multiplication.

Fractional PLL technology was not available for early quartz oscillators. The designers therefore had no alternative other than to develop a resonator technology that could support a range of frequencies, could be trimmed, and in some cases adjusted or pulled over temperature. On the other hand, we now have the circuit technology to program MEMS oscillator frequencies to their application requirements, trim them over production, and compensate them over temperature.

Using this circuit-centric approach simplifies the MEMS resonator development and production while also providing significant commercial benefits. For instance, inventories of oscillators with various pre-defined frequencies are not needed, and custom frequencies can be supplied quickly.

Figure 3.1 shows the circuit-centric oscillator topology. The architecture includes a PLL to translate the resonator frequency to the application requirements. The PLL is fractional in order to trim the resonator across production tolerances, and its multiplication value is variable in order to compensate for temperature. A state machine (in hardware or software) controls the PLL, and draws its parameters from non-volatile memory.

Since this architecture works well for MEMS oscillators then why is it not used for quartz too? Or in other words, why not apply this new circuit technology back to quartz oscillators? We have not seen this to any significant extent. There are likely many reasons, among them is that the incumbent industry has organizational inertia – most of the quartz suppliers are expert at crystal machining but not at circuit design. Developing a circuit infrastructure would be slow and expensive for them.

3.3 Key Circuits

The TDC (Temperature to Digital Converter) provides the temperature compensation data. The TDC in the oscillator detailed here is optimized for low temperature readout noise because noise in the temperature data becomes near phase noise in the output frequency. For this oscillator's target applications we require phase noise levels that necessitate particular attention to the TDC.

One has many options when designing temperature sensors. The default topology is a Delta-V_{BE} circuit using bipolar transistors in the CMOS die. These temperature sensors have many favorable characteristics: they are linear, easy to calibrate, well understood, highly evolved, use moderate die area, and are low power. In addition they are purely circuit-based.

However, in this case we have chosen to use thermistors as the detection elements because thermistors offer higher signal to noise.

Figure 3.2 shows the balanced bridge topology we use to digitize the thermistor resistance. The feedback loop works to keep the node between the Thermistor and the Trim reference resistor at mid value, the Temp output is a digital value.

To build this TDC one must have a stable reference resistance, but such a resistance is not available in common CMOS. We therefore construct an equivalent reference with a switched capacitor. Capacitors are inherently stable in CMOS and we have a stable reference frequency derived from the MEMS resonator to time the capacitor switching. The resonator frequency does change as a function of temperature, but two orders of magnitude less than the thermistor resistance, and also in a predictable manner, therefore it can be considered a fixed frequency. The resulting reference is thus highly stable.

Figure 3.3 shows further detail of the TDC converter. The TDC works by forcing the effective resistance of the switched capacitor to match the thermistor resistance. By closing Ø1 then Ø2 at a selectable rate, the switched reference capacitor C2 obeys Ohm's law with a selectable resistance. The rate is controlled by a sigma-delta feedback loop driving a fractionally derived frequency. Capacitor C1 averages current across the switching. In this way, the loop balances the Thermistor resistance and provides the digitized Temp output. The majority of the loop is digital, starting at the Quantizer.

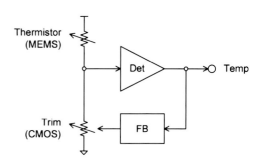

Fig. 3.2 Balanced bridge TDC

3 A MEMS TCXO with Sub-PPM Stability

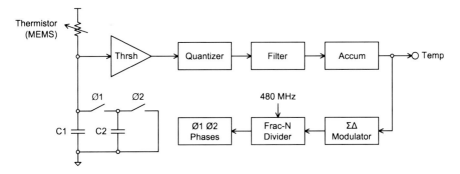

Fig. 3.3 Switched cap reference resistance

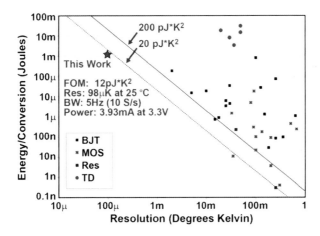

Fig. 3.4 Figure of merit comparing conversion energy and resolution (Comparison data compiled by K. Makinwa [3])

The TDC is more complex than what is described here in three ways: (1) the Thermistor is measured differentially with its polarity swapped and sensed by two sets of circuits, (2) the Thermistor is driven and sensed in a four-wire configuration with Kelvin connections, and (3) the implied mid-supply reference in the Quantizer is developed dynamically. A more detailed description of the TDC is published by M. Perrott [2].

Note that there is never a voltage or current from the Thermistor to an ADC. The Thermistor and TDC work in a feedback system to derive the temperature, there is never an analog temperature value that is digitized.

The TDC delivers a resolution of 98 uK at 5 Hz bandwidth while consuming 3.9 mA at 3.3 V. Figure 3.4 compares its energy efficiency and resolution against

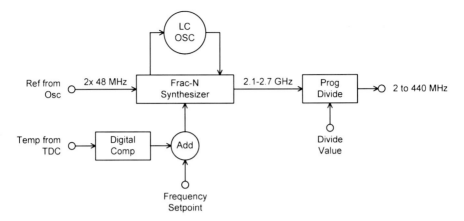

Fig. 3.5 Frac-N PLL

other integrated TDCs. This TDC provides over an order of magnitude lower noise than others in the comparison population and does so at under 20pJK^2. The comparison data in this plot is compiled by K. Makinwa [3].

The frac-N PLL provides an output frequency programmable from 1 to 220 MHz (after a final divide by 2). It is optimized for low phase noise and low power. Figure 3.5 shows its block diagram. Details of the PLL have been published by F. Lee [4].

A PLL normally is thought of as a frequency translation circuit, and for that function it consumes power. It is sometimes thought that this is a "waste" of power because if the resonator had been at the output frequency the PLL would not be needed. However, a key benefit of the PLL is that above its filter bandwidth it provides lower phase noise that its reference.

Figure 3.6 shows the phase noise of the MEMS output and the PLL output. The PLL output phase noise is 6 dB higher than the MEMS reference from 10 to 300 KHz, but its far phase noise is 12 dB lower. The far phase noise often matters more than the near phase noise. For instance in high speed serial links, the key specification is integrated jitter from 12 kHz to 20 MHz. The integration is linear across frequency, so the far phase noise dominates the near phase noise. The integrated jitter after the PLL is lower than before it.

For high speed serial links, at the common output frequency of 156.25 MHz, the oscillator show a 12 kHz to 20 MHz integrated phase jitter of 0.5 ps, which quartz suppliers consider extremely low jitter. Thus we say, "PLLs are our friends!" They do more than just frequency translation, trimming, and compensation. They decrease the output phase noise.

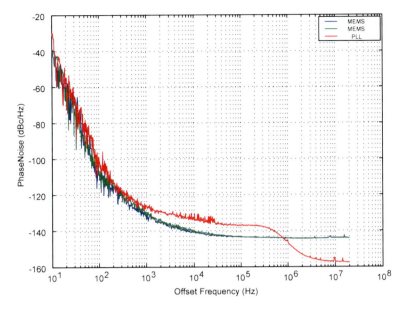

Fig. 3.6 MEMS and PLL phase noise, reference frequency is 48 MHz

3.4 MEMS Resonators and Thermistors

Development of MEMS resonators dates to the late-1960, with the first published results by Nathanson 1967 [5]. The early resonators were audio filters, and were not suitable for references. As references, they would not have provided the required stability or phase noise. Their limits can be traced to the resonator material and the packaging cleanliness.

In order for a resonator to have a stable frequency, one must build it from a stable material. The first resonators were built from metal, but were not sufficiently stable for frequency references as metal is subject to internal stresses, hysteresis, and aging. Material advances in the late 1970s at IBM [6, 7] and Berkeley in the 1990s showed silicon oxide and polysilicon respectively [8] as resonator materials. Further work in the 1990s and early 2000s developed single crystal silicon as an optimal resonator material. Presently both single crystal and polycrystalline silicon are used in MEMS resonators. Other silicon-centric materials, including Aluminum Nitride are in development, some for many years [9]. But to date none except silicon has seen commercial success. Research on more exotic materials such as polycrystalline diamond is underway [10, 11].

As resonators are reduced in size their volume to surface ratio decreases, consequentially their sensitivity to surface contamination increases. To maintain stabilities in the parts-per-million range expected of frequency references, one must minimize mass-loading the resonators with contamination. One finds that even a

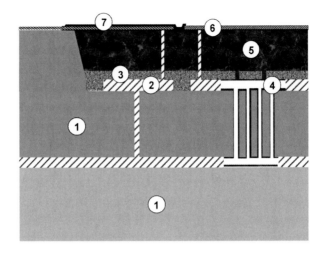

Fig. 3.7 MEMS encapsulation cross section

monolayer of surface contamination can shift a resonator's frequency out of specification. This drives a requirement for very clean packaging. Bonded covers have been made sufficiently clean to produce commercial resonators with XO-type accuracy. The epitaxially encapsulation used in the parts described here is likely cleaner and readily enabled TCXO to OCXO-level stability.

Figure 3.7 shows a cross-section diagram of the encapsulation used for the resonator and thermistor. This encapsulation is built as follows: (1) an SOI wafer is trenched to define the resonator, (2) protective oxide is deposited and patterned, (3) silicon is deposited and patterned with vents, (4) the resonator is released, (5) the vents are closed and thick encapsulation is deposited, (6) contact isolation trenches are etched and oxide filled, and (7) metal traces are fabricated and passivized in the normal way.

Figure 3.8 shows an isometric view of a 48 MHz resonator. The arrows indicate the mechanical motion in resonance with each ring expanding and contracting in phase. There are electrostatic drive and sense electrodes inside and outside of each ring. The rings are anchored proximal to the midpoint of the cross beams.

Figure 3.9 shows a network analyzer plot of a nominal resonator response. The resonant frequency is 48.016 MHz, and the quality factor is 147 k.

Figure 3.10 shows an example SEM cross section in this process. The SEM is taken of an edge of a wafer that has been cleaved and from which a resonator is protruding. The top of the SEM shows the top surface of the MEMS wafer.

The MEMS thermistor is built in single-crystal silicon. Like the resonator it is fully vacuum encapsulated and released. Because it is encapsulated with the resonator, it is protected from environmental contamination and therefore highly stable. The resistor is released from the substrate.

3 A MEMS TCXO with Sub-PPM Stability

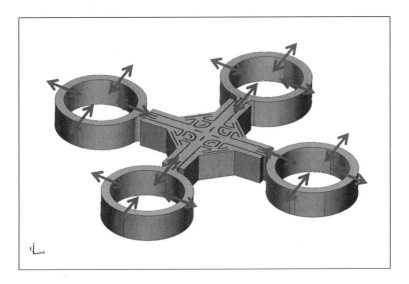

Fig. 3.8 Resonator isometric view

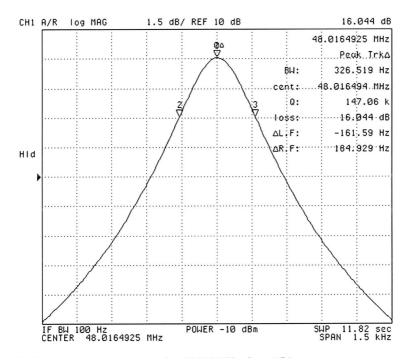

Fig. 3.9 Example resonator response, f = 48.016 MHz, Q = 147 k

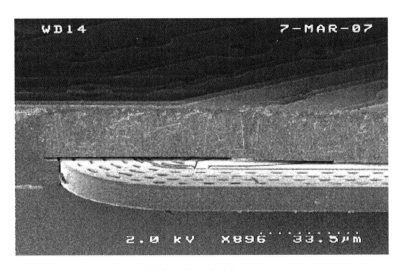

Fig. 3.10 Resonator cross section SEM of cleaved wafer

3.5 Performance

Figure 3.11 shows the output frequency as a function of temperature. The compensated output frequency of 68 parts over a temperature range of −40 C to +85 C is within ±0.2 ppm of the specified frequency.

Output phase noise is a function of the output frequency. Figure 3.12 shows the phase noise at 156.25 MHz, a common frequency for high speed serial link systems.

The integrated phase noise from 12KHz to 20 MHz in this case is 493 fs not including spurs and 689 fs including spurs. Generally the telecom application at this frequency accounts for spurs elsewhere and cares about the random integrated jitter, which is the 493 fs value.

Figure 3.13 shows the die photo, and Fig. 3.14 shows a diagram of the packaged MEMS and CMOS die. The MEMS die is mounted on top of the CMOS die which is molded into a QFN package.

3.6 Looking Toward TCOCXOs

When an oscillator is maintained at an elevated temperature it delivers better frequency stability than when its temperature is allowed to vary with the ambient. When a TCXO is ovenized it is called an OCTCXO.

3 A MEMS TCXO with Sub-PPM Stability

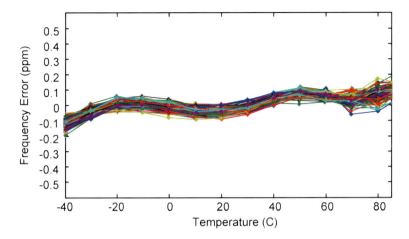

Fig. 3.11 Compensated output frequency vs. temperature

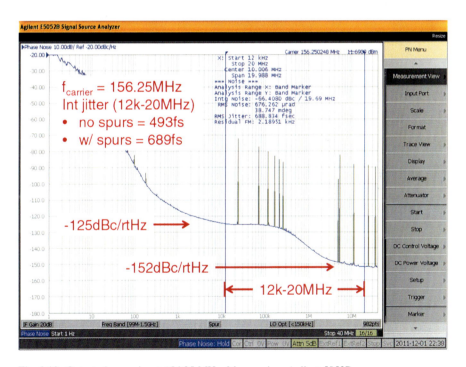

Fig. 3.12 Output phase noise at 156.25 MHz. Measured on Agilent 5052B

Fig. 3.13 Die photo

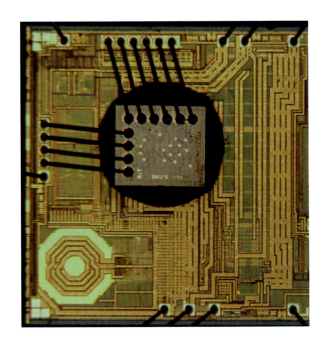

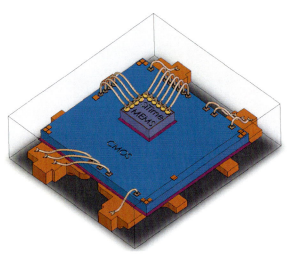

Fig. 3.14 Package diagram showing leadframe, CMOS, and MEMS die

Figure 3.15 shows the frequency vs. temperature of the MEMS oscillators built into an oven module. When operated as OCTCXOs their frequency can be stabilized to ±10 ppb over −45 C to +90 C. The brackets in Fig. 3.15 show common application temperature and accuracy requirements, including cell base stations and Stratum-3E oscillators for telecom networking. Accuracies from 10 to

3 A MEMS TCXO with Sub-PPM Stability

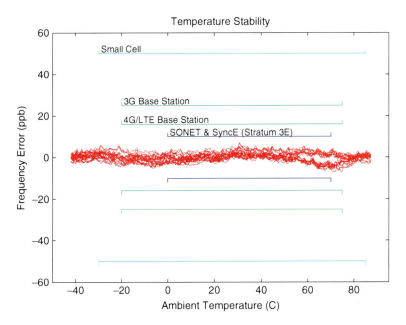

Fig. 3.15 Oven controlled output frequency vs. ambient temperature

100 ppb are normally provided by quartz OXCOs today, although most quartz OCXOs do not operate over the full −40 C to +85 C temperature range.

Note that the results shown in this paper are from production material, except for the data in Fig. 3.15 which is from a single part on a bench. Nonetheless, it is important to understand that MEMS oscillators will likely support these OCXO applications.

3.7 Conclusions

Looking forward, it is likely that this circuit-centric architecture will dominate timing generation. The advantages over the simple oscillator architecture are profound and the costs, in terms of both power and die area, are steadily decreasing.

Presently almost 200 million MEMS oscillators have been produced with this circuit-centric architecture. The growth rate of this segment is strong, at about 70 % per year. SiTime has been listed by Deloitte LLP in their Fast 500[TM] rankings as the fastest growing semiconductor company in North America [12]. The same is not true for the quartz suppliers, for whom growth is modest to negative. There are many reasons for this shift, among them are that MEMS shows better reliability, robustness, supply chain support, and cost structure that quartz. And a significant reason is that the MEMS oscillators are based on a programmable, flexible, and highly accurate, modern architecture.

References

1. R. Melamud, B. Kim, S.A. Chandorkar, M.A. Hopcroft, M. Agarwal, C.M. Jha, T.W. Kenny, Temperature-compensated high-stability silicon resonators. *Applied physics letters* **90**(24), pp. 244107–244107 (2007)
2. M. Perrott, J.C. Salvia, F.S. Lee, A. Partridge, S. Mukherjee, C. Arft, J. Jim, N. Arumugam, P. Gupta, S. Tabatabaei, S. Pamarti, H.C. Lee, F. Assaderaghi, A temperature-to-digital converter for a MEMS-based programmable oscillator with frequency stability and <1ps integrated jitter. IEEE J. Solid-State Circuit **48**(1), 276–291 (2013)
3. K.A.A. Makinwa, Smart temperature sensor survey, http://ei.ewi.tudelft.nl/docs/TSensor_survey.xls
4. F.S. Lee, J. Salvia, C. Lee, S. Mukherjee, R. Melamud, N. Arumugam, S. Pamarti, et al. A programmable MEMS-based clock generator with sub-ps jitter performance. In *VLSI Circuits (VLSIC), 2011 Symposium on*, pp. 158–159. IEEE, (2011)
5. H.C. Nathanson, W.E. Newell, R.A. Wickstrom, J.R. Davis Jr., The resonant gate transistor. IEEE Trans. Electron Devices **ED14**, 117–133 (1967)
6. K.E. Petersen, Silicon as a mechanical material. Proc. IEEE **70**(5), 420–457 (1982)
7. K.E. Petersen, private communication
8. C.T.-C. Nguyen, R.T. Howe, An integrated CMOS micromechanical resonator high-Q oscillator. IEEE J. Solid-State Circuit **34**(4), 440–445 (1999)
9. R. Ruby, P. Merchant, Micromachined thin film bulk acoustic resonators, In *Frequency Control Symposium, 1994. 48th., Proceedings of the 1994 IEEE International*, pp. 135–138. IEEE, 1994. Boston, MA, USA
10. J. Wang, J. E. Butler, T. Feygelson, C. T.-C. Nguyen, 1.51-GHz polydiamond micromechanical disk resonator with impedance mismatched isolating support, in *Proceedings of the seventeenth International. IEEE Micro Electro Mechanical Systems Conference*, Maastricht, 25–29 Jan 2004, pp. 641–644
11. T. Lin, T. beyazoglu, L. Wu, A. Lingqi, M. Akgul, Z. Ren, T.O. Rocheleau, CT-C. Nguyen, 2.98-GHz CVD diamond ring resonator with Q>40,000. In *Frequency Control Symposium (FCS), 2012 IEEE International*, Baltimore, MD, USA pp. 1–6, 2012.
12. Deloitte LLP, Deloitte's 2012 technology fast 500[TM] ranking, http://www.deloitte.com/assets/Dcom-UnitedStates/Local%20Assets/Documents/TMT_us_tmt/us_tmt_fast500_rankings_020713.pdf

Chapter 4
Dual Core Frequency Reference for Mobile Applications in 65-nm CMOS

Emmanuel Chataigner and Sébastien Dedieu

Abstract Wireless mobile devices require two reference clocks, a low-noise high-frequency one and a low-power low-frequency one.

The circuit described in this paper implements a single-crystal oscillator which generates a 26–52 MHz clock and a 32 kHz clock. Its reconfigurable architecture allows to re-use the crystal and the tank capacitances in two different core oscillators, each of them being optimized for one clock independently from the other one.

4.1 Introduction

Wireless mobile devices require two reference clocks to operate radio communications, digital processing or real time tracking. One clock, operating at a few tens of MHz, is used for cellular communications, WLAN and satellite localization and must be very stable. The other one, typically operating at 32.768 kHz, is used as real-time clock (RTC). The former is power-consumption hungry but only needed a small portion of the time. The latter is always on, dictating a very low power consumption.

The usual solution to generate these two clocks is to have two crystals working at two different frequencies, for example at 32 kHz and at 26 MHz. A cost effective solution would be to use only one crystal. Apart from the reduction of the bill of material and the PCB area, this solution would offer other advantages that will be presented hereafter.

E. Chataigner (✉) • S. Dedieu
STMicroelectronics, 850 rue Jean-Monnet, 38926 Crolles, France
e-mail: emmanuel.chataigner@st.com

A. Baschirotto et al. (eds.), *Frequency References, Power Management for SoC, and Smart Wireless Interfaces: Advances in Analog Circuit Design 2013*, DOI 10.1007/978-3-319-01080-9_4, © Springer International Publishing Switzerland 2014

In [1] a single-crystal solution is presented where the low power mode is obtained by changing the bias current and the supply voltage of the oscillator which remains unchanged. On the contrary, the circuit presented here implements a change of oscillator architecture while keeping the same crystal and the same capacitor tanks. Throughout this paper these two oscillators will be called cores. The High Power (HP) core is used during cellular communications to generate the High-Frequency (HF) clock while the Low Power (LP) core is used in stand-by mode to generate the low-power real-time clock.

This paper is organized as follows: Sect. 4.2 describes the dual core topology. In Sect. 4.3 the complete system is introduced and its advantages discussed. The measurement results are presented in Sect. 4.4. Finally, conclusions are drawn in Sect. 4.5.

4.2 Dual Core Oscillator Description

The circuit was implemented in STMicroelectronics 65 nm CMOS technology and was supplied by a 1.8 V voltage from a dedicated LDO.

4.2.1 Specific Requirements

The suppression of the 32 kHz crystal significantly reduces the cost of the final platform since this crystal is usually more expensive and has a larger footprint than a High-Frequency (HF) crystal, thus also reducing the area of the printed circuit board (PCB). The continuous push for low cost terminals also requests the oscillator to accommodate crystals from various providers and to be used with various packages. These requirements implied a larger frequency tuning range than usual. There was also a request to support multiple crystal frequencies from 26 to 52 MHz in order to address various platform needs.

The increased complexity of the silicon transceivers which support more and more RX/TX channels and integrate digitally-intensive blocks puts new constraints on the DCXO. To address this issue the following choices were made : (1) a dedicated LDO with high Power Supply Rejection Ratio (PSRR) is used to power the DCXO (2) the reference clocks are propagated as differential signals on the chip (3) a differential structure was chosen for the DCXO to ease the generation of those differential clocks.

Low phase noise specifications below 100-Hz and above 1-kHz frequency offsets were also required respectively for GPS and cellular standards.

Finally the re-use of the same crystal and the same tank capacitors was chosen to have a very good frequency matching over process and temperature variations. This feature is especially interesting for synchronizing both clocks since the temperature drift of a 32-kHZ crystal is ten times higher than a HF crystal.

Fig. 4.1 The HP core

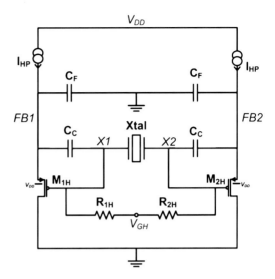

4.2.2 The HP Core

A Colpitts architecture was selected to address the low phase noise requirements while not consuming too much power. A pseudo-differential architecture (Fig. 4.1) was implemented for its simplicity and its ability to drive differential clocks.

The HP core has programmable current sources I_{HP} to guarantee fast and reliable start-up and to get the best trade-off between power consumption and phase noise performance. A peak detector measures the amplitude at the crystal and controls it through a feedback to the tunable current sources I_{HP}. The range of achievable currents guarantees the oscillation for 26–52 MHz crystals for all Equivalent Series Resistor (ESR) associated with 2,520 and 2,016 Surface Mounted (SMD) packages.

The load capacitance seen by the crystal is:

$$C_L = \frac{C_C \cdot C_F}{2(C_C + C_F)} \tag{4.1}$$

The core oscillates around the crystal parallel resonance frequency at:

$$\omega_p = \omega_s \cdot \sqrt{1 + \frac{C_1}{C_0 + C_L}} \tag{4.2}$$

with the series resonance frequency of the crystal given by:

$$\omega_s = \frac{1}{\sqrt{L_1 \cdot C_1}} \tag{4.3}$$

Fig. 4.2 The LP core

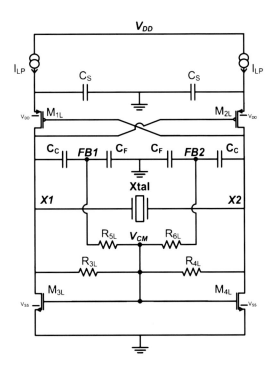

L_1 and C_1 are the electrical equivalent elements of the motional part of the crystal and C_0 its shunt capacitance associated with the package.

The coarse capacitance tank C_C consists of switchable Metal Insulator Metal (MIM) capacitors. The tuning range is ±40 ppm with a maximum frequency step of 3 ppm. This range will be used by the factory calibration to compensate for the crystal frequency tolerance, the PCB line variations and the silicon process variations.

The capacitance bank C_F consists of 8,191 unit varactors controlled in a thermometric way. Each PMOS-based varactor provides a 5 fF capacitance difference between the accumulation and depletion regions, giving an overall tuning range of ±25 ppm with a maximum frequency step of 0.03 ppm.

4.2.3 The LP Core

Re-using the same tank, a differential resonator-synchronized relaxation oscillator [2] was chosen for its low-noise, high-stability and low-consumption capability. The simplified circuit diagram is shown in Fig. 4.2.

As with the HP core, the load capacitance seen by the crystal is:

$$C_L = \frac{C_C.C_F}{2(C_C + C_F)} \qquad (4.4)$$

and the oscillation still occurs at ω_p.

In order to reduce the power consumption, C_L is set to its minimum value by changing C_C and C_F capacitance values.

Transistors M_{3L}/M_{4L} with biasing resistors R_{3L}/R_{4L} behave as current sources for differential signals and as diodes for common mode signals, thus attenuating the latter.

Applying the Barkhausen criteria on the impedance seen by the crystal resonator, it can be shown that its real part is negative and can sustain oscillations when :

$$\omega > \frac{(1+\eta).g_m}{C_s} \cdot \sqrt{\frac{1}{R.g_m - 1}} \qquad (4.5)$$

With η and g_m being respectively the body effect factor and the transconductance of M_{1L}/M_{2L} and R the resistance of R_{3L}/R_{4L}.

Once Eq. 4.5 is fulfilled we must guarantee that the oscillator will not oscillate in its original relaxation mode. This will occur when the imaginary part of the impedance seen by the crystal resonator is null. The frequencies at which this condition is fulfilled are:

$$\omega = \frac{(1+\eta).g_m}{C_s} \cdot \sqrt{\frac{C_s}{(1+\eta).(2.C_0 + C_L)} - 1} \qquad (4.6)$$

To avoid the relaxation mode one must guarantee that the Eq. 4.6 has no real solution. Then the only possible oscillation will occur at crystal parallel resonance ω_p.

From Eqs. 4.5 and 4.6 one can deduce that C_S must satisfy the following condition in order to get an oscillation at the crystal parallel resonance only:

$$\frac{(1+\eta).g_m}{\omega_p} \cdot \sqrt{\frac{1}{R.g_m - 1}} < C_s < (1+\eta).(2.C_0 + C_L) \qquad (4.7)$$

Depending on the crystal frequency and ESR and package shunt capacitance, it may be necessary to tune C_S in order to have Eq. 4.7 fulfilled for all g_m values. If C_S is too high the oscillator may oscillate in relaxation mode at a different frequency than the resonance of the crystal. If C_S is too low the real part of the impedance seen by the crystal will be positive at the crystal parallel resonance, preventing the circuit to oscillate.

The current consumption of the LP oscillator with its biasing is 30 µA. By changing to a single architecture and using N-type MOS transistors to create the negative-Gm, this current could be greatly reduced.

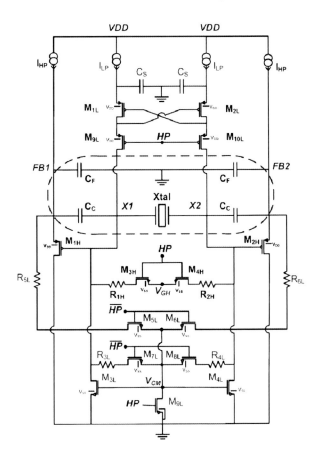

Fig. 4.3 The dual-core oscillator

4.2.4 The Dual Core

The dual-core architecture shown in Fig. 4.3 includes all the devices from HP core and the LP core shown in Figs. 4.1 and 4.2. Some additional transistors needed to efficiently switch between both cores have been added, namely M_{3H}/M_{4H}, M_{5L}/M_{6L}, M_{7L}/M_{8L} and M_{9L}/M_{10L}. Apart from the crystal and the tank capacitors C_C and C_F, each core has its own active devices and current sources allowing each of them to be independently optimized.

The detailed operation of the circuit will be described in the following sections.

When the dual core is configured in HP mode (Fig. 4.4) the two current sources I_{LP} are switched off, the biasing resistors R_{5L}/R_{6L} are disconnected and the active load transistors M_{3L}/M_{4L} are set to high impedance. As a result the cross-coupled pair M_{1L}/M_{2L} of the LP core should no longer be active. However, due to the oscillation, significant AC signals exist at the crystal nodes X_1/X_2 and may create currents across M_{1L}/M_{2L} which would in turn cause losses and degrade the low-frequency offset phase noise. Thus transistors M_{9L}/M_{10L} were added to fully isolate the HP core from the above devices allowing to meet the phase noise specifications for GPS applications.

Fig. 4.4 The dual core operated in HP mode

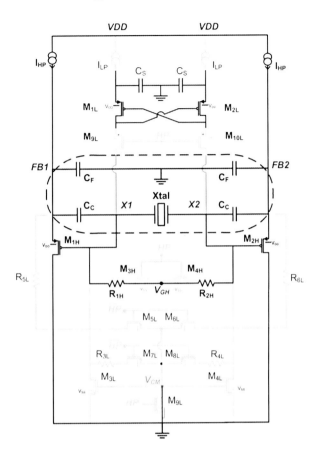

When the dual core is configured in LP mode (Fig. 4.5) the two current sources I_{HP} are switched off and transistors M_{3H}/M_{4H} disconnect the biasing voltage V_{GH} from the crystal nodes X_1/X_2. Transistors M_{1H}/M_{2H} are no longer active, leaving nodes FB_1/FB_2 at high impedance. In this way C_C and C_F are viewed by the crystal as connected in series. The common mode voltage V_{CM} is applied to nodes FB_1/FB_2 through resistors R_{5L}/R_{6L} in order to correctly bias the PMOS varactors of the fine tank C_F. As the amplitude of the voltages at nodes FB1/FB2 and X_1/X_2 is low, the threshold voltage of M_{1H}/M_{2H} is not reached and those transistors do not interfere with the oscillation.

4.3 System Description

The complete system entails the above dual-core oscillator, a fractional divider and a built-in temperature sensor as can be seen in Fig. 4.6. Also shown but not covered in this paper are the clock tree and the clock buffers towards the internal frequency synthesizers and towards external companion chips.

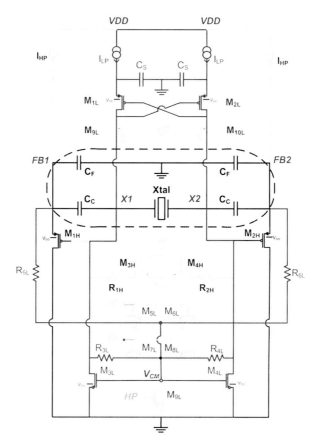

Fig. 4.5 The dual core operated in LP mode

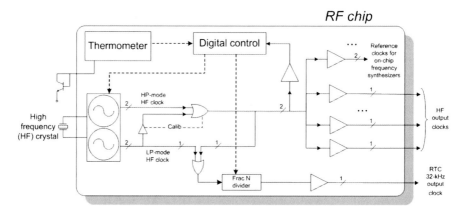

Fig. 4.6 The dual-core reference block diagram

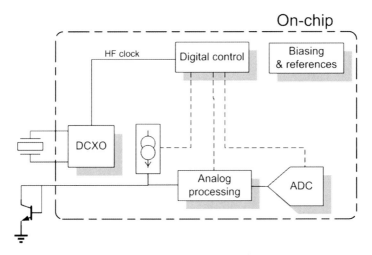

Fig. 4.7 Built-in thermometer diagram

4.3.1 The Fractional Divider

The 32-kHz clock is generated from the 26-MHz oscillator thanks to a fractional divider. It is used in both modes since the RTC clock is always present on the platform.

When switching to the LP mode the relaxation oscillator load capacitance is set to its minimum value. This load variation shifts the RTC frequency which must be re-tuned to the nominal value using the fractional divider ratio. The divider is supplied by a 1.2-V voltage generated from the 1.8-V supply through a diode-connected MOS. This supply voltage allows to use GO1 transistors to minimize capacitances and thus lower the consumption.

The main contributor to the divider consumption is the first divide-by-2 stage. The use of more advanced CMOS technologies will allow to further reduce this consumption.

A 14-bit fractional code allows to tune the frequency over a ± 600 ppm range with a resolution of 0.03 ppm.

4.3.2 Built-in Thermometer

The diagram of the built-in thermometer is shown in Fig. 4.7. It uses an external forward-biased silicon PN junction as temperature sensor. By feeding two different currents through this diode, the temperature can be calculated by measuring the V_{BE} voltage difference at both biasing.

The on-chip blocks used in the thermometer are:

- The reference current block which generates the two reference currents.
- A switched capacitor block which performs the analog processing of the two voltages taken from the PN diode.

- A digital control block which drives the reference currents and the switch capacitors sequencing.
- An 11-bit SAR ADC (Analog to Digital Converter) which converts the output voltage of the analog processor into a digital word.

As the temperature time constant is of the order of seconds, a measurement made each second is enough to have a good tracking of thermal variations. A few milliseconds are needed to complete a measurement, leading to an average current consumption of the thermometer of only a few μA.

4.3.3 Calibrations

The system requires three factory calibrations, one for the thermometer and two others for the two oscillator cores.

The thermometer requires a calibration at room temperature to cancel the offset of the signal processing chain. The gain of the thermometer being by design accurately known, this single-point calibration is enough to guarantee an accurate measurement over the ambient temperature range from $-30°C$ to $+85°C$. The offset is calculated from the measurements given by two thermometers. The first one is the internal one (to be calibrated) connected to an external diode sensor (see Fig. 4.7). The second one is an instrumentation thermometer whose sensor is placed close to the previous diode.

The two oscillator cores require two initial frequency calibrations to deal with the process variations of the crystal, of the tank capacitors and of the PCB lines:

1. First, in HP mode, the HF clock is tuned to 26 MHz by varying the coarse and fine capacitance words in an iterative manner.
2. Then, in LP mode, the coarse and fine capacitances are changed to their predefined LP values in order to lower the load capacitance. The HF clock frequency is measured and the fractional divider ratio is calculated in order to have the nominal 32,768 Hz on the RTC clock. This second frequency calibration is very fast since it does not use any iteration loop but only a frequency measurement.

4.4 Measurement Results

4.4.1 Process and Temperature Variations

One major benefit of the dual core is that ageing or temperature variations will produce similar effects on both oscillation modes due to common tank capacitors and crystal. This is especially true for the temperature-dependent frequency

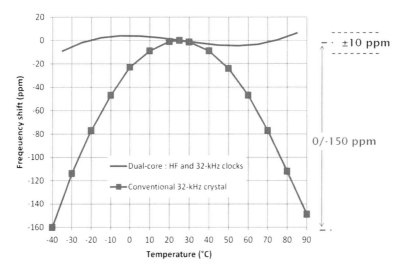

Fig. 4.8 Dual core temperature sensitivity

changes which lay mainly in the crystal. As shown in Fig. 4.8 the RTC clock temperature sensitivity is greatly improved since it inherits the sensitivity of the HF crystal. This feature can be used at the application level in conjunction with the temperature sensor to improve the tracking of clocks when switching from the RTC clock back to the HF clock.

4.4.2 The Thermometer Calibration

The Fig. 4.9 shows the measured temperature after calibration and the associated error after a temperature calibration was performed at room temperature and nominal supply voltage. The X-axis is the ambient temperature, the Y-axis the measured temperature.

The error after calibration does not exceed [−0.6°C, +0.6°C] over the full [−30°C, +85°C] ambient temperature range.

4.4.3 The Oscillator with Temperature Compensation

Using the thermometer with the off-chip PN diode sensor, the XO frequency has been corrected by changing the fine tuning capacitance. The output code of the on-chip ADC is read and used by the test program to tune the oscillator via a lookup table. Figure 4.10 shows that the frequency variations over the full temperature range have been reduced by a factor of 10 down to ±1.2 ppm.

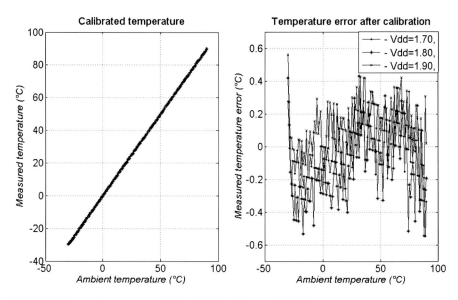

Fig. 4.9 Measured temperature (*left*) and calibration error (*right*)

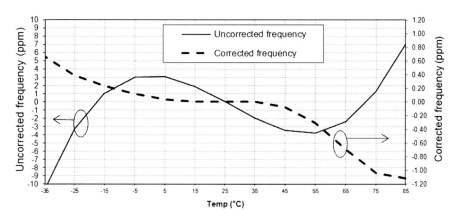

Fig. 4.10 Measured corrected frequency

4.4.4 Frequency Shift Between Oscillator Cores

First, the frequency calibrations are performed at room temperature (25°C) and nominal voltage (1.8 V) for the HP core and for the LP core. The coarse/fine capacitor bank and fractional divider settings found then are kept unchanged in all the subsequent switching between HP and LP mode. The ambient temperature and the supply voltage are varied and the output frequency is measured for each mode.

Fig. 4.11 Frequency shift error

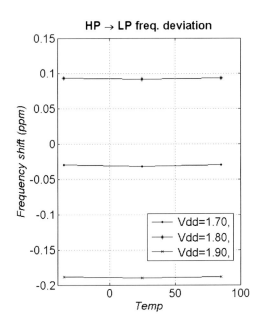

Figure 4.11 shows the frequency shift in ppm between LP-mode frequency and HP-mode frequency after the circuit has been switched to LP mode. As expected, the clock frequency is almost independent of the temperature since both cores undergo the same variations on the crystal and the tank capacitors. However, due to the different topologies between the two cores, the frequency shift varies with the supply voltage but this sensitivity is very small (< 0.2 ppm) and compatible with the application.

The measured transient behavior of the RTC clock when switching from HP mode to LP mode is shown in Fig. 4.12.

4.4.5 HP Mode Noise Quality

The Single Side Band (SSB) phase noise in HP mode with a 26-MHz crystal is shown in Fig. 4.13a. Cellular 2G/3G/4G as well as WLAN requirements are fulfilled with the −148.7 dBc/Hz phase noise at 100 kHz offset, while GPS requirement is also respected with −61 dBc/Hz @ 1 Hz offset. The significant margin permits to cope with process, voltage and temperature (PVT) variations.

The Single Side Band (SSB) phase noise measured in HP mode at 52 MHz with a 52 MHz crystal is shown in Fig. 4.13b. After a divider by 2 that will provide an improvement of a few dB, the phase noise at 26 MHz will be close to the one obtained with a 26 MHz crystal.

The oscillator demonstrates similar noise performance for both crystals frequencies giving more flexibility to address various platform needs.

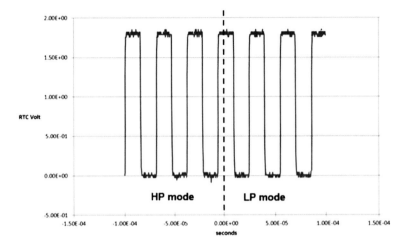

Fig. 4.12 Measured RTC clock transition

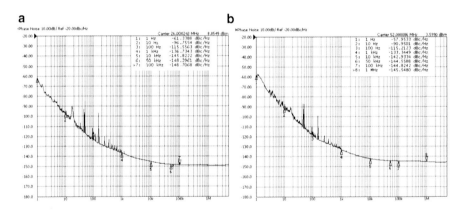

Fig. 4.13 HP mode phase noise (**a**) 26-MHz crystal (**b**) 52-MHz crystal

4.4.6 Performance Summary

The main performance of the circuit is summarized in Table 4.1. As previously mentioned one can see that the temperature stability is very good and is the same for both clocks.

An additional benefit of using a high-frequency crystal is the very fast start-up time allowing to get the RTC clock in a few milliseconds whereas 32 KHz crystals need about 1 s to be fully stabilized.

The chip, fabricated in ST-Microelectronics 65 nm high-performance CMOS process is shown in Fig. 4.14. The silicon active area of the dual-core oscillator with its tank, its biasing and the fractional divider is 0.17 mm^2.

Table 4.1 Performance summary

Parameter	HP mode	LP mode	Unit
Technology	65-nm CMOS		n/a
Supply voltage	1.8		V
Supply current	700[a]	65[b]	µA
Start-up time for 0.1 ppm accuracy	3		ms
Start-up time for 10 ppm accuracy	1.6		ms
Phase noise @ 1 Hz	−61.3	n/a	dBc/Hz
Phase noise @ 100 Hz	−115.5	n/a	dBc/Hz
Phase noise @ 1 kHz	−135.7	n/a	dBc/Hz
Phase noise @ 100 kHz	−148.7	n/a	dBc/Hz
Phase noise @ 1 MHz	−149.3	n/a	dBc/Hz
Phase noise in 10 kHz–10 MHz band	−78.0	n/a	dBc
Period jitter in 500 MHz bandwidth	± 0.18	± 40	ns
Uncompensated stability over −35/+85 °C	± 10		ppm
On-chip compensated stability over −35/+85 °C	± 1.5		ppm
Frequency step (coarse/fine)	3/0.03	0.08	ppm
Frequency tuning range (coarse/fine)	± 40/± 25	± 600	ppm
Power-supply sensitivity	± 0.2	± 1	ppm/V

[a]Includes the oscillator and its biasing
[b]All included: the oscillator, its biasing, the $\Sigma\Delta$ divider and the output buffer on a 10 pF load

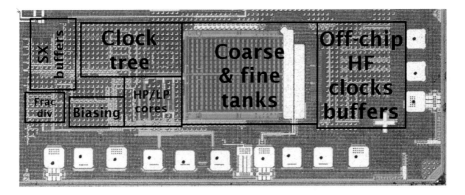

Fig. 4.14 Die photo of the dual core oscillator

4.5 Conclusion

A new oscillator architecture able to produce from a single crystal the two reference clocks needed on a mobile application platform has been presented. The design goal was to assess the feasibility of switching between two different oscillators architectures around the same crystal and the same tank capacitors. The behavior over temperature was also addressed.

The cost of the final application can be reduced thanks to the suppression of one crystal and its associated PCB area. The fully reconfigurable architecture of the two modes of operation allows to easily optimize each mode independently from the other one.

The RTC clock temperature sensitivity is greatly improved since it now depends on the HF crystal. The re-use of the same crystal and tank capacitors also allows to benefit from the HF clock factory calibration to get a calibrated RTC clock with a very small additional test time.

At the application level the better tracking of both clocks over temperature and process variations means an easier synchronization when switching between the stand-by and the high-power states. Finally, as the oscillator starts up in HP mode, the RTC clock start-up time is much faster than it would be when using a 32-kHz crystal.

It was also demonstrated that good design trade-offs allow to meet phase noise specifications for both GPS and cellular/WLAN applications.

Acknowledgements The authors wish to warmly thank Olivier Berchaud from ST-Ericsson for the efficient validation of the circuit.

References

1. D. Griffith, F. Dulger, G. Feygin, A.N. Mohieldin, P. Vallur, A 65 nm CMOS DCXO system for generating 38.4 MHz and a real time clock from a single crystal in 0.09 mm^2, in *Radio Frequency Integrated Circuits Symposium (RFIC), 2010* IEEE, (IEEE, 2010), Anaheim, CA, USA, pp. 321–324
2. D. Ruffieux, A high-stability, ultra-low-power quartz differential oscillator circuit for demanding radio applications, in *Proceedings of the 28th European Solid State Circuits Conference 2002*, (ESSCIRC, 2002), Firenze, Italy, pp. 85–88

Chapter 5
UHF Clocks Based on Ovenized AlN MEMS Resonators

Augusto Tazzoli and Gianluca Piazza

Abstract This work describes temperature compensated UHF oscillators based on AlN MEMS resonators with co-integrated heaters. Two approaches are proposed and experimentally verified. In the first, the heater is used to maintain the resonators temperature, and hence its frequency, constant. In the second, the heater is used as a tuning knob to compensate for both the temperature dependency of the resonator and its driving circuitry. With this latter approach, a 586 MHz oscillator was shown to exhibit a temperature stability of 1.7 ppm from $-45°C$ to $85°C$, with phase noise better than 91 and -160 dBc/Hz at 1 kHz and 40 MHz offsets, respectively.

5.1 Introduction

Microelectromechanical system (MEMS) resonators have emerged as a promising alternative to bulky and unintegrable quartz-crystal and SAW resonators. Small-form factor, high frequency of operation, and co-integration with CMOS circuits are some of the main features making MEMS resonators the best candidate for the implementation of compact and multi-frequency banks of high-quality-factor mechanical elements for reconfigurable local oscillators and frequency synthesizers [1, 2].

Aluminum Nitride (AlN) Contour-Mode Resonators (CMRs) [3, 5] have emerged as one of the most promising solutions in enabling the fabrication of multiple frequencies (10 MHz–10 GHz) and high-performance resonators on the same silicon chip. This is made possible by the high transduction efficiency of the piezoelectric film, which translates to low values of achievable motional resistance (tens of ohms), and the ability to be integrated with conventional electronics.

A. Tazzoli (✉) • G. Piazza
Carnegie Mellon University, 5000 Forbes Avenue, Pittsburgh, PA, USA
e-mail: atazzoli@andrew.cmu.edu; piazza@ece.cmu.edu

A. Baschirotto et al. (eds.), *Frequency References, Power Management for SoC, and Smart Wireless Interfaces: Advances in Analog Circuit Design 2013*, DOI 10.1007/978-3-319-01080-9_5, © Springer International Publishing Switzerland 2014

Nonetheless, modern telecommunication systems require oscillators that are stable over a wide range of parameters and especially versus temperature [5]. In high precision commercial oscillators either temperature compensated crystals (TCXO) [6, 7] or oven stabilized devices (OCXO) [8] are used. Uncompensated AlN MEMS based oscillators suffer instead from large temperature dependence and their frequency exhibits a linear dependence on temperature of about 28 ppm/K [9]. Prior works have shown that at the MEMS scale [10, 11] low power heaters can be co-integrated with low frequency resonators and few mW (instead of Ws used in OCXO) can be used to thermally stabilize the resonator. A 623 MHz thermally tunable AlN resonator that uses a separate portion of the device for heating has been described in [12].

In this work we demonstrate temperature stable Ultra High Frequency (UHF) oscillators based on UHF ovenized AlN CMRs based on two different controlling techniques: the resonator ovenization, and the pulling of the resonator center frequency through the integrated heater. With the first method the resonator temperature is kept constant despite changes in the environment temperature, whereas in the second case the oscillator frequency is kept stable by pulling the resonator center frequency and flowing a pre-determined (look-up table) current value into the integrated heater. Pros and cons of the two techniques are discussed.

5.2 Device Operation and Fabrication

A CMR is composed of an AlN film sandwiched between two metal electrodes, see Fig. 5.1. When an alternating-current signal is applied across the thickness T of the AlN film, a contour extensional mode of vibration is excited through the equivalent d_{31} piezoelectric coefficient of AlN.

Given the equivalent mass density, ρ_{eq}, and Young's modulus, E_{eq}, of the material stack that forms the resonator, the center frequency, f_0, of the laterally vibrating mechanical structure is set by the period, W, of the metal electrode patterned on the AlN plate and can be approximately expressed as:

$$f_0 = \frac{1}{2W}\sqrt{\frac{E_{eq}}{\rho_{eq}}}. \quad (5.1)$$

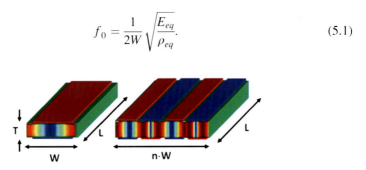

Fig. 5.1 Schematic of the contour extensional mode of vibration excited through the equivalent d_{31} piezoelectric coefficient of AlN of a single resonant element (*left*) and a multifinger device (*right*). The resonator geometrical dimensions are indicated in the figure

5 UHF Clocks Based on Ovenized AlN MEMS Resonators

Fig. 5.2 SEM pictures of fabricated ovenized AlN contour mode resonators: 250 MHz (**a**), 590 MHz (**b**), 1.1 GHz (**c**)

The other three geometrical parameters, *i.e.*, thickness, T, length, L, and number of electrodes, n, set the equivalent electrical impedance of the resonator and can be designed independently of the desired resonance frequency. Further details on the resonators operating principle can be found in [2].

The resonators of this work are formed by a 1 μm thick Aluminum Nitride film sandwiched between two metal layers and integrate microscale heaters in the bottom or top layer. In the first design the bottom plate of traditional LFE-F excited resonators [13] was modified in a serpentine (the heater) [14], which is made of 50 nm thick and 2 μm wide series-connected lines of Platinum (Pt) separated by a 2 μm gap. SEM pictures of the fabricated resonators with a resonant frequency of 250 MHz, 590 MHz, and 1.1 GHz are shown in Fig. 5.2.

In this design, the serpentine itself acts as the floating bottom electrode in the LFE-F excitation technique. The presence of this bottom metal layer serves the dual purpose of enhancing the electromechanical coupling in the resonator and heating the body of the device when a current is flown through it. Furthermore, the same serpentine will be used as a temperature sensor to control and keep the resonator temperature constant, as shown in Sect. 5.1. The top metal layer (100 nm thick Aluminum) constitutes the multi-fingered top electrode, whose pitch is used to set the frequency of vibration of the device and effectively excite the contour-extensional mode of vibration in the piezoelectric resonator.

In the second design the heater is integrated on the same layer of the top RF electrodes (see Fig. 5.3). The heater is designed all around the resonator, and the small tethers help keeping the generated heat confined within the resonator body improving the heating efficiency.

The AlN LFE contour-mode resonators with heaters integrated on the bottom layer were fabricated in a four-mask post-CMOS-compatible microfabrication process (see Fig. 5.4). The Pt serpentine/floating electrode was first patterned by lift-off on top of a high resistivity silicon wafer. The 1 μm thick AlN film was then sputtered deposited using a Tegal AMS 2004 SMT (OEM Group) and its quality was optimized to achieve rocking curve values as good as 1.2°. In order to access the heater on the bottom layer, VIAs were opened in the AlN film by wet etching in phosphoric acid (H_3PO_4). Optical lithography and lift-off were performed for the

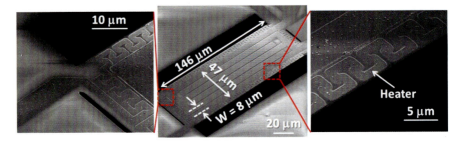

Fig. 5.3 SEM images of a 586 MHz resonator. The serpentine pattern placed all around the resonator body is the high-efficiency heater used to pull the resonator frequency

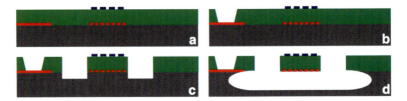

Fig. 5.4 Four mask fabrication process: (**a**) sputter deposition of Pt (50 nm) serpentine, AlN layer and Al top electrodes; (**b**) via opening to the serpentine pads in H3PO4; (**c**) dry etching of AlN in Cl2-based chemistry; (**d**) XeF$_2$ dry release of the AlN resonator

definition of the top Aluminum (Al) electrodes. The in-plane dimensions of the resonators were defined by dry etching (Trion Inductively-Coupled Plasma Phantom III) of the AlN film in Cl$_2$-based chemistry using photoresist as a mask. Finally, the device was released from the silicon substrate by isotropic dry etching in XeF$_2$.

A similar process was adopted for the fabrication of resonators integrating the heater on the top layer, but this new layout allowed simplifying the manufacturing process since only three masks are required since there is no need to open VIAs to access the bottom-layer as in the previous design.

5.3 Resonator Ovenization: Design and Characterization

As anticipated, uncompensated AlN-MEMS-based oscillators suffer from large temperature dependence and their frequency exhibits a linear dependence on temperature of about −28 ppm/K. This shift is primarily induced by softening of the AlN and metal electrodes (Al) and the TCE of Al with temperature. An example of the resonant frequency shift of a 1 GHz resonator with temperature is shown in Fig. 5.5a.

5 UHF Clocks Based on Ovenized AlN MEMS Resonators

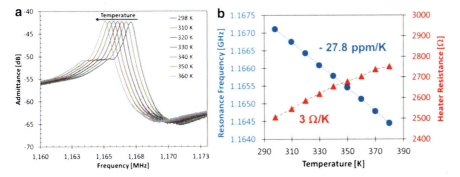

Fig. 5.5 (**a**) $Y_{11}(f)$ (*admittance in dB*) measured in a Lakeshore probe station at different temperatures for the 1.1 GHz ovenized resonator with heater on the bottom layer. (**b**) Resonance frequency and heater resistance shift vs. temperature of uncompensated 1.1 GHz AlN MEMS resonator vs. temperature. A TCF of about −28 ppm/K and a TCR of about 3 Ω/K were recorded

Such dependence with temperature is clearly unacceptable for modern oscillators and a way to keep the resonance frequency within few ppm from −40 °C to +85 °C is needed. It has been shown that at the MEMS scale [10], low power heaters can be co integrated with low frequency resonators and few mW (instead of the hundreds of mWs used in traditional OCXO) can be used to thermally stabilize the resonator. To address the issue of temperature shifts in the CMR technology, we integrated a heater directly into the body of the resonator and kept the oscillator frequency constant by either adopting a resonator ovenization technique similar to what is done with OCXO (analog temperature controller), or pulling the resonator frequency using a microcontroller (digital temperature controller, see next section).

Different variations of the serpentine, in terms of line width, spacing and orientation with respect to the RF electrodes, have been designed, fabricated, and tested. In order to prove the effectiveness of the concept, this ovenization technique was successfully verified in devices operating at 250 MHz, 580 MHz, and 1.1 GHz (see Fig. 5.2) integrating the heater on the bottom layer. Thanks to this innovative design, it has been possible to obtain ovenized resonators with higher performances, *i.e.* high quality factors (Q up to 2,000 @ 1.1 GHz) and k_t^2 (electro-mechanical coupling up to 0.6 % @ 1.1 GHz), than any prior implementation [12].

Resonators with heaters patterned around the top RF electrodes showed a resonant frequency of ~586 MHz at room temperature, with an unloaded $Q > 4,000$, a $k_t^2 > 0.8$ %, and a series resistance of 100 Ω. Further results on these devices can be found in [15].

The efficacy of the proposed solutions was characterized by first using the resonator itself as a thermometer and monitoring its frequency shifts when subjected to external temperature variations induced in a controlled Lakeshore probe station (Fig. 5.5a). The temperature dependence of the serpentine resistance was simultaneously measured with an Agilent B1500 Semiconductor Parameter Analyzer (Fig. 5.5b).

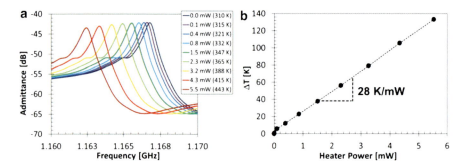

Fig. 5.6 (a) $Y_{11}(f)$ measured at different supplied heater powers (i.e. temperatures) for the 1.1 GHz resonator with the heater on the bottom layer. Minimal change in the device Q and k_t^2 ensures that the resonator will operate in an oscillator even when heated at the max temperature of 443 K. (b) Measurement of the temperature increment (from $T_{amb} = 310$ K) as a function of the power dissipated by the integrated heater

A representative 1.1 GHz resonator (heater on the bottom layer) is reported here, but all devices exhibited similar temperature dependence. A temperature coefficient of frequency (TCF) of -27.9 ppm/K and a temperature coefficient of the heater resistance (TCR) of 3 Ω/K were extracted (bottom layer Pt heaters).

The effectiveness of the integrated micro-oven was then tested in air (at 310 K) by applying different power levels to the heater, and recording the resonator admittance (Fig. 5.6a, b). According to the calibration curve shown in Fig. 5.6b, it was possible to extract the temperature increase *vs.* the power supplied to the heater and obtain a temperature rise factor of 28 K/mW. This translates to the ability of operating a resonator at around 100 °C with few mW. Further investigations on FEM transient simulations and comparison with measured results have been shown in [14] by the same authors.

5.4 Temperature Compensated AlN MEMS Oscillator

5.4.1 Analog Temperature Controller

The good performance in terms of quality factor and coupling coefficient of the fabricated micro-ovenized resonators permitted us to build self-sustained oscillators based on AlN MEMS resonators, temperature compensated through a simple analog feedback circuit. A Pierce-like oscillator was made with a commercial GaAs p-HEMT (ATF551M4) (see Fig. 5.7a), whereas a simple feedback temperature controller (Fig. 5.7b) based on a Wheatstone bridge (R1-R2-R3-R4) was designed to keep the resonator at a constant temperature exploiting the same serpentine (R3) as a heater and sensor. The control circuitry is based on using the operational amplifier to measure the imbalance in the bridge and output a

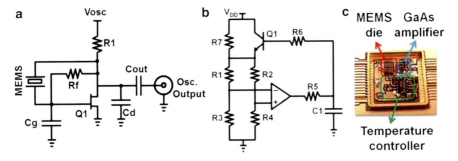

Fig. 5.7 Pierce oscillator (**a**) and temperature controller (**b**) schematics. The packaged oscillator is shown in (**c**)

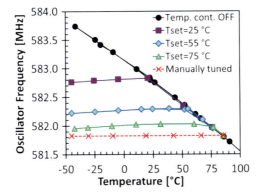

Fig. 5.8 Oscillator frequency vs. temperature at different oven set points (Tset)

proportional voltage to self-balance the bridge (*i.e.*, $V_+ = V_-$). The output voltage is then filtered by an RC low pass filter (R5-C1) and used to control the BJT Q1. The amount of current flowing through the BJT (and then partially through the serpentine) is used to heat the device so that a stable resonator temperature is attained. The circuitry and the MEMS die were packaged in a 25 × 25 mm^2 package (Fig. 5.7c) and hermetically sealed by Vectron International before testing inside a Tenney TPS climatic chamber.

The temperature stability of the proposed solution is shown in Fig. 5.8, which compares the oscillator frequency *vs.* temperature for different oven set points (R4). A temperature stability of ~125 ppm was recorded in the whole temperature range with power dissipation lower than 10 mW. A decrease of the oscillator frequency was observed for temperatures lower than −20°C, due to the temperature sensitivity of the GaAs transistor, as confirmed by ADS simulations. The RF performance, short term stability, and temperature sensitivity of the packaged oscillator were tested placing the packaged oscillator connected on an ad-hoc test fixture inside a Tenney TPS climatic chamber. The oven set point was set in order to keep the resonator temperature constant at +85°C (the maximum of the temperature range of interest).

The oscillator short-term stability was measured with an Agilent E5052B Signal Source Analyzer. Stable Phase Noise (PN) was recorded over the entire temperature range. PN values better than -93 dBc/Hz were shown at 1 kHz offset frequency, with a floor better than -160 dBc/Hz at an output RF power level of -6.2 dBm. Floor values better than -170 dBc/Hz were obtained increasing the oscillator bias voltage. However, a slight degradation of the close in was observed, likely due to the increase of the GaAs transistor noise up-converted to the carrier frequency. Further experimental results can be found in [16].

Despite the low power consumption of this approach compared to traditional OCXO, the ovenization of only the MEMS resonator is clearly a limit of this technique leaving the oscillator sensitive to any temperature dependence of the sustaining circuitry. However, temperature stability as good as 2 ppm over the entire temperature range was attained by manually tuning the temperature controller set point (see Fig. 5.8, "manually tuned" curve). This result demonstrates the possibility of achieving good temperature stability in high frequency AlN CMR based oscillators by ovenizing the resonator, consuming low power, and still attaining good short term stability, and inspired the development of the digital temperature controller described in the next section.

5.4.2 Digital Temperature Controller

A digital temperature controller circuitry based on a microcontroller circuit was then implemented to form a Microcontroller Compensated MEMS Oscillator (MCMO), which yields a tighter control on the overall circuit temperature stability. The sustaining oscillator was realized with the same Pierce circuit shown in Fig. 5.7a and it was integrated with a microcontroller based temperature controller circuit (Fig. 5.9a) to form the MCMO. The use of a MCMO permits full temperature compensation of the oscillator over the temperature range of interest, eliminating also the shifts induced by the GaAs transistor and the other discrete components. The MCMO stores in a look-up table the corresponding amount of current needed by the heater to keep the oscillator frequency constant in a temperature range from $-25\,^{\circ}$C to $+85\,^{\circ}$C and therefore can compensate for frequency shifts of any origin.

The ambient temperature needed as a reference for the microcontroller is sensed by a digital thermometer placed with the microcontroller inside a 20×20 mm^2 package (Fig. 5.9b). The microcontroller calculates from a 501-entry look-up table the current needed by the heater. The digital value is then transferred to a 12-bit DAC and converted into a current by OP1-Q2-R2. The communication between the microcontroller, the temperature sensor, and the DAC relies on an I^{2}C bus. A digital solution was chosen to avoid problems with ADC converters sensitivity to temperature variations. The packaged MCMO was tested inside a Tenney climatic chamber. The look-up table was calibrated to keep the oscillator frequency around 585.920 MHz (Fig. 5.10) and a temperature stability of 1.7 ppm was recorded in the whole temperature range with a total power dissipation of about 17 mW.

5 UHF Clocks Based on Ovenized AlN MEMS Resonators

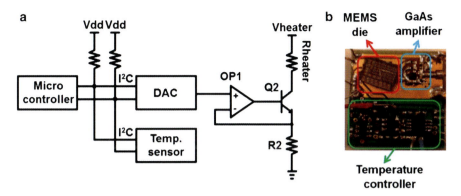

Fig. 5.9 Schematic of the microcontroller based temperature controller (**a**). The PCB realization is shown on the right (**b**). The PCB size is 20 × 20 mm^2

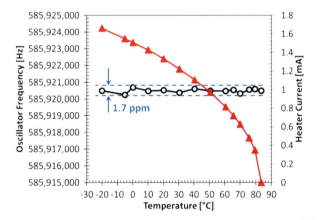

Fig. 5.10 Oscillator frequency and heater current vs. temperature measured during the microcontroller calibration procedure

The oscillator short-term stability was measured with an Agilent E5052B SSA. Phase noise values better than −91 and −160 dBc/Hz were measured over the entire temperature range at offset frequencies of 1 kHz and 40 MHz, respectively (see Fig. 5.11).

5.5 Conclusions

UHF oscillators temperature compensated based on ovenized CMRs were presented. We proposed two different layouts to integrate efficient heaters on the resonator body (bottom or top layer) and we demonstrated two different techniques to stabilize the oscillator frequency based on analog and digital implementations.

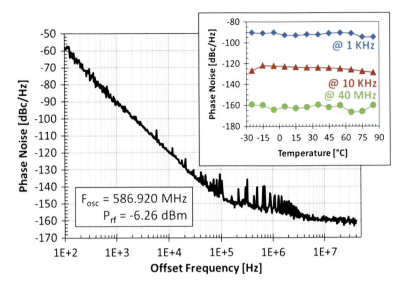

Fig. 5.11 Oscillator phase noise at $P_{RF} = -6.26$ dBm. Phase noise values better than -91 dBc/Hz @ 1 kHz and -160 dBc/Hz at 40 MHz offsets were recorded. Inset: Phase noise evolution at offset frequency of 1 KHz, 10 KHz, and 40 MHz vs. temperature

The analog implementation exploited the patterned serpentine as a high efficiency heater and temperature sensor with the aim of ovenizing the MEMS resonator and keeping its temperature constant. A value of ~125 ppm was recorded for the oscillator frequency stability over the temperature range $-40 \div +85$°C with a power consumption of less than 10 mW. Even if power efficient, this technique leaves the oscillator sensitive to any temperature dependence coming from the electronic circuitry.

This problem was overcome in the 586 MHz Microcontroller Compensated MEMS Oscillator (MCMO) where the oscillator frequency is kept constant pulling the resonator resonant frequency through the calibrated flow of current into the MEMS heater. The implementation of a MCMO permits full temperature compensation of the oscillator over the temperature range of interest, eliminating also the shifts induced by the sustaining electronic circuitry.

A complete characterization of the MCMO frequency and short-term stability was carried out over the entire temperature range, recording a frequency stability of 1.7 ppm with a maximum power consumption of about 17 mW, phase noise values better than -91 and -160 dBc/Hz at offset frequencies of respectively 1 kHz and 40 MHz.

These ultra-stable oscillators with low jitter and phase noise will ultimately benefit defense as well as commercial communication systems.

Acknowledgments This work was supported by the DEFYS DARPA Award # FA86501217264, "PiezoElectric Non Linear Nanomechanical Temperature and Acceleration Insensitive Clocks" (PENNTAC).

Authors would like to acknowledge Matteo Rinaldi and Nai-Kuei Kuo for devices fabrication and Vectron International, PA, USA, for packaging the oscillator.

References

1. C.Y.C. Nguyen, MEMS technology for timing and frequency control. IEEE Trans. Ultrason. Ferroelectr. Freq. Control **54**(2), 251–270 (2007)
2. G. Piazza, P.J. Stephanou, A.P. Pisano, Piezoelectric aluminum nitride vibrating contour-mode MEMS resonators. J. Microelectromech. Syst. **15**(6), 1406–1418 (Dec. 2006)
3. M. Rinaldi, C. Zuo, J. Van der Spiegel, G. Piazza, Reconfigurable CMOS oscillator based on multi-frequency AlN contour-mode resonators. IEEE Trans. Electron Devices **58**(5), 1281–1286 (2011)
4. M. Rinaldi, C. Zuniga, G. Piazza, 5 – 10 GHz AlN contour-mode nanoelectromechanical resonators. Micro Electro Mechanical Systems (MEMS), IEEE Hilton Sorrento Palace, Sorrento, Italy, 25–29 January 2009, pp. 916–919
5. J.R. Vig, Military applications of high accuracy frequency standards and clocks. Ultrason. Ferroelectr. Freq. Control IEEE Trans. **50**(5), 522–527 (1993)
6. Datasheets from Vectron International web-site: http://www.vectron.com/products/tcxo/tcxo_index.htm
7. Datasheet from SiTime Corporation web-site: http://www.sitime.com/products/datasheets/sit5000/SiT5000-datasheet.pdf
8. Datasheets from Vectron International web-site: http://www.vectron.com/products/ocxo/ocxo_index.htm
9. C.-M. Lin et al., Thermally compensated aluminum nitride Lamb wave resonators for high temperature applications. Appl. Phys. Lett. **97**(8), 083501 (2010)
10. C.M. Jha, M.A. Hopcroft, S.A. Chandorkar, J.C. Salvia, M. Agarwal, R.N. Candler, R. Melamud, B. Kim, T.W. Kenny, Thermal isolation of encapsulated MEMS resonators. J. Microelectromech. Syst. **17**, 175–184 (Feb 2008)
11. J. Salvia, R. Melamud, S. Chandorkar, S.F. Lord, T.W. Kenny, Real-time temperature compensation of MEMS oscillators using an integrated micro-oven and a phase lock loop. J. Microelectromech. Syst. **19**(1), 192–201 (2010)
12. B. Kim, R.H. Olsson, K.E. Wojciechowski, Ovenized and thermally tunable aluminum nitride microresonators, in *Ultrasonics Symposium (IUS)*, 2010 IEEE, 11–14 October 2010, Town & Country Inn & Convention Center, San Diego, CA, USA, pp. 974–978
13. C. Zuo, J. Van der Spiegel, G. Piazza, 1.05-GHz CMOS oscillator based on lateral-field-excited piezoelectric AlN contour-mode MEMS resonators. IEEE Trans. Ultrason. Ferroelectr. Freq. Control **57**(1), 82 (2010)
14. A. Tazzoli, M. Rinaldi, G. Piazza, Ovenized high frequency oscillators based on aluminum nitride contour mode MEMS resonators, in *IEEE International Electron Device Meeting (IEDM)*, December 2011, Hilton Washington and Towers, Washington, DC, USA, pp. 481–484
15. A. Tazzoli, N.-K. Kuo, M. Rinaldi, H. Pak, D. Fry, D. Bail, D. Stevens, Gianluca Piazza, A 586 MHz microcontroller compensated MEMS oscillator based on ovenized aluminum nitride contour-mode resonators. in *Proceedings of the IEEE International Ultrasonics Symposium (IUS)* 2012, Dresden, 7–10 October 2012, 1055–1058
16. A. Tazzoli, M. Rinaldi, G. Piazza, Ultra high frequency temperature compensated oscillators based on ovenized AlN contour-mode MEMS resonators. in *IEEE International Frequency Control Symposium* 2012, Renaissance Baltimore Harborplace Hotel, Baltimore, MD, USA, 21–24 May 2012, pp. 1–5

Chapter 6
Towards Portable Miniature Atomic Clocks

David Ruffieux, Jacques Haesler, Laurent Balet, Thomas Overstolz, Jörg Pierer, Rony Jose James, and Steve Lecomte

Abstract Accurate clocks play a fundamental role in modern communication systems, especially given the trend towards ever increasing data rates. Atomic clocks easily achieve sub-ppb stability, orders of magnitude better than even the best quartz and MEMS-based references. Driven by recent developments in photonics and MEMS technology, they now dissipate several tens of mW and are matchbox size. The challenge is how to improve on this and realize clocks that can be used in mobile phones and tablets. This paper will present recent progress towards this goal with an emphasis on the required signal processing and RF circuitry.

6.1 Introduction

The field of atomic clocks is currently undergoing a scientific and technical revolution. Technology based on MEMS components manufacturing shows the promise of having miniature (<1 cm^3) and low-power (<100 mW) atomic clocks (MACs). They are typically based on the coherent population trapping (CPT) scheme where the microwave interrogation is directly coupled to the laser such that the dimensional constrains linked to the microwave cavity are avoided [1]. MACs have the potential to be integrated in future battery-operated portable devices for communication, navigation, signal processing and many other mobile applications requiring ultra-stable timekeeping/frequency references.

The paper is organized as follow: the CPT interrogation principle is first discussed to derive the corresponding clock architecture and the high level circuit building blocks. The RF lock loop constituting circuits are then discussed, followed

---

D. Ruffieux (✉) • J. Haesler • L. Balet • T. Overstolz • J. Pierer • R.J. James • S. Lecomte
CSEM, Neuchâtel, Switzerland
e-mail: david.ruffieux@csem.ch

A. Baschirotto et al. (eds.), *Frequency References, Power Management for SoC, and Smart Wireless Interfaces: Advances in Analog Circuit Design 2013*, DOI 10.1007/978-3-319-01080-9_6, © Springer International Publishing Switzerland 2014

6.2 CPT Interrogated Atomic Clock Operating Principle

The coherent population trapping (CPT) interrogation [1] consists in modulating the bias current of a wavelength tuned VCSEL at an RF frequency (ν_{RF}) that is half the ground state hyperfine splitting of ^{87}Rb ($\nu_{HF}/2 = 3.417$ GHz). Figure 6.1 illustrates the principle. Let's consider that a ramp on the laser diode (LD) bias current sweeps the pumping photon energy (E = hν) producing two peaks of absorption corresponding to transitions between each of the two ground states toward the excited states (top plot on the right of Fig. 6.1). When amplitude modulation of the LD is added, the absorption spectrum is composed of each individual laser wavelength (carrier plus sidebands). For $\nu_{RF} = \nu_{HF}/2$, five absorption peaks are produced (middle plot on the right of Fig. 6.1). Within the central one, a transmission peak hundreds of Hertz wide and with a ~1 % contrast (CPT signal, shown on bottom plot in Fig. 6.1) is observed and this resonance provides the high-Q (>10^6) frequency reference of the MAC. To maintain the system locked around that peak and to provide a highly stable clock, the frequencies of both the RF modulating signal and VCSEL light must be controlled with feedback loops. In particular, locking of the RF frequency onto the atoms is obtained by frequency modulation of the RF carrier, which allows deriving a signed error signal used to form the RF frequency lock loop.

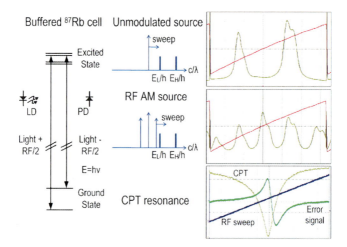

Fig. 6.1 Principle of CPT interrogation

6.3 Clock and ASIC Main Block Diagrams

Figure 6.2 shows a simplified schematic of the CPT atomic clock. The core of the system is the physics package (PP) comprising the Rb-cell, the laser (VCSEL) and photodetector (PD). The 3rd generation of the control ASIC hosts three main different control loops: the RF, laser and temperature lock loops.

The RF lock loop generates the 3.417 GHz RF signal used to modulate the amplitude (AM) of the VCSEL current. As mentioned previously, the RF signal is further frequency modulated (FM) so that the loop can be maintained locked and an error signal be generated from that obtained on the PD followed by proper demodulation. The laser lock loop provides the laser DC biasing current to ensure that its optical frequency is accurately tuned. A tiny amplitude modulated current is further added to adjust dynamically the DC biasing current with high resolution after demodulation of the corresponding signal obtained on the PD. The two loops output signals are combined with a bias-T on the laser side while they share a single PD and hence input signal. To maintain the above loop locked, the temperature of both the VCSEL and Rb-cell need to be tightly controlled, necessitating two additional thermostat loops. A controlled biasing current used to generate a magnetic field (B) within the Rb-cell is also implemented. An additional loop controlling the laser output power (P) might be added in a near future.

Figure 6.3 shows a detailed block diagram of the RF (blue) and laser (green) lock loops while Fig. 6.4 illustrates their working principles. The PD is followed by a transimpedance amplifier (TIA) and two coherent demodulators formed with a chopper, an integrator and a low-pass filter to retrieve the error signals generated with the help of the AM and FM modulations. A dynamic tracking current is generated for the laser loop with an OTA and applied in parallel to the laser diode DC bias and tiny amplitude modulated current. The demodulator output of

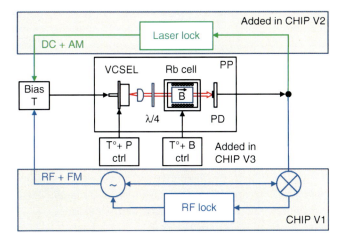

Fig. 6.2 High level block diagram of a CPT atomic clock

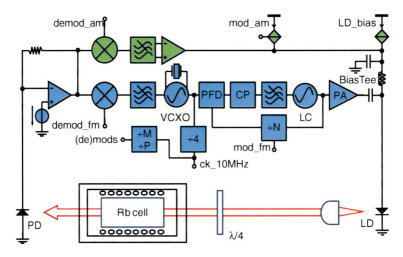

Fig. 6.3 Detailed block diagram of the RF and laser lock loops

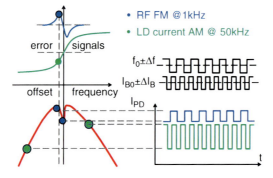

Fig. 6.4 Illustration of the two loops interrogation signals obtained with AM/FM

the RF loop fine tunes dynamically the frequency of a 40 MHz VCXO which is used as the reference of a fractional-N PLL generating the slightly frequency modulated 3.417 GHz signal that is buffered with a power amplifier before modulating the LD current. A 10 MHz output signal obtained after division of the VCXO signal by 4 is delivered externally.

The effect of the AM and FM signals is to probe alternatively on either side of the ^{87}Rb absorption and CPT transmission peaks respectively so as to derive a signed error signal allowing to lock both loops with the help of two integrators. The laser optical and the RF frequencies are hence adjusted so that they match the three energy level transitions of ^{87}Rb when the two probed signals are equal thus nulling the error signal. This is illustrated in Fig. 6.4.

6.4 RF Lock Loop Implementation

The left part of Fig. 6.5 shows a schematic of the transimpedance amplifier (TIA) while the coherent demodulator is depicted to the right. At low frequencies, the TIA works as a current conveyor (CCI) with transistor M_1 sourcing a current equivalent to the photocurrent generated on the photodiode (PD). This is obtained with the arrangement of M_1, M_2 and the biasing current I_B that impose a constant reverse biasing voltage across the PD whatever the photocurrent level. Above a frequency defined by the output conductance of M_2 and the large off-chip filtering capacitor C_F, ($\omega > g_{DS,2}/C_F$) the source conductance of transistor M_1 sets the transimpedance ($1/g_{mS,1}$) of the TIA. A modulated photocurrent above that frequency and up to that of the low pass pole defined by $g_{mS,1}/C_P$, with C_P, the junction and total parasitic capacitance present at the PD cathode and circuit interface node, will generate a voltage swing on that same node as the CCI feedback via M_1 is eliminated. In order to maximize the TIA gain and lower its noise figure, M_1 is implemented with fingered native near 0-V_T transistors that can be switched on/off in a binary weighted way to maximize the transimpedance for a given photocurrent level. It can be adjusted automatically by measuring a copy of the current circulating in M_2 (equal to I_B) that can be fed externally. As the transimpedance is increased, the gate to source voltage of M_1 is increased until the biasing current source I_B leaves saturation lowering in turn the mirrored current. The TIA added noise is negligible as M_1 is operated with a large overdrive voltage resulting in a much lower noise contribution (PSD $= 4 \cdot \gamma \cdot k \cdot T \cdot g_{mS}$) than the photodiode shot noise whose PSD is equal to $2 \cdot q \cdot I_{PD}$. The white noise of M_2 (PSD $= 2 \cdot q \cdot I_B$), which together with M_3 are operated in weak inversion (sub-threshold regime) to maximize the transconductance of M_3, is low pass filtered. The latter converts the voltage swing appearing at the PD cathode to a modulated current that is fed to the coherent demodulator implemented with N and PMOS switching pairs driven by the modulation signal as in a conventional active mixer or chopper with the

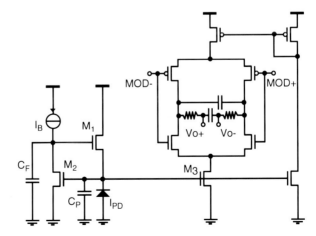

Fig. 6.5 TIA and demodulator transistor-level implementation

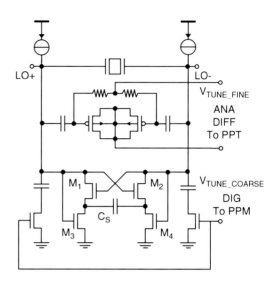

Fig. 6.6 VCXO transistor-level schematic

exception that the signal is injected in a complementary way before being chopped and integrated on an off-chip capacitor. The resulting saw-tooth waveform is further low pass filtered with off-chip passives implementing a cut-off at a few Hz delivering a slowly varying differential tracking signal (Vo+;Vo−) used to lock the VCXO PLL reference to the atomic transitions. The band-pass filter characteristic of the front-end avoids folding dominant noise components at harmonics of the modulating frequency despite the facts that the chopper is hard switched.

Figure 6.6 shows the VCXO implementation which is using a differential topology proposed in [2]. Cross-coupled transistor pair M_1 and M_2 forms the gain stage providing the negative differential transconductance $1/g_{m1} + 1/g_{m2} = 2/g_m$ similarly to LC oscillator implementations. In order to cope with the high DC impedance of the XTAL and prevent the circuit from latching at DC, a capacitor, C_S, should be added between the two sources of M_1, M_2 to cancel the positive feedback at low frequencies. Transistors M_3 and M_4 are used to set the common mode of the oscillator and provide negative feedback at low frequency to minimize any offset voltage across the resonator. Since all gate voltages are equal in the absence of oscillation, M_1, M_3 and M_2, M_4 are essentially diode connected in a cross-coupled configuration. This is exploited to implement an amplitude regulating loop in a way similar to the single-ended version proposed in [3]. The VCXO is tuned within ppm-level accuracy to 40.0 MHz with a bank of digitally controlled switched capacitor. Then a low gain differential analog varactor implemented with accumulation MOS devices connected to the filtered output of the demodulator is used for fine frequency tracking down to ppt (1E-12) level. A simple 3-bits differential ADC is used to sense the loop filter output voltage level so that the DCXO control can be modified to re-center the analogue varactor characteristic in case of e.g. XTAL ageing. The 40 MHz frequency was chosen since it is an integer multiple of the standard 10 MHz reference used in accurate

Fig. 6.7 RF VCO transistor-level schematic

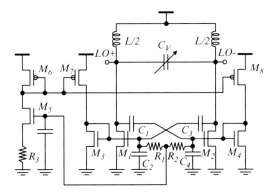

timing system. Choosing a higher frequency yields two advantages: the XTAL is much smaller (footprints available down to 1.2×1.0 mm^2); the divider ratio between RF and reference frequencies is reduced improving the in-band noise performance of the synthesizer.

The VCXO reference then drives a conventional fully integrated fractional-N PLL locked near 3.417 GHz. A prescaler by three is used to relax the multi-modulus divider switching speed while a 40-bits, 2nd order $\Delta\Sigma$ mash modulator clocked at 40 MHz is used for ppt-level frequency adjustment. A differential charge pump and loop filter is used for better noise immunity. Figure 6.7 shows a schematic of the 3.4 GHz VCO using an analog varactor topology similar to that of the VCXO. Capacitively cross - coupled transistors M_1, M_2 form a linearized negative transconductor that has to compensate for the tank losses. Decoupling gates and drains at DC yields LO signals with a constant common mode voltage, which can conveniently be set through the biasing of the integrated differential inductor mid-point. The choice of the capacitive attenuation factor ($1 + C2/C1$) allows an easy adjustment of the LO amplitude depending on how one wants to address the phase noise/power consumption trade-off. Amplitude regulation is implemented with the help of the two current mirrors M1, M3 and M2, M4 so that the average current drawn by the oscillator is precisely set by the current sources M7, M8. The latter currents are both copies of that of M6, which is set itself by the degenerated current mirror formed by M3-M4, M5, R3 and differential resistors R1, R2. This arrangement forms a PTAT reference current source setting the start-up current of the oscillator well above its critical current, leading to a quick start-up. Owing to the non-linearity of the two diode-connected NMOS transistors M3, M4, the common mode voltage at the gates of M1 to M4 has to decrease as the oscillations grow. This voltage is sensed with differential resistors R1, R2, filtered by C5 and fed back to the biasing loop, reducing the oscillator current until equilibrium is reached. The principle used for the amplitude regulation has long been proposed for quartz oscillator in a single - ended version [2] but proves very attractive for VCOs, since it minimize the number of noisy transistors (bias and amplitude regulation are both obtained with M3, M4, the noise of the other transistors can be minimized if they are operated in the strong inversion regime).

Fig. 6.8 PA and pre-PA transistor-level schematic

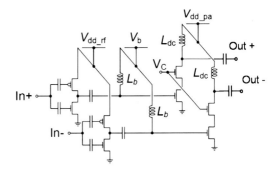

The power amplifier chain, depicted in Fig. 6.8, was designed in view of reaching an output power that can be varied between 0 and 10 dBm. A preamplifier with inductive loading is used to achieve sufficient voltage gain to drive correctly the power stage, still preserving the power consumption and the transmitter overall efficiency at low output power levels. The preamplifier is made of an inductively loaded complementary push-pull class AB stage. Its load inductor is used to bias the gate of the PA at various voltages with the help of a current DAC to sweep the output power over a range of 10 dB. The PA, operating in class-C, makes use of cascode transistors to increase its reliability by splitting the drain voltage swing across the stacked transistors. Additionally, a reduction of the amplifier feedback capacitance simultaneously improves the amplifier stability. On-chip differential inductors with an estimated Q in the range of 7–8 have been used.

6.5 Laser Lock Loop Implementation

Figure 6.9 shows a transistor level implement of the laser biasing loop. The leftmost current sources, OTA and resistor implement a low noise, linear voltage controlled current source used to bias the laser diode. The control voltage is generated externally with a DAC embedded in the controller. A small square wave modulated current at ~50 kHz, whose two levels are controlled with a 4-bits current switching DAC is impinged in parallel to interrogate the ^{87}Rb atoms as depicted in Fig. 6.4. Eventually the error signal integrated and low pass filtered on the loop filter (V_{TRACK}) is converted to a tracking current with a resistively degenerated linearized two stages OTA that fine tunes the laser diode optical frequency dynamically to maintain the lock state. The three low frequency currents are fed to the low frequency input port of an off-chip balun, while the 3.417 GHz RF amplitude modulation signal generated by the PA is superimposed on the high frequency port of the balun. A three bit differential ADC sense the loop filter output signal so that the controller can adjust the DC laser biasing current via the DAC to maintain the V_{TRACK} signal within its linear range should the laser lock loop drift with time. The monitoring is handled externally on the micro-controller.

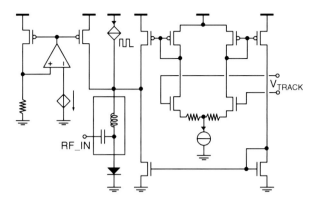

Fig. 6.9 Transistor-level implementation of the laser biasing loop

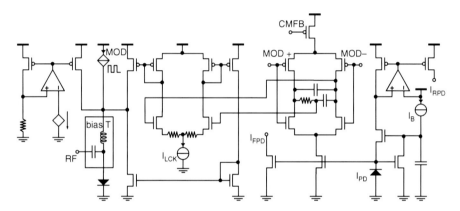

Fig. 6.10 Complete laser lock loop schematic

Figure 6.10 shows a transistor level schematic of the complete laser lock loop. The coherent demodulator topology is similar to that of the RF lock loop discussed previously. The TIA which is shared among both loops is powered at 1.8 V and topped with a current mirror whose gates are driven by an OTA to extract a copy of the unfiltered raw photodiode current (I_{RPD}). It is helpful to perform the optical path alignment and set the laser diode biasing current. The latter is first swept to tune the laser optical frequency and measure the corresponding absorption spectra signal detected by the photodiode. A band-pass filtered copy of the photocurrent (I_{FPD}) is also available externally to characterize the TIA performance and optimize the internal gain settings. All blocks but the TIA are powered at 3.3 V so that the laser can be driven directly (Vbias ~ 2 V) and the transconductance value reduced thanks to the increased dynamic range for maximum noise immunity.

6.6 Laser and Rb-Cell Temperature Control

Since both the laser optical frequency and the atomic transitions energy levels in the ^{87}Rb-cell exhibit a temperature dependency, the two devices need to be thermostated with mK-level stability. To avoid having to cool the devices in high temperature environment, the oven temperatures are chosen between 80°C and 90°C furthermore ensuring enough ^{87}Rb is in the vapor state to increase the interactions with the interrogating light. Heating sources such as power MOS, combined to a NTC resistor both mounted on the element to be heated are usually used to build the thermostat. Placing the whole setup suspended with very small tethers of low thermal conductivity in a vacuum chamber eliminates conduction and convection losses, minimizing the required heating energy. The thermistor is typically placed within a Wien bridge and a high resolution ADC, digital signal processing and DAC combination is used to control the oven temperature.

In the proposed implementation, voltage controlled current source reusing the topology proposed for the laser bias shown in the left of Fig. 6.10 and controlled by an external DAC are fed to heating resistors laid out directly on the sidewalls of the Rb-cell or on the laser heater. As a significant amount of power drawn by the ASIC will be converted to heat, the chip includes, as an alternative, a site to assemble the laser directly on top of an embedded heating meander resistor realized with the top metal. The controlled heater is to be used for residual temperature adjustment while benefitting from the readily available heat dissipated in the die to lower the overall power consumption.

The temperature sensor implementation is based on a ring oscillator that is fed with a temperature dependent current obtained with a resistor having a strong temperature dependency as in [4]. The arrangement, detailed in the left part of Fig. 6.11, is improved by eliminating most of the supply dependency. The sensor is made of a ring oscillator with nine stages and a highly temperature sensitive resistor, where both are fed with identical currents obtained by matched current sources. The operational amplifier driving the current sources whose transistor implementation is detailed in the right part of Fig. 6.11, further imposes that the voltages across the ring and resistor be equal.

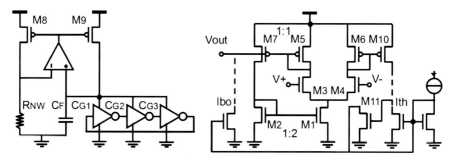

Fig. 6.11 Temperature sensor schematic

To minimize the sensitivity of the oscillator to a supply voltage variation, an adaptive self-biasing amplifier topology [5] was used. By matching all PMOS transistors including M8 and M9, the current biasing the amplifier is a fraction of that consumed by the loads owing to the factor two gain in the positive feedback loop formed by the current mirrors M5, M7 and M2, M1. In this way, a very high DC gain is obtained almost zeroing the equivalent amplifier input impedance. Any supply dependant offset voltage variation between the amplifier inputs is hence minimized. A filtering capacitor C_F is used to reduce the ripple at the positive ring supply. Using Ohm's law and the dynamic equation for the current in digital circuit ($I = f \cdot C_G \cdot V$), one shows that the oscillator frequency is given by $1/RC_G$. As the thermal oxide forming the gate capacitance of the ring stages exhibits a much lower temperature dependency, the oscillator TCF is hence mostly controlled by the resistor TC_R. The sensing resistor are interleaved directly in-between the heating resistors meanders on the Rb-cell, the laser off or on-chip heaters. An additional temperature sensor using a N-well resistor was placed at a reasonable distance from the blocks draining a large current to measure the die temperature and evaluate whether temperature gradients between the oxide insulated upper on-chip laser heater and the substrate appear.

Another current source is used to circulate a controlled current through an Helmoltz coil so as to generate an internal magnetic field to split the hyperfine structure of ^{87}Rb (Zeeman effect [6]) and make the clock less sensitive to external magnetic field variations.

6.7 Miniature Batch-Fabricated ^{87}Rb-Cells

Fabricating small leak free atomic cavities with reproducible filling is still challenging. An important effort has been done to develop wafer level fabrication of millimetre size atomic MEMS cells made of silicon and glass by using alkali azide as starting material.

After full evaporation of the solvent, the cavities are sealed by anodic bonding under controlled atmosphere. Metallic rubidium and nitrogen are obtained by UV decomposition of the crystallized rubidium azide.

The first cells (10×10 mm^2) fabricated in this way showed promising performances for a CPT miniature atomic clock and optimization of the buffer gas mixture is still on-going. The cells were miniaturized down to 1×1 mm^2. Much of the work has nevertheless been pursued on 4×4 mm^2 cells in order to add functionalities to the glass windows as shown in Fig. 6.12. Both faces have integrated heaters, temperature sensors and Helmholtz coils.

The functionalized cells are currently being characterized. Heating the cells up to 100°C has been achieved without problems and first 0–0 CPT signal have been measured in a laboratory setup. Integrating the functionalized cells in the prototype is ongoing.

Fig. 6.12 Functionalized 4 × 4 × 1.6 mm3 atomic MEMS cell

6.8 Physics Package, Atomic Clock Assembly and Control

The core physics package is realized by a stacking of PCB layers. It measures 11 × 11 × 8.5 mm^3 (1 cm^3), including the functionalized atomic vapor cell with dimensions downsized to 4.0 × 4.0 × 1.6 mm^3 (26 mm^3). Figure 6.13 illustrates the core physics packages (middle), as well as the laser (VCSEL) PCB layer (left) and the optical PCB layer with two photodetectors (right).

The core physics package is mounted in a commercial ceramic package shown in Fig. 6.14, for subsequent vacuum encapsulation. The resulting assembly is surrounded by an external $\emptyset = 42$ mm magnetic shielding, the overall volume of the physics package reaching 22 cm^3. The right part of Fig. 6.14 shows a photograph of the ASIC implemented in a standard digital 0.18 μm CMOS. It is pad-limited and measures 2.3 × 2.1 mm. It is not thermostated in the actual clock PCB implementation.

The MAC prototype is controlled by means of a MSP430 microcontroller embedding the required DACs to drive all the voltage controlled current sources. A LabVIEW® interface is used to perform the initial locking of all loops but such functions could later be implemented directly on the microcontroller. In operation mode the controller only has to deal with the voltage controlled current sources used for the laser bias, laser and cell heating and temperature measurement and the magnetic field generation. Only the temperature control loops require slightly more demanding controller resources. The others have much more relaxed time constants since they are activated to compensate for long term drift using coarse tracking, the fine, fast dynamic tracking being implemented in the analogue domain.

6 Towards Portable Miniature Atomic Clocks

Fig. 6.13 MAC physics package realized by stacking functionalized PCB layers (laser, optics and atomic vapour cell layers)

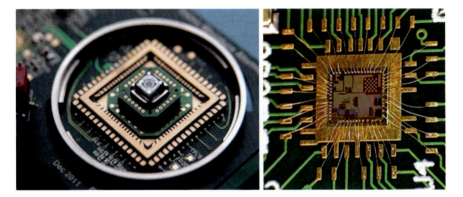

Fig. 6.14 Photographs of clock package (*left*) and ASIC (*right*)

6.9 Measurements

The described MAC prototype is currently in its integration and test phase. The main lock loops could already be closed and preliminary frequency stability measurements could successfully be realized using an external miniature glass atomic vapor cell (100 mm^3). Three generations of ICs with increased functionalities were designed and tested over the recent years.

Figure 6.15 shows a measurement of the synthesizer phase noise of the 1st ASIC version at the output of the power amplifier (black), reaching −85.6 dBc/Hz at 1 kHz offsets from the 3.417 GHz carrier. At the PLL cutoff frequency of 250 kHz, the phase noise is −98 dBc/Hz. The noise of the RF signal is compared with that achievable with typical laboratory equipment. The dark gray curve depicts the noise measured using a bulky low-noise oven controlled crystal oscillator (OCXO) in combination with a N5181A RF analog signal generator. The light gray curve shows the phase noise resulting from the use of the N5181A synthesis together with the integrated 40 MHz VCXO output after division by 4. The biggest difference is seen from 100 Hz offset and above with up to 30 dB noise degradation using

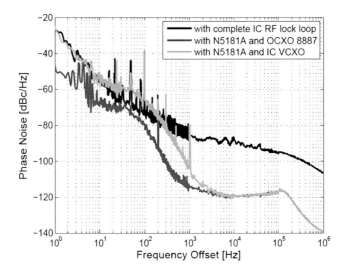

Fig. 6.15 Phase noise performance comparison

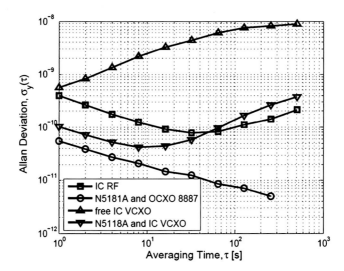

Fig. 6.16 Allan deviation performance comparison

the traditional synthesizer approach. Let's now see how such a difference affects the clock stability.

Figure 6.16 depicts the Allan deviation measured on the 10 MHz clock when locked through the IC RF loop to the physics package (□). The clock exhibits a 1 s intercept point of $\sigma_y(\tau=1) = 4 \times 10^{-10}$, improving as $\sigma_y(\tau) \propto \tau^{-1/2}$ up to some tens of seconds. The poor medium term stability observed for integration time $\tau > 20s$ is also evidenced when using the IC VCXO in combination with the

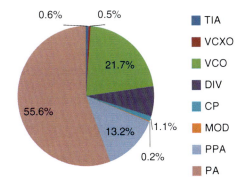

Fig. 6.17 ASIC power consumption breakdown

N5181A synthesizer (∇). The free VCXO, whose frequency stability is also plotted (Δ), also exhibits the same behaviour. The last curve (o) represents the Allan deviation measured when locking the laboratory equipment to the ^{87}Rb cell. By comparing phase noise and Allan deviation one can gauge how the former affects the clock stability. The most important synthesizer noise contribution affecting the clock stability is that present near the FM modulation frequency and first harmonics at a few 100 Hz.

The power consumption of the circuit is summarized in Fig. 6.17. The TIA and VCXO contribute to the overall consumption with only 0.3 mW, while the PLL including the fully integrated LC VCO consumes 7.9 mW. The overall consumption of 26.3 mW at an output power of 0 dBm suffers from the low efficiency of the power stage, which has been initially designed to reach output power up to 10 dBm. Such figures could greatly be lowered using a deeper sub-micron CMOS node as demonstrated in [7] for a 2.4 GHz transmitter consuming only 5.4 mW at 0 dBm output power. The LD typically draws 1–2 mA depending on the VCSEL type. The power consumption needed to heat the laser and Rb-cell is not yet known, the clock prototype being still at the assembly stage.

Nonetheless, a very impressive physics package with 10 mW consumption was demonstrated in [8]. High performance miniature atomic clocks with <20 mW power consumption will hence soon become a reality.

As a final measurement, Fig. 6.18 shows the Allan deviation measured with the third ASIC version using the RF and laser lock loops. It demonstrate a frequency stability very close to the telecom specifications, reaching $\sigma_y = 3 \cdot 10^{-11}$ at 1 day integration time and an impressive $\sigma_y = 6 \cdot 10^{-11}$ at 1 s. This demonstrates that that the integrated electronics supports very good short term frequency stability now rivaling with that achievable with laboratory equipment. Further characterization and analyses are however required to improve the clock long term stability. The complete MAC prototype using the miniature MEMS Rb-cell will be fully characterized in 2013.

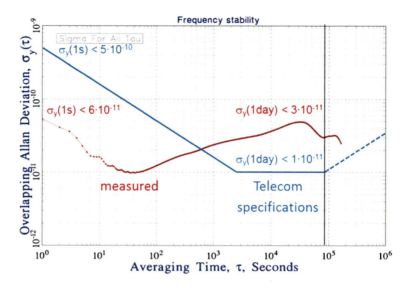

Fig. 6.18 Allan deviation measurement of the atomic clock with a glass Rb-cell

6.10 Conclusions and Perspectives

This paper has discussed the implementation of an ASIC designed to build compact CPT-interrogated atomic clocks. Several loops have to be implemented so that a VCXO can be locked to the atomic transitions of e.g. ^{87}Rb so as derive an accurate 10 MHz reference. This is obtained by modulating the biasing current of a VCSEL at RF with an FM sub-modulated 3.417 GHz signal and at low frequency to stabilize the laser optical frequency. Additional thermostat loops controlling the laser and Rb-cell temperature with mK accuracy have also been implemented but not yet tested. Allan deviation of $\sigma_y = 6 \cdot 10^{-11}$ at 1 s was demonstrated at a power dissipation of 30 mW excluding the thermostats using a 0.18 μm CMOS technology. The clock long term stability has to be improved to reach telecom specifications.

One can nonetheless reasonably project that an overall power dissipation of 20 mW to implement a complete MAC could be achieved in a near future allowing its integration into mass-market higher performances portable devices to satisfy their wireless communication demand for ever increasing data-rates.

References

1. R. Lutwak, D. Emmons, W. Riley, R.M Garvey, The chip-scale atomic clock – coherent population trapping vs. conventional interrogation, in *34th Annual Precise Time and Time Interval (PPTI) Meeting*, 2002, Reston, Virginia, pp. 539–550
2. D. Ruffieux, A high-stability, ultra-low power differential oscillator circuit for demanding radio applications, in *Proceeding Europeans Solid-State Circuit Conference*, 2002, Firenze, Italy, pp. 85–88

3. E.A. Vittoz, M.G.R. Degrauwe, S. Bitz, High performance crystal oscillators: theory and applications. IEEE J. Solid-State Circuit **23**(3), 774–783 (1988)
4. C.-K. Kim, J.-G. Lee, Y.-H. Jun, C.-G. Lee, B.-S. Kong, CMOS temperature sensor with ring oscillator for mobile DRAM self-refresh control. Micorelectron. J. **38**(10–11), 1042–1049 (2007)
5. M.G. Degrauwe, J. Rijmenants, E.A. Vittoz, H. DeMan, Adaptive biasing CMOS amplifiers. IEEE J. Solid-State Circuit **17**(3), 522–528 (1982)
6. Online: http://en.wikipedia.org/wiki/Zeeman_effect
7. Y. Liu et al., A 1.9nJ/b 2.4GHz multistandard transceiver for personal/Body-Area networks, in *proceeding IEEE ISSCC*, 2013, San Francisco, California, USA, pp. 446–447
8. R. Lutwak et al., The chip-scale atomic clock – recent developments, in *IEEE International Frequency Control Symposium*, 2009, Besançon, France, pp. 573–577

Part II
Power Management for System-on-Chip

Andrea Baschirotto

This second part of the book is dealing with '*Power Management for System-on-Chip*' and deals with the design aspects relative to managing the power within specific application devices. The case of cellular is phone, due to its large popularity, is taken as a benchmark example. The topic is addressed with two papers coming from Universities and discussing advanced solution, while four papers coming from industries discuss about the most advanced solutions already present in actual products.

In the seventh Chapter Hans Meyvaert discusses advanced aspects regarding AC-to-DC and DC-to-AC conversion. The AC-to-DC conversion allows to interface the mains voltage while achieving a high level of integration and compatibility to low voltage CMOS circuits, e.g. power supply straight from the wall socket. Alternatively the DC-to-AC interaction targets inversion of low voltage DC values to higher AC values for driving purposes.

The eighth Chapter deals with innovative switched-capacitor techniques to realize fully integrated DC-DC converters with maximum efficiency and power density. Elad Alon introduces circuit design methods to develop multiple topologies (and hence output voltages). These techniques are verified by a proof-of-concept converter prototype.

Francesco Rezzi, in the ninth Chapter, describes the operation of the Li-ion batteries to demonstrate that multiple charges reduce battery life-time. Thus proper battery handling is a hot topic in the engineering community, in consideration also that more and more devices moved or are moving to non-replaceable batteries. The historical trend of battery technology and address battery and power management techniques aimed to increase battery life and safety with particular focus on smartphone and tablets are reviewed.

The tenth Chapter from Sebastien Cliquennois reviews the challenges which integrated voltage regulators have and will have to tackle for power management of portable applications while focusing on how the Digital Switched-Mode Power Supplies (SMPS) technology, already widely used for medium and high power systems, is able (or not) to challenge the classical analog loops.

Jay Ackermann in the eleventh Chapter discusses the overall mobile phone charging system and some key factors that make battery chargers unique. The design of a switch-mode battery charger is presented, including subsystem circuit architecture, stability analysis, and sequencing logic, plus the key performance parameters of the design are summarized.

David Flynn in the twelfth Chapter discusses promising approaches to enhance Power Gating (PG) and State Retention Power Gating (SRPG) techniques, which are appropriate to digital designers without the need to resort to full-custom design techniques. The aim is also to increase designer understanding of how the essentially analog circuit challenges can be abstracted for a richer set of standby power management schemes.

Chapter 7
From AC to DC and Reverse, the Next Fully Integrated Power Management Challenge

Michiel Steyaert, Hans Meyvaert, and Piet Callemeyn

Abstract This chapter discusses the advances of the next leap in integrated power management beyond the scope of today's DC-DC converters: the interaction with AC. Both AC to DC and DC to AC conversion with the intention of eventually interfacing mains voltages are investigated and reported on. The AC-DC conversion research aims to interface the mains voltage while achieving a high level of integration and compatibility to low voltage CMOS circuits, enabling power supply straight from the wall socket. Alternatively the reverse interaction from DC to AC is investigated similarly and targets inversion of low voltage DC values to higher AC values for driving purposes. Because of the integrated approach, the bill of materials is drastically reduced.

7.1 Introduction

DC-DC Converters currently are a subject of intense study: not only for discrete realizations [1], but also for integrated solutions this is true [2]. Higher operating frequencies, made possible through technological advances, led to the decrease in size of the passive components. Discrete realizations, with switching frequencies up to 5 MHz have seen a tremendous increase in power density as a result of this [3]. Moreover, fully integrated solutions have appeared taking advantage of the even higher switching speeds, 70 MHz [4] up to 1 GHz [5] and even 3 GHz [6] implementations have been reported, that can be achieved with CMOS transistors. As more and more fully integrated designs are being presented, it is becoming evident that history is repeating itself. Similar to the CMOS breakthrough in RF circuits enabling monolithic telecommunication circuits, CMOS is slowly increasing its market share in the field of power management circuits [7].

M. Steyaert (✉) • H. Meyvaert • P. Callemeyn
ESAT MICAS – KU Leuven, Kasteelpark Arenberg 10, 3001 Heverlee, Belgium
e-mail: steyaert@esat.kuleuven.be

A. Baschirotto et al. (eds.), *Frequency References, Power Management for SoC, and Smart Wireless Interfaces: Advances in Analog Circuit Design 2013,*
DOI 10.1007/978-3-319-01080-9_7, © Springer International Publishing Switzerland 2014

A similar trend of increased switching frequency has also appeared in switched mode AC-DC converters and DC-AC inverters. However, full integration in these cases has not nearly evolved as it has with DC-DC conversion. Nonetheless integration is a logical next step as it is the major driving force in decreasing the cost in Bill of Materials (BOM), reducing the volume and enable higher performance at lower power consumption.

For power supplies that step the mains AC down to typical DC bus voltages of 12 V or lower, progress has resulted in the miniaturization of the isolating step-down transformer in the popular flyback converter design on one hand and a smaller inductor in a buck converter approach on the other. But still a costly transformer is required in the former case and a complex high side switch along with a narrow duty cycle in the latter. Herein lays a strong motivation to adopt a solution that removes these components. It is therefore that the approach taken in this paper prefers capacitors above inductors for AC-DC conversion.

Previous work on circuits with the mains as input concentrated on the feasibility of interfacing the high input voltage by integrating the power supply in a process such as silicon on sapphire (SOS) [8]. Hereby taking benefit from the technology's high voltage capability and thus the ability of implementing active circuits that can withstand the mains input. To circumvent the need for high voltage active circuits, a capacitive division approach has been suggested in [9]. On one hand dividing the input voltage to a lower level shifts the high voltage requirement from the active circuit to the passive components and enables regular CMOS processes to be used for the subsequent power management. On the other hand the power throughput is deteriorated as discussed in Sect. 7.2. Therefore the research in this paper targets the benefit of low voltage operation, but at a sustained power throughput.

The main feature of a DC-AC converter, or inverter, exists in converting a DC source into an AC source. The frequency and amplitude of the converter output can be adjusted. Inverters are extensively used in uninterrupted power supplies (UPS), motor drivers, cold cathode fluorescent lamps (CCFL) and photovoltaic (PV) panels. The common output frequency range of these inverters is from 0 to 10 kHz. Present-day commercial inverters are using external components, increasing the BOM.

By taking the next leap towards a monolithic integration of inverters, a less expensive solution is achieved. Moreover, this will result in a flexible inverter that can achieve a much wider frequency range. This enables the possibility to control micro scale piezoelectric motors and magnetic machines which require high driving frequencies above 100 kHz.

On the other hand, the generation of low-frequency signals should remain possible on-chip as well. This is a major challenge as the on-chip passives are inherently small, thus requiring techniques to overcome this issue.

As one refers to DC to AC conversion, an interfacing of the output with the mains is one of the major applications. This means that the on-chip solution should be adapted to comply with these requirements. More specifically, one will be interested in both a high output voltage and high output power. This implies that state-of-the-art circuit techniques will be needed to achieve these requirements.

This paper is structured into two parts, each discussing a mains interface direction. Firstly, the AC-DC interaction and its challenges are reviewed in Sect. 7.2. The selected solution approach is presented and discussed in detail in Sect. 7.3. Secondly, Sect. 7.4 explores the reverse interaction of DC to AC. Finally, conclusions are drawn in Sect. 7.5.

7.2   Coping with High Voltage Inputs

Interfacing voltages beyond the nominal rated device voltage generally requires special circuit techniques to prevent overvoltage from destroying the devices. Successful techniques to do so include device stacking [10, 11], where cascaded devices each share a portion of the total voltage, and voltage domain stacking [12] in which multiple nominal voltage rails are serialized. However, even with these techniques the maximal achievable interface voltage is still limited to a few times the nominal rated supply voltage as the complexity to implement these techniques increases substantially for each added level of stacking.

When considering the mains voltage with a nominal peak voltages of 375 V in the $265V_{RMS}$ case, it is clear that these techniques are inadequate and alternative approaches are needed. With the mains voltage input exceeding the rated voltage of the active circuitry by two orders of magnitude, it is required to create a voltage gap between the mains input and the active circuit. This can be achieved by placing an impedance in series [8] over which the voltage is dropped, e.g. a resistor. But such an approach would suffer from an unacceptable low efficiency due to the very large voltage drop and is therefore undesirable. Another possibility is to use a capacitor, which is lossless in the ideal case and thus a better choice.

The series capacitor approach is taken in the work of [9] in the form of a capacitive voltage divider, as shown in Fig. 7.1a. The mains input voltage V_{AC} is divided by the combination of capacitors C_{in} and C_{div} to a lower value V_X, which can be withstood by the rectifier and the rest of the active circuit. The divided voltage V_X is then rectified onto a smoothing capacitor C_{DC} and supplies a load current. V_{DC} in worst case condition (no load) is $\sqrt{2}\ V_{X,RMS}$. Therefore, the capacitive voltage division ratio r_{div} must fulfill $r_{div}\ \sqrt{2}\ v_{AC,RMS} < V_{rated}$ in order to mitigate a V_{DC} overvoltage occurrence. But r_{div} reduces power throughput considerably in all other load conditions as the rectifier diodes are only turned on when $V_X > V_{DC}$ (Fig. 7.1b), which only occurs for a short time near the peak of V_X. Afterwards, V_X decreases below V_{DC} and the rectifier diodes turn off until V_X goes below $-V_{DC}$. During this time C_{DC} buffers V_{DC}.

The approach in this work proposes to use a series capacitor as a capacitive step-down due to its interaction with the load and the power management regulation circuits located behind the rectifier, showing similarity to [8]. But other than

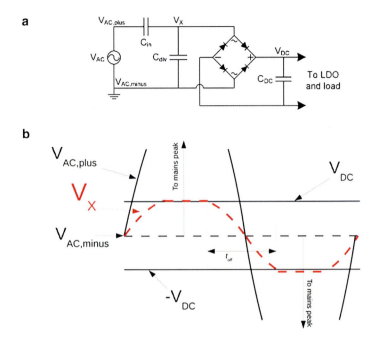

Fig. 7.1 (**a**) Capacitive division topology: the mains amplitude is divided down to V_X before rectification. (**b**) Operational waveforms of the capacitive divider: power throughput is limited due to increased t_{off}

in [8], this work aims to use a cheap CMOS process by moving the high voltage towards the integrated passive components. And unlike [9], overvoltage is mitigated by providing proper current sinking after rectification in the form of a shunt regulation path. This approach maximizes the rectifier diode on time as $V_{AC,low}$ floats at the rate of the mains when the rectifier is off, keeping t_{off} to a minimum. Hereby power throughput is optimal for any given amount of series capacitance C_{in}, reducing the necessary capacitor size and cost in comparison to other approaches such as the capacitive division. This concept is demonstrated in Fig. 7.2.

7.3 Capacitive Step-Down Explored

Continuing upon the reasoning of the previous Section, a system topology is proposed in Fig. 7.3 and differentiates between two cases. The first case follows a fully integrated approach, integrating all components on chip. The system is composed of a capacitive AC-DC step-down and a regulation stage, which will be discussed in Sects. 7.3.1.1 and 7.3.1.2 respectively. A second case allows the use of external passive components to scale up the total design and power level.

7 From AC to DC and Reverse, the Next Fully Integrated Power Management Challenge

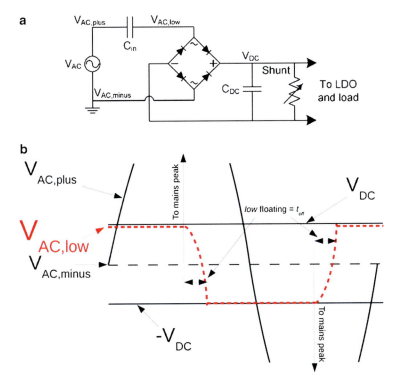

Fig. 7.2 (**a**) Proposed capacitive step-down: V_{DC} is set by load and shunt. (**b**) Capacitive step-down waveforms of operation

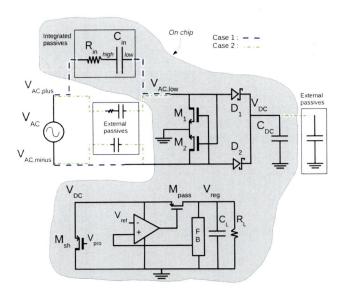

Fig. 7.3 System topology of the proposed AC-DC converter showing the two implemented options: the fully integrated case 1 and the partially integrated case 2

7.3.1 Operation

7.3.1.1 Capacitive AC-DC Step-Down

The AC-DC step-down stage, shown in Fig. 7.3, consists of the high voltage passive components C_{in}, R_{in} together with a rectifier [13] and a smoothing capacitor C_{DC}. To understand the operation, case 1 is discussed and the assumptions for simplicity that $R_{in} = 0$, the rectifier is ideal ($V_{th,M1-M2} = 0$, $V_{D,D1-D2} = 0$), C_{DC} is infinite and charged to a voltage V_{DC} are made. With a mains RMS voltage (Fig. 7.2b) V_{AC} present at the input terminals and $V_{AC,plus}$, referred to $V_{AC,minus}$, increases from 0 V up to V_{DC}: devices M1, D1 and D2 remain off while M2 is on. Next for $V_{DC} \leq V_{AC,plus} \leq \sqrt{2}\, V_{AC}$, D1 turns on and current flows to C_{DC}. Immediately after the mains peak, D1 turns off followed soon after by M2 since the low terminal of C_{in} starts to float and decreases at the same rate as $V_{AC,plus}$ until a drop of $2V_{DC}$ has taken place. At that time $V_{AC,minus} - V_{AC,plus} = V_{DC}$ and D2 turns on, while M1 has already turned on just before, providing another current towards C_{DC}. This continues until the negative peak at which D2 turns off. The above operation continues to alternate.

Input series capacitor C_{in} separates the active circuit from the high input mains voltage. While the high terminal of C_{in} is subjected to the full mains voltage, meaning a peak-to-peak voltage $V_{ptp,high}$ of $2\sqrt{2}\, V_{AC}$, this is not true for the low terminal ($V_{AC,low}$). The low terminal is bound by the rectified voltage V_{DC} resulting in a $V_{ptp,low} = V_{DC}$. An input series resistor is added, in addition to C_{in}, to protect the circuit against inrush current that occurs when the system is connected to the mains at the time of a high voltage or peak while C_{in} is not charged. Without resistor R_{in} a potentially destructive current charges C_{in}, only limited by the parasitic series resistance located between $V_{AC,plus}$ and $V_{AC,minus}$.

7.3.1.2 Shunt and Series Regulation

C_{DC} is assumed infinite until now and fixed at V_{DC}, limiting $V_{AC,low}$ with respect to ground during both positive and negative mains half cycle. In practice this is done by the parallel combination of a shunt regulation path and a low dropout (LDO) series regulator passing the current to the load. At nominal load the shunt path is inactive and all power is passed on by the rectifier to be consumed in the load, satisfying both $<|i_{Cin,nom}|> = i_{load,nom}$ and $V_{DC,nom} = V_{reg}$ (aside from the minimal dropout voltage). The resulting equilibrium of $V_{DC,nom}$ is given by V_{out} of Eq. 7.6, in which $<|i_{Cin,nom}|>$ equals the load current $i_{load,nom}$ for that nominal case.

When load power decreases to a lower level $i_{load,low}$, V_{reg} will be kept constant by the series regulator. This is not true for V_{DC} which will settle at a new equilibrium $V_{DC,low}$ in order to satisfy $<|i_{Cin,low}|> = i_{load,low}$. From Eq. 7.6 it can be seen that, for a given set of fixed parameters V_{in}, C_{in} and f_{mains}, this can only occur by increasing V_{out} (i.e. V_{DC}). The new $V_{DC,low}$ equilibrium can be calculated

7 From AC to DC and Reverse, the Next Fully Integrated Power Management Challenge   109

according to Eqs. 7.1 and 7.2. In conclusion, this means that for a lower than nominal load current V_{DC} will easily exceed the safe nominal voltage limit. For this reason a shunt path was included through M_{sh} in parallel with the series regulator in order to limit V_{DC} to a maximum of $V_{pro} + V_{th,Msh}$ at less than nominal loads as it allows $< |i_{Cin}| >$ to remain constant throughout any load current variation.

$$\frac{< \left| i_{C_{in,low}} \right| >}{< \left| i_{C_{in,nom}} \right| >} = \frac{4f_{mains}C_{in}\left(\sqrt{2}V_{in} - V_{DC,low}\right)}{4f_{mains}C_{in}\left(\sqrt{2}V_{in} - V_{reg}\right)} \qquad (7.1)$$

$$V_{DC,low} = \sqrt{2}V_{in} - \frac{< \left| i_{C_{in,low}} \right| >}{< \left| i_{C_{in,nom}} \right| >}\left(\sqrt{2}V_{in} - V_{reg}\right) \qquad (7.2)$$

7.3.2   AC-DC Step-Down Modeling

A representation of an ideal capacitive AC to DC step-down conversion is shown in Fig. 7.4. This subsection analyses the power throughput of such a circuit with respect to its parameters: V_{in}, f_{mains}, C_{in} and V_{out}.

$$V_{in}(t) = \sqrt{2}V_{in}\sin\left(2\pi f_{mains}t\right) \qquad (7.3)$$

$$V_{C_{in}}(t) \approx \left(\sqrt{2}V_{in} - V_{out}\right)\sin\left(2\pi f_{mains}t\right) \qquad (7.4)$$

$$i_{C_{in}}(t) = C_{in}\frac{dV_{C_{in}}}{dt} = C_{in}\left(\sqrt{2}V_{in} - V_{out}\right)\cos\left(2\pi f_{mains}t\right)2\pi f_{mains} \qquad (7.5)$$

$$< \left| i_{C_{in,low}} \right| > = 4f_{mains}C_{in}\left(\sqrt{2}V_{in} - V_{out}\right) \qquad (7.6)$$

$$P_{out} = < \left| i_{C_{in}} \right| > V_{out} \qquad (7.7)$$

The input voltage as function of time is given by Eq. 7.3 and is present at the high terminal of the capacitor C_{in}. On the other hand the low terminal of C_{in} exhibits a square wave pattern with amplitude V_{DC}. As a result of these voltages present at the capacitor terminals, the voltage over C_{in} can be approximated by Eq. 7.4. The capacitor current as function of time is then given by Eq. 7.5. Averaging this over time consequently leads to the average capacitor current $<|i_{Cin}|>$ in Eq. 7.6, which can be combined with the output voltage V_{out} to calculate the output power P_{out} according to Eq. 7.7.

A power throughput bottleneck is introduced as result of the low mains frequency and a low capacitance value for C_{in}. Only low values can be integrated due to the high voltage nature of this component, which require sufficient spatial separation, leading to a low capacitance density. Because this bottleneck pushes the absolute power levels downwards, the available capacitance becomes a valuable

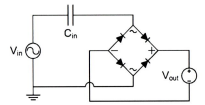

Fig. 7.4 Representation of an ideal capacitive AC to DC step-down

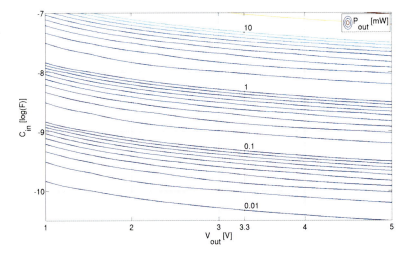

Fig. 7.5 Output power trade-off calculated with the compact model in Eqs. 7.3, 7.4, 7.5, 7.6 and 7.7

commodity and the necessity of using it efficiently is strengthened. To achieve efficient utilization of C_{in}, the rectifier diode on time should be maximal. It can be seen in Fig. 7.2b that the proposed architecture operation constitutes voltages $V_{AC,low}$ and $V_{AC,minus}$ to exhibit block pulse like behavior approaching the ideal case, i.e. an AC square wave output of the capacitive AC-DC step-down topology that is fed into the rectifier. This is opposed to the waveforms of the capacitive division approach of Fig. 7.1b [9]. A diode on time of 91 % and 93 % were achieved in the proposed demonstrator for the US and EU mains cases respectively.

Since the mains voltage is standardized, both system parameters V_{AC} and f_{mains} are already fixed. Figure 7.5 shows the output power capability of an ideal AC-DC stage as function of the two remaining degrees of freedom. A trade-off between input capacitance C_{in} and the output voltage V_{out} is observed. Equations 7.6 and 7.7 demonstrate the linear influence of the input capacitor C_{in} the output power. Alternatively, when keeping the series input capacitor constant, a higher output voltage V_{out} increases P_{out}, even though increasing V_{out} results in a lower average

7 From AC to DC and Reverse, the Next Fully Integrated Power Management Challenge   111

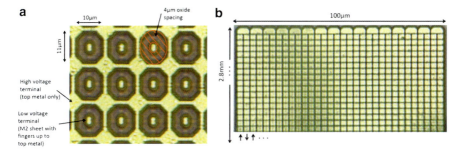

Fig. 7.6 (a) Top view of custom high voltage capacitor layout. (b) High voltage input resistor comprised of meandering top metals and vias

input capacitor current (Eq. 7.5). This can be explained by the fact that this effect is negligible for V_{out} voltages below 50 V and therefore the output power relation as function of V_{out} scales linearly in this region, as given by Eq. 7.7.

7.3.3 Implementation

Figure 7.3 introduced two separate integration cases. Both are implemented in a 0.35 µm CMOS technology, extended with a set of DMOS devices up to 25 V. The first case implements a fully integrated design, including all high voltage components. A similar second case implements a scaled up version of the integrated design and requires external passive components to achieve higher power levels. The implemented circuits are now discussed.

7.3.3.1 High Voltage Passive Components

Capacitor C_{in} bridges the high voltage gap between the high voltage mains input and the low voltages on-chip, as discussed in Sect. 7.3.1.1. While the active circuits do not come in contact with high voltage, the input capacitor C_{in} and input resistor R_{in} are subjected to a maximum voltage of $\sqrt{2}\,V_{AC}$, up to 375 V in the case of $V_{AC, RMS} = 265V$. With the oxide in the metal stack having a breakdown of at least 1 MV cm [14], a minimal spacing of 3.75 µm in needed to ensure breakdown will not occur. To this end the input capacitor was implemented as a metal-metal fringe capacitor with at least 4 µm of oxide between the capacitor plates. On top of that, metal corners were rounded to avoid the point effect. In Fig. 7.6a, a top view of the custom layout of the capacitor is depicted. The high voltage plate is located solely in the top metal as to ensure sufficient spacing (>4 µm) to the low voltage terminal and the substrate. Voids are left in this high voltage plate through which the low voltage plate, mainly located lower in the metal stack, rises up to the top metal. This structure was found to maximize fringing while considering metal density reliability rules. Nevertheless capacitance density suffers from the widely spaced capacitor plates and 12.5 pF/mm^2 is achieved for this structure, resulting in a total of 50 pF. The input resistor (Fig. 7.6b) implements 36 kΩ using a series connection of vias and the top two metals in the stack, ensuring a large spacing to the substrate. Oxide spacing exceeds 6 μm to ground to be able to withstand even higher voltages such as short spikes in the mains input.

7.3.3.2   Regulation Circuits

The rectifier [13] transistors M1 and M2 are implemented with available thick oxide DMOS devices, as part of a set power devices rated up to 25 V available in the technology, in order to extend the rectifier's safe operation slightly. D1 and D2 are implemented as Schottky diodes to reduce the forward voltage drop.

After rectification onto smoothing capacitor C_{DC} post regulation is required to deliver a regulated output voltage V_{reg} of 3.3 V. In order to save area C_{DC} is placed underneath C_{in} and implements more than 10 nF in NMOS gate capacitors. The shunt and series path perform a dual function. First, the shunt path limits V_{DC} by providing a current sinking when the load current decreases and secondly a series regulator chops off the ripple voltage that remains after the previous rectification.

1. *Shunt regulation:* Transistor M_{sh} is a thick oxide PDMOS device biased with an overvoltage protection control signal V_{pro}. When load power is decreased and V_{DC} increases above $V_{pro} + V_{th,Msh}$ the PDMOS will start to conduct and will limit the maximum of the rectified voltage to a safe value.
2. *Series regulation:* The LDO regulates the rectified voltage into a ripple free output voltage V_{reg}. Considering the limited power budget available at the output from the AC-DC stage, power consumption in the regulator is minimized in order to limit the impact on system efficiency. Due to very low frequency time constants in the system it is possible to achieve sufficient performance with very low static current. The error amplifier and feedback path together consume only 150 nA. A Gain-bandwidth of 100 kHz was achieved when loaded by the gate capacitance of M_{pass}, which does not need to be large at the expected current levels. To reduce the area of the resistive feedback, sub threshold biased devices were used to create large resistance density.

7.3.4   *Measurement Results*

Both prototypes, case 1 and case 2, are measured for various mains voltage, frequency specifications. Figure 7.7a shows the maximum achievable output power of the fully integrated case as function of the input mains RMS voltage ranging from 85 up to 265 V, 50–60 Hz. Along with an input RMS voltage increases, achievable load power is scaled linearly. The deviation between measurement and

7 From AC to DC and Reverse, the Next Fully Integrated Power Management Challenge 113

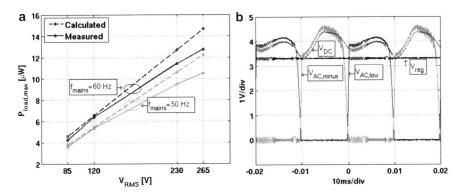

Fig. 7.7 (a) Calculated and measured output power for prototype case 1 with $C_{in} = 50$ pF. (b) Operational waveforms

calculated value increases towards higher input voltage as the limit of what the integrated smoothing capacitor C_{DC} can buffer are reached. For a 50 Hz condition, the load power scales from 3.6 µW for a 85V_{RMS} input up to a maximum of 10.5 µW at 265V_{RMS}. Load power increases from 4.2 µW up to 12.7 µW similarly when input frequency is 60 Hz. Figure 7.7b shows the system voltage waveforms of both rectifier inputs, the rectifier output V_{DC} and the regulated output voltage V_{reg} for the typical EU mains input case.

Series regulation removes the ripple in V_{DC} to a voltage variation smaller than 5 % of the regulated output V_{reg}. The waveform inputs of the rectifier show the presence of a parasitic coupling in the measurement setup. Signal $V_{AC,minus}$ contributes more input current than its complementary signal $V_{AC,low}$. This imbalance is due to the fact that the generated mains signal in the test setup is not only AC coupled as it should be, but also exhibits DC coupling to ground. Since the fully integrated case 1 of Fig. 7.3 only employs one series capacitor, the parasitic DC coupling to ground can propagate into the measurement via the path with no series capacitor. This issue is resolved when a series capacitor is inserted in both connections to the mains as is done while measuring the case 2 prototype. Two series capacitors are now used as Fig. 7.3 demonstrates. Figure 7.8b shows the waveforms in the case of two external series input capacitors. As predicted, the imbalance has disappeared. The according power for a set of several series input capacitor measurements is combined in Fig. 7.8b to demonstrate the increased output power capability enabled by larger input capacitors. Figure 7.9 shows the die measuring 6 mm^2. Most of the area is occupied by the fully integrated prototype with its high voltage capable passive components C_{in} and R_{in}, which is typical for monolithic power management circuits. The second prototype case can be found in the upper right corner. Finally, a comparison of the proposed converter with a prior state of the art is given in Table 7.1. The measurement results of this converter increase upon the previously achieved power density, demonstrating the enhanced capacitor utilization architecture used in this work. On top of that the input voltage range has been extended from 120V_{RMS} up to the maximum of 265V_{RMS}.

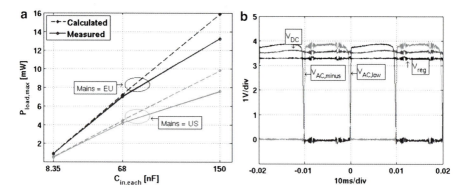

Fig. 7.8 (a) Calculated and measured output power for prototype case 2 for multiple C_{in} capacitors and the two most prominent mains configurations. (b) Operational waveforms

Table 7.1 Comparison of the demonstrated cases to the prior art

Reference	[9]	Case 1		Case 2 (2 × 68 nF)	
Tech node	0.13 μm	0.35 μm			
V_{RMS}	120 V	120 V	230 V	120 V	230 V
f_{mains}	60 Hz	60 Hz	50 Hz	60 Hz	50 Hz
Power/area	0.43 $\mu W/mm^2$	1.06 $\mu W/mm^2$	1.58 $\mu W/mm^2$	–	–
V_{reg}	4 V	3.3 V			
$t_{on,diode}$	48 %	91 %	93.5 %	91 %	93.5 %
$P_{out,max}$	1.5 μW	6.4 μW	9.5 μW	4.2 mW	7 mW

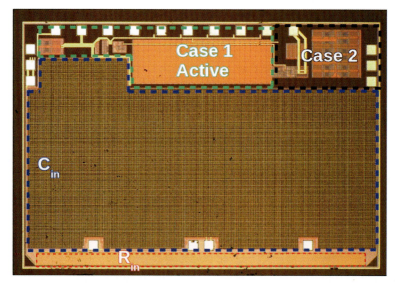

Fig. 7.9 Die photo showing case 1, comprised of the case 1 active circuits, C_{in} and R_{in}. Case 2 is located in the upper right corner

7.4 DC-AC Conversion

The integration paradigm for power supplies has already been discussed in [2]. These recent research efforts have cleared the path to develop fully-integrated DC-DC converters in standard CMOS technology. The next leap in this trend is the integration of even more complex power conversion blocks. AC-DC converters were already discussed in Sects. 7.2 and 7.3, DC-AC conversion will be discussed in this section.

7.4.1 DC to AC Conversion: Applications and On-Chip Challenges

DC-AC converters are commonly used in uninterruptible power supplies, motor drivers, cold cathode fluorescent lamps and photovoltaic panels. The common output frequency of these devices is from 0 to 10 kHz. On the other hand, when going towards small-scale applications such as piezoelectric motors and micro magnetic machines, higher driving frequencies are needed up to 100 kHz.

An on-chip realization of this DC-AC conversion, alleviating the need for external components, would reduce the bill of materials drastically. Moreover, a more flexible inverter can be made to achieve a much wider frequency range.

The generation of very low frequency signals on-chip (e.g. 50 Hz AC) should remain possible as well. Moreover, often a high output voltage is required. These are two major challenges for an on-chip realization and will be addressed hereafter.

7.4.2 Possible Circuit Topologies for On-Chip DC-AC Converters

To achieve high efficiency, one will need a switching converter topology. Due to the small size of on-chip passives, a high frequency will be needed to transfer energy from the input to the output.

In this section, single-input single-output converters containing a single inductor will be discussed. Topologies using multiple inductors exist, but this leads to an increased use of chip-area. This is why only the single inductor topologies are discussed here. An inductor can be connected in a switching circuit in a limited number of ways. One can consider two intervals during which the inductor is either connected to the load or the source. This means that the inductor is connected in the circuit in two different ways during the first and the second interval.

By elimination of redundant circuits, one comes to eight possible converter topologies. These are fully described in [15]. Of these eight, the most appropriate topologies to achieve high efficient inversion will be touched briefly in this chapter.

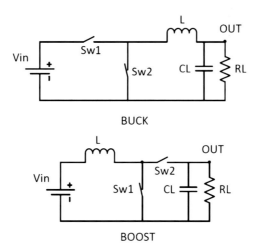

Fig. 7.10 Buck and boost topology

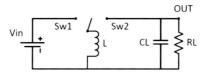

Fig. 7.11 Buck-boost topology

Buck, boost and buck-boost: the basic buck and boost topologies are given in Fig. 7.10. Energy transfer from the input to the output is made possible by the combination of inductor and capacitor. The buck and boost converters produce a positive unipolar output voltage. The buck-boost topology in Fig. 7.11 combines the possibility to achieve both a higher or lower output voltage. This topology produces a negative unipolar output voltage.

With these converters it is possible to increase, decrease or invert a dc voltage. The control is done using PWM (pulse width modulation).

Bridge: the buck, boost and buck-boost converters produce a unipolar output voltage. For inverter applications, one might be interested in bipolar output voltages. A technique to achieve this, is the differential connection of a load over two buck converters. Figure 7.12 clarifies this approach. If converter one produces voltage V_1 and converter 2 produces voltage V_2, the load voltage will be given by:

$$V = V_1 - V_2 \qquad (7.8)$$

Both V_1 and V_2 are individually positive, but the load voltage V can either be positive or negative.

If one simplifies the circuit topology of Fig. 7.12, the resulting circuit on the right hand side is commonly known as an H-bridge or full bridge inverter. The advantage of this approach is the bipolar output voltage. The drawback of this approach are the

7 From AC to DC and Reverse, the Next Fully Integrated Power Management Challenge   117

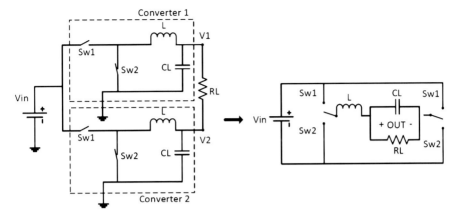

Fig. 7.12 Bridge topology

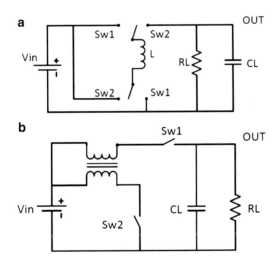

Fig. 7.13 (a) Watkins-Johnson topology. (b) Watkins-Johnson using transformer

extra switches compared to basic buck or boost converters, leading to added switching losses.

Watkins-Johnson: the combination of two boost converters, analogous to the previously tackled bridge converter, yields the topology given in Fig. 7.13a. The number of switches can be reduced by using a two-winding inductor as shown in Fig. 7.13b. The advantages of this converter are its ground-referenced load and the ability to produce a bipolar output voltage using only two switches. There are however some drawbacks for on-chip realization. In the case of the first topology (Fig. 7.13a) four switches are needed, leading to increased switching losses. In the latter topology (Fig. 7.13b) an on-chip transformer must be developed. The added parasitic capacitive coupling will inevitably lead to increased power losses in the structure.

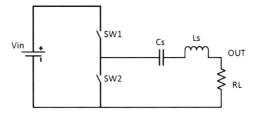

Fig. 7.14 Resonant converter topology

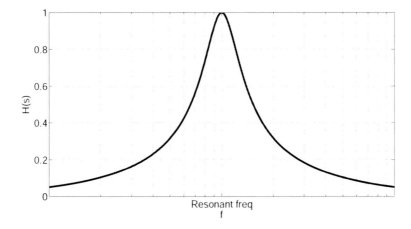

Fig. 7.15 Bode diagram for a tank response network.

Resonant converters: this converter consists of the combination of a switch network and a resonant tank network. An example topology can be seen in Fig. 7.14. This topology will deliver a high frequency ac output. Because of the resonant network, the output voltage amplitude will change according to the switching frequency. Figure 7.15 shows the Bode diagram for the tank response network. It can be seen that the output amplitude is highly dependent of the switching frequency. Using PFM (pulse frequency modulation), one is now able to change the output amplitude by changing the switching frequency around the resonant frequency.

The advantage of this topology is the high switching frequency. This allows the on-chip passives to be smaller. Moreover, switching loss reduction is inherent for resonant converter circuits. The drawback of this topology is the tuning of the tank, which means that it can only be optimized for a small range of loads.

Out of all these different topologies, the most promising circuit to be integrated is the resonant conversion circuit. It enables the on-chip passives to become small. Due to the added inherent switching loss reduction, the efficiency can be higher compared to non-resonant circuits although there is a higher switching frequency.

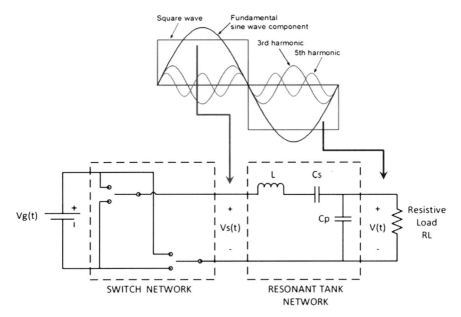

Fig. 7.16 Basic resonant converter circuit and waveforms

7.4.3 Resonant Conversion

A resonant power converter has some advantages over switching PWM converters. These PWM converters will suffer from switching losses and low efficiency at high frequencies [16]. The main advantage of resonant converters is their reduced switching loss. This is achieved via mechanisms known as zero-current switching (ZCS) and zero-voltage switching (ZVS). The turn-on and turn-off transitions of the various converter switches can occur at zero crossings of the resonant converter quasi-sinusoidal waveforms. This will reduce switching losses, meaning that resonant converters can be operated at higher switching frequencies than PWM converters. Zero-voltage switching can also eliminate some of the sources of converter-generated electromagnetic interference.

The basics for a resonant conversion are explained in Fig. 7.16. This circuit depicts the two basic elements in a resonant converter: the switch network followed by the resonant tank. The switch network produces a square wave voltage $V_S(t)$ at frequency f_{RES}. The frequency f_{RES} is the resonant frequency of the tank network.

By its nature, a square waveform consists of several harmonics. The fundamental component is at f_{RES}. The third, fifth, ... harmonic are filtered by the resonant tank network. Essentially, these harmonics will be negligible at the output, meaning that the load current i_{OUT} and voltage v_{OUT} will be sinusoidal waveforms of frequency f_{RES}. This is depicted in Fig. 7.16.

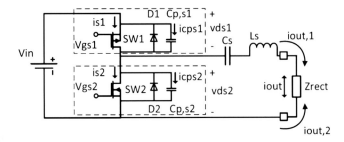

Fig. 7.17 Equivalent circuit used for the design of the series resonant class DE inverter

By changing the switching frequency f_S closer to or further from f_{RES}, the output amplitude can be modulated. This yields a modulated envelope that can be filtered and used as a low frequent AC waveform. It can be seen as a low frequent envelope that is modulated on a high frequency carrier.

Examples of resonant converter circuits are class E inverters [17] and class DE inverters [18]. In a class E inverter, high peak switching voltages are present, up to four times the supply voltage. However, the breakdown voltage for transistors in deep-submicron technologies is low, requiring measures to prevent degradation or breakdown of the switches. For a class DE inverter on the other hand, the peak switch voltage will be limited to the supply voltage. It is thus more suitable for integration in deep-submicron technologies.

7.4.3.1 Series Resonant Class DE Inverter

Figure 7.17 illustrates the schematic of a series resonant class DE inverter with ideal components. It consists of two MOS transistors SW_1 and SW_2, a MIM-capacitor C_S, an inductor L_S and a resistive output load Z_{RECT}. The capacitor, inductor and resistor form the series resonant tank. The capacitors $C_{P,S1}$ and $C_{P,S2}$ are the parasitic output capacitances of the MOS transistors. Diodes D_1 and D_2 are the intrinsic body-drain pn-junction diodes. These will be used as anti-parallel diodes.

The switches turn on and off periodically. This is controlled by a non-overlapping clock: there is a dead-time between the switch on-times to prevent short-circuit currents flowing through both transistors. The operation is described as follows. During the first switching stage SW_1 is closed and SW_2 is open. The current $i_{OUT,1}$ through the inductor and capacitor now starts to flow. During the second stage SW_1 is open and SW_2 is closed. The current $i_{OUT,2}$ due to the stored magnetic and electric energy in the resonant tank now flows in the opposite direction of $i_{OUT,1}$. Over one complete period, the current i_{OUT} through this resonant tank is nearly a sine wave. The output load Z_{RECT} sees this sinusoidal current that changes direction every clock cycle, this yields an AC voltage across the load resistor. This principle of operation is also explained by the waveforms, sketched in Fig. 7.18.

7 From AC to DC and Reverse, the Next Fully Integrated Power Management Challenge   121

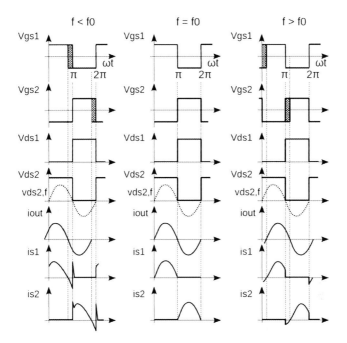

Fig. 7.18 Waveforms for operation at f < f0, f = f0 and f > f0

7.4.3.2 Inherent Reduced Switching Losses

An interesting feature of this topology is the inherent reduced switching loss compared to a classic pulse-width modulated (PWM) switch-mode power supply. Figure 7.18 shows the waveforms for the operation at, below and above the resonant frequency of the series resonant converter. For operation at the resonant frequency, $f = f_0$, the transistors turn on and off at zero current, resulting in low switching losses and high efficiency. In many cases, the operating frequency f is not equal to the resonant frequency f_0 since the output voltage and power can be controlled by varying the operating frequency f. Each transistor should be turned off for $f < f_0$ and turned on for $f > f_0$ during the time interval when the switch current is negative. During this time, the current can flow through the anti-parallel diode. To prevent short-circuit currents, a non-overlapping clock with sufficient dead-time must be used.

When the inverter is operated below resonance, zero-current switching (ZCS) will occur. The series resonant circuit will represent a capacitive load [19], this means that the current i through the resonant tank will lead the fundamental component $v_{DS2,F}$ of the voltage v_{DS2}. In this case, the transistor current goes to zero before the transistor is turned off. The circuit inherently causes the turn-off transition to be lossless. However, when the transistor is turned on, its parasitic output capacitance is discharged through its on-resistance, causing a switching loss.

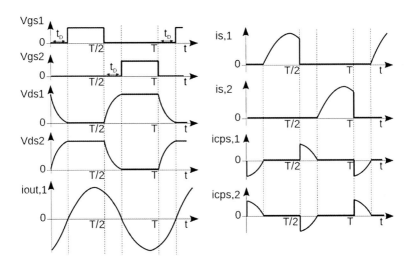

Fig. 7.19 Waveforms for the series resonant class DE inverter

When the inverter is operated above resonance, zero-voltage switching (ZVS) will occur. The series resonant circuit now represents an inductive load. The circuit naturally causes the transistor voltage to become zero before the transistor is turned on. The current i lags $v_{DS2,F}$. The turn-on transition is now lossless. Both the switch voltage and current waveforms overlap during turn-off, causing a switching loss.

7.4.3.3 Use of the Parasitic Output Capacitances

To assist the turn-off process above resonance, small shunt capacitors can be introduced in parallel with the transistors. The transistors used in this circuit are sufficiently large to use the output capacitances $C_{P,S1}$ and $C_{P,S2}$ as shunt capacitors. These shunt capacitors eliminate the turn-off switching loss during operation above the resonant frequency. This principle is depicted in Fig. 7.19. There is again a dead time in the gate-to-source voltages, during which both transistors are off. During the dead time, shunt capacitors become part of the resonant circuit. One shunt capacitor is charged and the other is discharged during the dead time. These capacitors introduce commutation intervals at transistor turn-off. When SW_1 is turned off, the tank current flows through capacitance $C_{P,S1}$ instead of SW_1 itself and the voltage across SW_1 and $C_{P,S1}$ increases. If the turn-off time is sufficiently fast, the transistor is turned off before the drain voltage rises too much above zero. A negligible switching loss is now incurred. This series resonant converter achieves the class E switching conditions [17], which means zero-voltage switching and zero-voltage slope switching. This topology enables higher efficiency since it eliminates switching power losses during the transition time of the switches.

7 From AC to DC and Reverse, the Next Fully Integrated Power Management Challenge   123

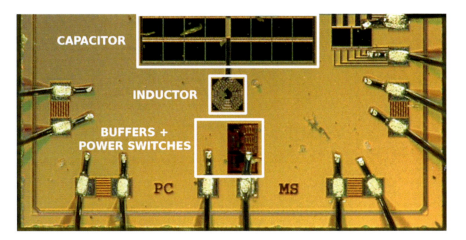

Fig. 7.20 Die photograph of the chip

7.4.4 Monolithic Integration of a Class DE Series Resonant Inverter

7.4.4.1 Implementation

The design and layout of the different on-chip components in the series resonant class DE inverter are discussed in this section. An external load resistor is used to have freedom during the measurements. This corresponds to the load Z_{RECT} in Fig. 7.17.

The inductor is implemented as an octagonal metal track, hollow spiral inductor. A standard metal layer, a thick top metal layer of 2 μm, and an aluminium layer of 1.2 μm are used to reduce the parasitic series resistance. It consists of two windings of 20 μm, resulting in a total on-chip inductance of 1.7 nH. The load is connected off-chip using a bondwire, this introduces extra inductance. The total measured extracted inductance equals 5.1 nH. The series resistance at a frequency of 500 MHz is 0.5 Ω, and is taken as an upper limit since the inverter is designed to be used around 100 MHz. The inductor area is 125 μm by 125 μm. This can be seen in the middle of Fig. 7.20.

The capacitor in the presented class DE inverter is implemented by means of a MIMCAP. In the used technology this yields a capacitance density of 2 fF/μm2. A total capacitance of 145.6 pF is realised. The electrical series resistance was estimated using the sheet resistances of the two metal layers, resulting in 0.8. The capacitor can be seen in the top part of Fig. 7.20.

The power switches, SW_1 and SW_2 are implemented using a fingered layout. The width of the power switches is 1.8 and 0.75 mm respectively. This width yields

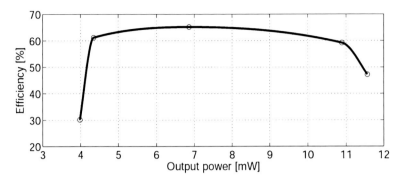

Fig. 7.21 The efficiency of the series resonant class DE inverter for varying loads

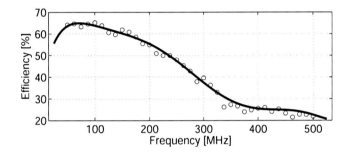

Fig. 7.22 The efficiency at increasing frequency for a 25 Ω load

an optimal tradeoff between losses in the buffers driving the switches and the conduction losses. The switches are driven by two buffer trains. The buffers apply the non-overlapping clock to the pMOS (SW_1) and nMOS (SW_2) transistor. These buffers and power switches are located in the bottom part of Fig. 7.20.

7.4.4.2 Measurements

The series resonant class DE inverter is implemented in a 130 nm 1.2 V CMOS technology. It measures 1.5 × 0.75 mm. Figure 7.20 shows a die photograph. Figure 7.21 shows the efficiency of the inverter as a function of the output power at varying resistive loads. The maximum efficiency is 65.2 % at a load of 6.9 mW with an output peak amplitude of 585 mV at a switching frequency of 100 MHz. A maximal output power of 11.6 mW is achieved at an efficiency of 47 % with an output peak amplitude of 312 mV.

Figure 7.22 shows the efficiency of the inverter in function of the switching frequency for a load of 25 Ω. An efficiency of 65.2 % is achieved. At a higher switching frequency, the efficiency decreases. It is paramount to control the switching frequency as to achieve the highest efficiency. This point is shifted depending on

7 From AC to DC and Reverse, the Next Fully Integrated Power Management Challenge   125

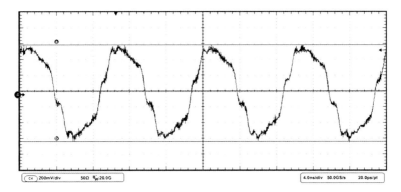

Fig. 7.23 Measured output voltage waveform at 100 MHz for 25 Ω

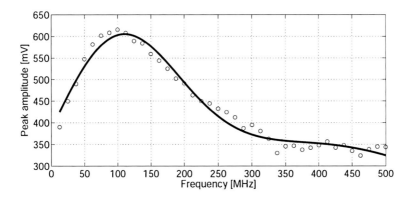

Fig. 7.24 Peak output amplitude as a function of switching frequency

the output load. For this integrated circuit, the efficient switching frequency will be around 100 MHz, depending on the load used.

In Fig. 7.23 the output voltage sine wave at 100 MHz is presented for a load of 25 Ω. The peak amplitude of the output waveforms is given in Fig. 7.24 in function of the switching frequency. There is a peak around 100 MHz.

The switching frequency is now varied between 100 and 400 MHz using a sinusoidal pulse frequency modulation at 20 kHz (period of 50 μs). This is depicted in Fig. 7.25. Using this PFM modulation, a modulated high-frequency carrier is generated at the output. The result is a high frequency carrier with a sinusoidal envelope at 20 kHz. This is shown in Fig. 7.26. The output peak amplitude varies between 326 and 648 mV.

Table 7.2 summarizes the measurement results and makes a comparison with existing simulations of an integrated series resonant class DE inverter [18]. The measurements of the discussed monolithic series resonant class DE inverter confirm these simulation results. Moreover, the performance is increased compared to the simulated results.

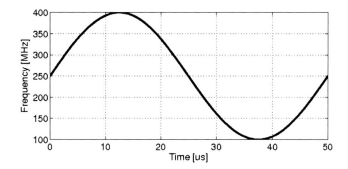

Fig. 7.25 Frequency sweep, one period of modulated 20 kHz sine wave shown

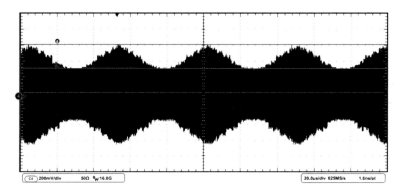

Fig. 7.26 Measured PFM modulated output waveform

Table 7.2 Summary of the measurement results

Reference	[18]	This work
Process (µm)	1.2 µm	0.13 µm
Input voltage	2 V	1.2 V
Maximum output peak amplitude	320 mV	600 mV
Maximum efficiency	48 %	65.2 %
Power at maximum efficiency	6.4 mW	6.9 mW
Switching frequency	500 MHz	100 MHz
Maximum output power @ efficiency	Not available	11.6 mW @ 47 %
Notes	Simulation only	Fully-integrated

7.5 Conclusions

This paper has discussed the advances of the next leap in integrated power management beyond the scope of today's DC-DC converters: the interaction with AC. Both AC to DC and DC to AC conversion were tackled with the intention of eventually interfacing mains.

An AC to DC conversion was implemented in CMOS. The topology was selected to combine both low voltage compatibility while maximizing the power throughput for a given set of component resources by an increased rectifier diode on time than in the prior art. A compact calculation model describing an ideal AC-DC capacitive step-down is presented and was found suitable due to the close match in operation of the selected topology and the ideal model, demonstrating its efficiency. Two separate prototypes of this design were implemented to show the practical limits of a fully integrated AC-DC converter and demonstrate the possibilities of allowing small external components to be present.

A DC-AC conversion was realized in a 130 nm 1.2 V CMOS technology using a series resonant class DE inverter topology. An on-chip spiral inductor and an integrated MIMcap were used. The inherent soft switching yields high conversion efficiency at high switching frequencies. The use of on-chip passives reduces the bill of materials considerably. The measurement results confirm and improve previous simulations.

References

1. A. Radić, Z. Lukić, A. Prodić, R.H. de Nie, Minimum-deviation digital controller IC for DC–DC switch-mode power supplies. Power Electron. IEEE Trans. **28**(9), 4281–4298 (2013)
2. M. Steyaert, T. Van Breussegem, H. Meyvaert, P. Callemeyn, M. Wens, DC-DC converters: from discrete towards fully integrated CMOS, in *Proceedings of the ESSCIRC (ESSCIRC)*, Helsinki, 2011, 12–16 Sept 2011, pp. 42–49
3. F. Waldron, R. Foley, J. Slowey, A.N. Alderman, B.C. Narveson, S.C. Ó Mathúna, Technology roadmapping for power supply in package (PSiP) and power supply on chip (PwrSoC). Power Electron. IEEE Trans. **28**(9), 4137–4145 (2013)
4. T.M. Van Breussegem, M.S.J. Steyaert, Monolithic capacitive DC-DC converter with single boundary–multiphase control and voltage domain stacking in 90 nm CMOS. Solid-State Circuit IEEE J. **46**(7), 1715–1727 (July 2011)
5. D. Somasekhar, B. Srinivasan, G. Pandya, F. Hamzaoglu, M. Khellah, T. Karnik, K. Zhang, Multi-phase 1 GHz voltage doubler charge pump in 32 nm logic process. Solid-State Circuit IEEE J. **45**(4), 751–758 (2010)
6. S. Sheikhaei, M. Alimadadi, G.G.F. Lemieux, S. Mirabbasi, W.G. Dunford, P.R. Palmer, Energy recycling from multigigahertz clocks using fully integrated switching converters. Power Electron. IEEE Trans. **28**(9), 4227–4239 (2013)
7. M. Steyaert, P. Vancorenland, CMOS: A paradigm for low power wireless? in *Proceedings of the 39th Design Automation Conference*, New Orleans, LA, USA, 2002, pp. 836–841
8. M. Pomper, L. Leipold, R. Muller, R. Weidlich, On-chip power supply for 110 V line input. Solid-State Circuit IEEE J. **13**(6), 882–886 (1978)
9. A.A. Tamez, J.A. Fredenburg, M.P. Flynn, An integrated 120 volt AC mains voltage interface in standard 130 nm CMOS, in *Proceedings of the ESSCIRC*, Sevilla, 2010, 14–16 Sept 2010, pp. 238–241
10. A.-J. Annema, G.J.G.M. Geelen, P.C. de Jong, 5.5-V I/O in a 2.5-V 0.25-μm CMOS technology. Solid-State Circuit IEEE J. **36**(3), 528–538 (2001)
11. B. Serneels, T. Piessens, M. Steyaert, W. Dehaene, A high-voltage output driver in a 2.5-V 0.25-μm CMOS technology. Solid-State Circuit IEEE J. **40**(3), 576–583 (2005)
12. V.W. Ng, S.R. Sanders, A high-efficiency wide-input-voltage range switched capacitor point-of-load DC–DC converter. Power Electron. IEEE Trans. **28**(9), 4335–4341 (2013)

13. M. Ghovanloo, K. Najafi, Fully integrated wideband high-current rectifiers for inductively powered devices. Solid-State Circuit IEEE J. **39**(11), 1976–1984 (2004)
14. Timedomain CVD Inc. silicon dioxide: Properties and applications. [Online]. Available: http://www.timedomaincvd.com/CVD/Fundamentals/films/SiO2 properties.html/
15. R. Erickson, D. Maksimovic, *Fundamentals of Power Electronics*, New York, (Springer, 2004)
16. B. Sahu, G. Rincon-Mora, An accurate, low-voltage, cmos switching power supply with adaptive on-time pulse-frequency modulation (pfm) control. Circuit Syst. I: Regul. Pap. IEEE Trans. **54**(2), 312–321 (2007)
17. N. Sokal, Class e high-efficiency switching-mode tuned power amplifier with only one inductor and one capacitor in load network-approximate analysis. J. Solid-State Circuit **16**(4), 380–384 (1981)
18. T. Suetsugu, M. Kazimierczuk, Integration of class de inverter for on-chip dc-dc power supplies, in *Circuits and Systems, 2006. ISCAS 2006. Proceedings. 2006 I.E. International Symposium on*, Island of Kos, May 2006, p. 4
19. M. Kazimierczuk, D. Czarkowski, *Resonant Power Converters*, Hoboken, New Jersey, USA, (Wiley, 2010)

Chapter 8
Fully Integrated Switched-Capacitor DC-DC Conversion

Elad Alon, Hanh-Phuc Le, John Crossley, and Seth R. Sanders

Abstract This chapter describes techniques to maximize the achievable efficiency and power density of fully integrated switched-capacitor (SC) DC-DC converters. Circuit design methods to support multiple topologies (and hence output voltages) are described. These techniques are verified by a proof-of-concept converter prototype implemented in 0.374 mm^2 of a 32 nm SOI process. The 32-phase interleaved converter can be configured into three topologies to support output voltages of 0.5–1.2 V from a 2 V input supply, and achieves ~80 % efficiency at an output power density of 0.86 W/mm^2.

8.1 Introduction

As parallelism is now the dominant mechanism by which integrated circuit designers improve the computing performance of their chips while remaining within strict power budgets, there is increasing need and potential benefit to utilizing an independent power supply for each processing core. Simply adding off-chip supplies not only incurs significant degradation of supply impedance due to e.g. split package power planes, but also additional cost due to increased motherboard size and package complexity. Therefore, there is strong motivation to fully integrate voltage conversion on the die, as shown in Fig. 8.1.

Although on-die DC-DC converters are currently almost always implemented as linear regulators, achieving high efficiency across a broad range of output voltages necessitates the use of switching converters. Inductor-based switching converters are dominant in off-chip converters, and recent efforts to co-package and reduce the inductor size [1, 2] have brought them closer to complete integration. However, fully integrated DC-DC converters based on CMOS inductors either require costly

---

E. Alon (✉) • H.-P. Le • J. Crossley • S.R. Sanders
University of California, Berkeley, USA
e-mail: elad@eecs.berkeley.edu

A. Baschirotto et al. (eds.), *Frequency References, Power Management for SoC, and Smart Wireless Interfaces: Advances in Analog Circuit Design 2013*, DOI 10.1007/978-3-319-01080-9_8, © Springer International Publishing Switzerland 2014

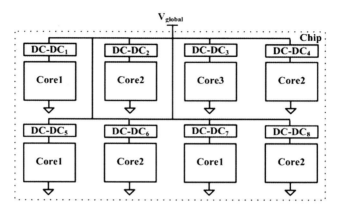

Fig. 8.1 Chip with multiple, local, on-die DC-DC converters

extra fabrication steps [3, 4] (e.g., thick metals or integrated magnetics), or suffer from the high series resistance (and hence low efficiency [5]) of standard on-die inductors. Integrated capacitors on the other hand can achieve low series resistance and high capacitance density, and most importantly, can be used to implement DC-DC converters in completely standard CMOS processes.

Given these advantages, fully integrated switched-capacitor (SC) converters have recently received significant attention from multiple researchers. For example, [6] and [7] both investigated multiphase interleaving for fully integrated SC voltage doublers, with [6] demonstrating high efficiency (82 %) but low power density (0.67 mW/mm^2), and [7] achieving high power density (1.123 W/mm^2), but low efficiency (60 %). The need for high efficiency is self-evident, but high power density is also critical since it sets the area overhead of the converter relative to the on-chip circuitry it is supplying power to.

In order to explore the boundaries of their capabilities, in this chapter we describe a methodology to achieve the optimal tradeoff between efficiency and power density for fully integrated SC converters. Section 8.2 therefore presents an analysis and optimization of SC converter losses as a function of power density, and discusses the use of topology reconfiguration and output impedance control to enable wide output voltage range. Section 8.3 then describes a converter prototype with reconfigurable topology. Measurement results from the prototype converter verifying the predicted performance and proposed techniques are presented in Sect. 8.4, and the chapter is finally concluded in Sect. 8.5.

8.2 SC Converter Analysis and Optimization

To achieve the optimal tradeoff between power density and efficiency, in this section we will analyze the operation and loss mechanisms of SC converters. This analysis will lead to design equations for switching frequency and switch width that

8 Fully Integrated Switched-Capacitor DC-DC Conversion

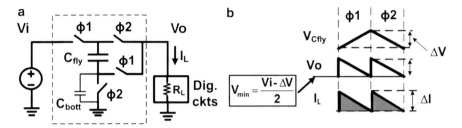

Fig. 8.2 (a) 2:1 step-down SC DC-DC converter and (b) its operating waveforms

minimize losses in a given technology and power density. Since a single-topology SC converter is only efficient when generating an output voltage within a limited range, this section also describes a simple design strategy for enabling reconfigurable topologies as well as predicting the overall efficiency vs. output voltage.

8.2.1 Operation of a Sample SC Converter

In order to highlight the key loss mechanisms that will set the tradeoff between converter efficiency and area (i.e., power density), we will begin by examining the operation of the 2:1 step-down converter shown in Fig. 8.2a.

Switched-capacitor DC-DC converters typically operate in two phases, each of which ideally has 50 % duty cycle. While it is possible to operate SC DC-DC converters at a fixed switching frequency and use variable duty cycle to adjust the output impedance [8, 9], maximum efficiency can only be achieved by optimizing switching frequency and operating with 50 % duty cycle.

Under this condition (shown in Fig. 8.2), during phase $\phi 1$, the flying capacitor C_{fly} is connected between the input node Vi and the output node Vo. The charge drawn from Vi though C_{fly} charges up this capacitor and flows to the load. In phase $\phi 2$, C_{fly} is connected between Vo and GND, and thus the charge previously stored on the flying capacitor is transferred to the output. Since the switching cycle is often much smaller than the charge/discharge time constant (which is set by $R_L C_{fly}$), the ramp rate of the voltage across the capacitor is relatively constant, and hence the load can be treated as a current source.

As will be described later, in order to maximize efficiency it is desirable to utilize all available capacitance within the converter itself. Therefore, we will assume that there is no explicit output filtering capacitor, which in the case of the simple SC converter described so far, makes the peak-to-peak voltage ripple across the capacitor and the converter's output equal, as shown in Fig. 8.2b. This voltage ripple has a direct implication on the loss – and hence the achievable efficiency – of the converter.

8.2.2 Loss Analysis

The voltage ripple across the capacitors scales with the load current, and will therefore act as a form of series loss similar to the switch conduction losses. In addition, any SC converter will also have shunt losses that are independent of the load current, including gate and bottom plate capacitor switching losses. Note that the control circuitry for an SC converter will also contribute to shunt loss, but will be neglected here since this loss can usually be made relatively small. These losses can be modeled as shown in Fig. 8.6, where the series losses are represented by an equivalent output resistance R_o [10, 11], the shunt losses by the parallel resistor R_p, and the transformer represents the ideal voltage conversion ratio.

In order to show the relationship between voltage ripple across the capacitor and loss, we should recall that most fully integrated switched-capacitor converters will be delivering power to synchronous digital circuitry. The performance of synchronous digital systems is determined by the operating frequency, which in turn is set by the minimum average voltage over a clock period. Since the clock period of most digital circuits will be short in comparison to the converter's switching period, the performance of these circuits is typically simply set by the minimum voltage V_{min} of the supply [12]. In this case, the efficiency of the converter should be calculated relative to the power that would have been consumed by the load if it was constantly operating at exactly V_{min} [12]. In other words, the ideal power consumed by the load is:

$$P_{L-min} = V_{min}I_L \tag{8.1}$$

where $I_L = \frac{V_{min}}{R_L}$. However, due to the voltage ripple from the converter, and assuming that this ripple is relatively small compared to the nominal voltage, the average power dissipated by the load is approximately:

$$P_{L-tot} \approx \left(V_{min} + \frac{\Delta V}{2}\right)\left(I_L + \frac{\Delta I}{2}\right) \tag{8.2}$$

where ΔV is the output voltage ripple (due to the operation of the converter) and $\Delta I = \Delta V/R_L$.

Although P_{L-tot} is indeed dissipated by the load, any power consumed beyond P_{L-min} should be counted as loss since this extra power does not contribute to an increase in performance. In order to quantify this loss, we need to calculate V_{min} and ΔV; as shown in Fig. 8.2b, for the 2:1 converter considered here, V_{min} is lower than the ideal output voltage $Vi/2$ by $\Delta V/2$:

$$V_{min} = \frac{Vi}{2} - \frac{\Delta V}{2}, \tag{8.3}$$

8 Fully Integrated Switched-Capacitor DC-DC Conversion 133

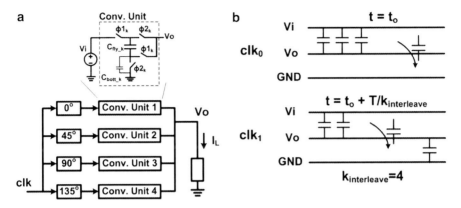

Fig. 8.3 (a) Sample four-phase interleaved SC converter and (b) operation of its flying capacitors

and the voltage ripple ΔV is set by:

$$\Delta V = \frac{I_L}{C_{fly}} \cdot \frac{T}{2} = \frac{I_L}{2C_{fly}f_{sw}}, \qquad (8.4)$$

where T is the switching period and $f_{sw} = 1/T$ is the switching frequency.

As should be clear from Eq. 8.2, the loss caused by the operation of the converter is due to both the voltage ripple ΔV as well as due to the excess current flowing in the load ΔI. The loss due to the voltage ripple ΔV is unavoidable because the voltage drop $\Delta V/2$ in Eq. 8.3 is inherent to the fact that charge (power) is being delivered through a capacitor. However, the current ripple ΔI can be eliminated if the ripple in the output voltage above V_{min} is minimized.

Fortunately, the ripple in the output voltage and hence the load current ripple can be reduced by multiphase interleaving. As described in [6], [7], and [13], multiphase interleaving is implemented by partitioning the converter into sub-units and switching each one of these units on a different clock phase. Figure 8.3 depicts a sample four-phase interleaved design and the operation of the flying capacitors in clock phase 0 (clk$_0$) and clock phase 1 (clk$_1$). Each unit in this converter uses 1/4 of the total capacitance and a clock that is 45° phase-shifted from its neighbor.

The total charge (per switching cycle) required by the output is the same as that in the converter without interleaving, but is equally divided among each unit. Thus, the charge flowing through each unit flying capacitor in the interleaved design is the same as it would be in the original design. As illustrated in Fig. 8.4a, the voltage ripple on each unit capacitor required to deliver that charge is therefore essentially identical to the previous ΔV from Eq. 8.4. As a result, V_{min} is unchanged. However, because the charge delivered to the output is divided more finely, the output voltage and current ripple are reduced by the interleaving factor ($k_{interleave} = 4$), as shown in Fig. 8.4b. This leads to a reduction in the loss associated with the current ripple:

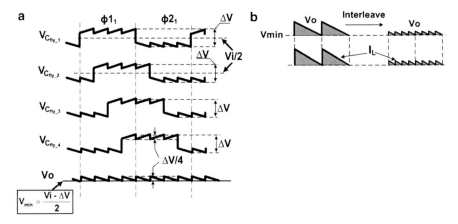

Fig. 8.4 (a) Flying capacitor voltages and (b) their effect on output voltage and current ripple of the four-phase interleaved converter in Fig. 8.3

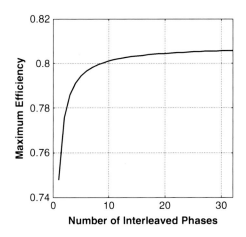

Fig. 8.5 Efficiency of an example 2:1 SC converter as a function of $k_{interleave}$

$$P_{L-tot} \approx \left(V_{min} + \frac{\Delta V}{2}\right)\left(I_L + \frac{1}{k_{interieave}} \cdot \frac{\Delta I}{2}\right) \qquad (8.5)$$

As shown in Fig. 8.5, interleaving the converter by roughly by a factor of 10 (which is relatively simple in an integrated design) is sufficient to essentially eliminate the efficiency penalty due to load current ripple. In other words, extreme levels of interleaving are generally not necessary – especially if they would result in significant control overhead. Assuming sufficient interleaving (i.e., $k_{interleave}$ > ~10) we can generally ignore the loss caused by the current ripple, resulting in the classic SC loss [10] given by:

Fig. 8.6 Simplified SC converter model for calculation of losses

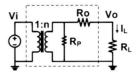

$$P_{C_{fly}} = I_L \cdot \frac{\Delta V}{2} = \frac{I_L^2}{M_{cap} C_{fly} f_{sw}} \tag{8.6}$$

where M_{cap} is a constant related to the converter's output resistance and is determined by the converter's topology (e.g., $M_{cap} = 4$ for a 2:1 SC converter).

Beyond the intrinsic SC loss, the finite conductance of the switches leads to another series loss term. To simplify the equations, we will assume here that all of the switches have identical characteristics (regardless of type or gate overdrive), but it is straightforward to extend the analysis to handle differences between each switch. The switch conductance loss $P_{R_{sw}}$ is therefore set by:

$$P_{R_{sw}} = I_L^2 \frac{R_{on}}{W_{sw}} M_{sw} \tag{8.7}$$

where R_{on} is the switch resistance density measured in $\Omega \cdot m$, W_{sw} (m) is the total width of all switches, and M_{sw} is a constant determined by the converter's topology. For the 2:1 converter in Fig. 8.2, there are four switches, and each occupies 1/4 of the total switch area. During each half of a switching period, two of the four switches conduct the current flowing into to the output, resulting in:

$$M_{sw} = N_{switches, tot} \times \left(\frac{T_{ph1}}{T} \times N_{sw, on, ph1} + \frac{T_{ph2}}{T} \times N_{sw, on, ph2} \right) \tag{8.8}$$

$$= 4 \times \left(\frac{1}{2} \times 2 + \frac{1}{2} \times 2 \right) = 8$$

As shown in Eqs. 8.6 and 8.7, the intrinsic switched-capacitor loss and switch conductance loss are both set by the load current, and can hence be modeled by the equivalent output resistance R_o in Fig. 8.6. The total series loss is therefore approximately set by:

$$P_s = I_L^2 R_o = P_{R_{sw}} + P_{C_{fly}} \tag{8.9}$$

The other key portion of an SC converter's losses stems from shunt losses due to switching the parasitic capacitance of the flying capacitors and power switches. Any flying capacitor – particularly fully integrated ones – will have parasitic capacitance associated with both its top plate and its bottom plate. In steady-state operation, both of these plates experience approximately equal voltage swings.

Therefore, we will group both losses caused by the top-plate capacitor $C_{top-plate}$ and the bottom-plate capacitor $C_{bottom-plate}$ into one parasitic capacitor switching loss $P_{bott-cap}$, given by:

$$P_{bott-cap} = M_{bott}V_o^2 C_{bott}f_{sw} \qquad (8.10)$$

where M_{bott} is a constant determined by the converter's topology (e.g., $M_{bott} = 1$ for a 2:1 SC converter) and $C_{bott} = C_{bottom-plate} + C_{top-plate}$. For simplicity, once again assuming that all of the switches have identical characteristics, the gate parasitic capacitance switching loss $P_{gate-cap}$ is given by:

$$P_{gate-cap} = V_{sw}^2 W_{sw} C_{gate}f_{sw} \qquad (8.11)$$

where V_{sw} is the gate voltage swing and C_{gate} is the gate capacitance density (F/m) of the switches.

8.2.3 Loss Optimization

In order for the converter to achieve the highest overall efficiency at a given power density we must minimize the total loss, which is set by the combination of the four previously discussed components:

$$\begin{aligned}
P_{loss} &= \left(P_{C_{fly}} + P_{R_{sw}}\right) + \left(P_{bott-cap} + P_{gate-cap}\right) \\
&= \left(\frac{I_L^2}{M_{cap}C_{fly}f_{sw}} + I_L^2 \frac{R_{on}}{W_{sw}}M_{sw}\right) + \left(M_{bott}V_o^2 C_{bott}f_{sw} + V_{sw}^2 W_{sw} C_{gate}f_{sw}\right)
\end{aligned}$$

$$(8.12)$$

For a given technology, R_{on} and C_{gate} are set by the available transistors, and hence are essentially fixed. Similarly, the intrinsic switched-capacitor loss $P_{C_{fly}}$ (and hence the overall loss) will always be minimized by utilizing as large of a flying capacitor C_{fly} as possible given the chip area constraint. Therefore, at a given power density, the only two variables that can be optimized to minimize the total losses are switch width W_{sw} and switching frequency f_{sw}.

Increasing either switch width or switching frequency decreases the series losses at the cost of increasing the shunt loss. Minimizing the converter's total losses therefore boils down to setting the values of W_{sw} and f_{sw} to balance series and shunt losses. As we will detailed next, the power density required of the converter plays an important role in determining the most dominant loss components, and hence how W_{sw} and f_{sw} should be set to minimize loss.

At high power densities (i.e., large I_L or equivalently small R_L, where $R_L = V_o/I_L$), W_{sw} and f_{sw} must both increase with the load current in order to suppress the series losses. Since $P_{gate-cap}$ is proportional to both width and frequency while $P_{bott-cap}$ scales only with switching frequency, beyond a certain load current the bottom plate loss becomes the least significant term. To arrive at simple analytical equations for the optimal f_{sw} and W_{sw} in this regime, we can thus ignore the bottom plate portion of the shunt losses. In this case, the optimal f_{sw} and W_{sw} will be:

$$f_{sw_{opt}} = \frac{1}{\sqrt[3]{M_{cap}^2 M_{sw}}} \cdot \sqrt[3]{\frac{V_o^2}{V_{sw}^2}} \times \frac{1}{R_{on} C_{gate} \left(R_L C_{fly}\right)^2} \qquad (8.13)$$

$$W_{sw_{opt}} = \sqrt[3]{M_{sw}^2 M_{cap}} \cdot \sqrt[3]{\frac{V_o^2}{V_{sw}^2} \frac{R_{on}^2 C_{fly}}{R_L^2 C_{gate}}} \qquad (8.14)$$

Under these conditions and with the optimal f_{sw} and W_{sw}, the minimum normalized loss (which sets the efficiency $\eta = (1 + P_{loss}/P_L)^{-1}$) is approximately:

$$\frac{P_{loss}}{P_L} = 3 \sqrt[3]{\frac{M_{sw}}{M_{cap}}} \cdot \sqrt[3]{\frac{V_{sw}^2}{V_o^2} \frac{R_{on} C_{gate}}{R_L C_{fly}}} \qquad (8.15)$$

This relative loss expression highlights the tradeoff between power density and efficiency (Fig. 8.7). For a given technology and converter topology, increasing the power density by a factor of x at a given output voltage implies that R_L also decreases by a factor of x, leading to an increase in the minimum normalized loss by a factor of $\sqrt[3]{x}$.

This relative loss expression also highlights that the most important technology metric guiding the selection of the switches is the product of gate voltage swing squared and intrinsic time constant (i.e., $V_{sw}^2 R_{on} C_{gate}$). Similarly, since it is the ratio of this switch metric to the load voltage squared multiplied by the effective time constant for charging/discharging the flying capacitors (i.e., $V_o^2 R_L C_{fly}$), increasing the density of the capacitors also directly improves efficiency at a given power density.

Although the previous analysis provides a clear intuitive picture of the relationship between power density and efficiency, it is only accurate at high power densities where the loss due to switching the "bottom-plate" parasitics of the flying capacitors is negligible compared to the other losses. Both the optimal switching frequency and the switch area scale down as output power drops, and hence at low power densities the losses due to driving the switch parasitic capacitors become much smaller than the all of the other losses. Therefore, in this regime we can approximately find the optimum loss by ignoring the switch gate loss and finding the optimum switching frequency $f_{sw,opt}$:

$$f_{sw,opt} = \frac{1}{\sqrt{M_{cap} M_{bott} k_{bott}}} \frac{1}{C_{fly} R_L} \qquad (8.16)$$

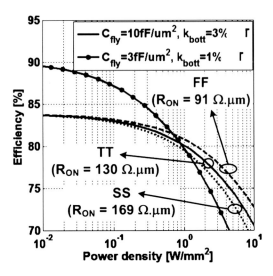

Fig. 8.7 Analytical predictions of optimized power density vs. efficiency for a 2:1 SC converter. The switch characteristics of a 32 nm CMOS technology (i.e., $R_{ON} = 130\ \Omega \cdot \mu m$, $C_{gate} = 3\ fF/\mu m$, $V_{sw} = 1\ V$) were used to generate these curves, which also highlight the impact of +/−30 % variations in R_{ON}

where $k_{bott} = C_{bott}/C_{fly}$ is the parasitic to flying capacitance ratio. Although the switch gate losses were assumed to be small, we can still size the switches to minimize the total loss in Eq. 8.12 with the frequency found in Eq. 8.16:

$$W_{sw} = \sqrt{\frac{V_o^2}{V_{sw}^2} \cdot \frac{R_{on}C_{fly}}{R_L C_{gats}}} \cdot \sqrt{M_{sw}^2 M_{cap} M_{bott} k_{bott}} \qquad (8.17)$$

Combining the results from Eqs. 8.12, 8.16 and 8.17, the normalized loss in the low power density regime is:

$$\frac{P_{loss,opt}}{P_L} = 2\sqrt{\frac{M_{bott}}{M_{cap}} k_{bott}} + 2\sqrt{\frac{M_{sw}}{\sqrt{M_{cap} M_{bott}}} \cdot \frac{1}{\sqrt{k_{bott}}}} \cdot \sqrt{\frac{V_{sw}^2}{V_o^2} \cdot \frac{R_{on}C_{gate}}{R_L C_{fly}}} \qquad (8.18)$$

This result highlights a key intrinsic limit on the efficiency of a switched-capacitor DC-DC converter. Even in very light load conditions (i.e., $R_L = \infty$), the maximum efficiency of the converter is limited by the bottom-plate capacitance ratio k_{bott} and the converter's topology – i.e., by the first term in Eq. 8.18. For example, with a bottom-plate capacitor ratio of 1 %, the efficiency of a 2:1 converter is limited to 90.9 %. Of course, any non-zero load will decrease the efficiency of the converter, but for sufficiently light loads the efficiency will still be dominated by bottom-plate losses.

To illustrate these effects, Fig. 8.7 shows the efficiency vs. power density curves of two optimized converter designs with different flying capacitor characteristics. One converter employs capacitors with high capacitance density but also a higher k_{bott} (e.g., MOS capacitor), while the other employs capacitors with lower density but also lower parasitics (e.g., a MIM or MOM capacitor). At high power densities (where Eq. 8.15 accurately predicts the minimum loss), high capacitance density directly translates into higher efficiency. However, at low power densities (where Eq. 8.18 is more accurate) the flying capacitors should have as low parasitic capacitance as possible in order to maximize peak efficiency.

Beyond illustrating the importance of selecting an appropriate capacitor given the target power density, Fig. 8.7 also predicts that a 2:1 SC converter using currently available CMOS technology can achieve ~80 % efficiency at a power density of ~1 W/mm^2. While this performance is substantially better than previous predictions or demonstrations of fully integrated DC-DC converters [3, 5–7, 14], it is only achievable at a single output voltage.

8.2.4 Output Voltage Range Considerations

Unlike in inductor-based converters where charge is saved and transferred in the form of current in the inductors – which enables efficient control of the output voltage by modulating the DC voltage applied to one side of the inductor – SC converters save and transfer charge as a voltage on the flying capacitors. The output voltage of a SC converter is thus determined by its topology.

To efficiently achieve a wider output voltage range, SC converters require reconfigurable topologies that can support multiple conversion ratios [15, 16]. By using a given number of reconfigurable topologies, an SC converter can support the same number of discrete open-circuit voltage levels. Intermediate voltages between these discrete levels can then be obtained by controlling R_o, which is equivalent to linear regulation off of the open-circuit voltages.

As discussed in [10] and shown here in Eqs. 8.9 and 8.12, the converter's output impedance R_o can be adjusted by controlling one or the combination of switching frequency f_{sw} [16], switch sizing W_{sw}, and effective flying capacitance C_{fly} [17]. Figure 8.8 shows the resulting efficiency vs. output voltage for a converter operating off a 2 V input and allowing reconfiguration into one of three possible topologies with conversion ratios of 1/3, 1/2, and 2/3. Even with linear regulation performed only by adjusting f_{sw} (which is slightly sub-optimal), Fig. 8.8 predicts that such a converter could achieve above 70 % efficiency for most output voltages spanning from ~0.5 up to ~1.2 V.

8.3 SC Converter Circuit Design

In order to verify the previously described optimization strategy as well as the predicted performance, a prototype SC converter was designed and fabricated in a 32 nm SOI test-chip in collaboration with AMD [16]. Although proper selection of the flying capacitor, switch width, and switching frequency (as outlined in the previous section) are critical to achieving a converter with high efficiency and

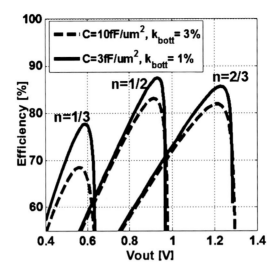

Fig. 8.8 Predicted efficiency vs. Vo with three reconfigurable topologies for two capacitor implementations. For both types of capacitors, the load is adjusted so that the converter is supplying 0.1 W/mm^2 with a Vo of 0.95 V. These curves assume that R_L varies along with the output voltage in the same manner as a CMOS ring oscillator

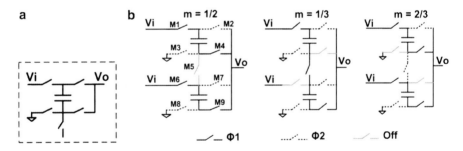

Fig. 8.9 (a) Standard cell and (b) reconfigurable converter unit

minimal area overhead, the need to support reconfigurable topologies (in this design, 2/3, 1/2, and 1/3) and multiple output voltage results in several circuit design challenges which must be overcome as well.

As with any custom designed VLSI structure, a physical design strategy that enables one to construct larger SC converters blocks by arraying identical sub-converter unit cells is highly desirable. In order to achieve this goal while supporting topology reconfiguration, we therefore propose to partition the converter into a unit cell consisting of one flying capacitor and five switches, as shown in Fig. 8.9a. Conceptually, each standard cell can be configured to operate in series or in parallel with the rest of the cells, leading to a simple physical design strategy that supports multiple conversion ratios. As shown in Fig. 8.9b, for this prototype

8 Fully Integrated Switched-Capacitor DC-DC Conversion 141

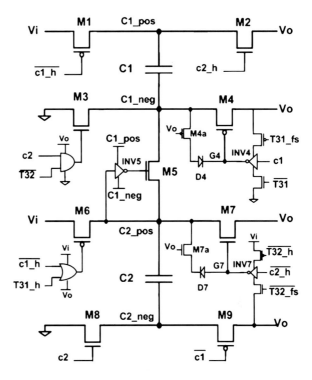

Fig. 8.10 Transistor-level implementation of the converter unit cell. The converter operates off of two non-overlapping clocks c1 and c2

converter we have grouped together two standard cells to form a converter unit supporting three topologies with conversion ratios of 1/3, 1/2, and 2/3 ($V_o = 0.66$, 1, or 1.33 V with a 2 V input). For simplicity, the intermediate voltage levels are generated by controlling f_{sw}. Figure 8.10 highlights the complete transistor-level implementation of this converter unit.

8.4 Experimental Verification

A die photo of the implemented SC converter employing the previously described design techniques is shown in Fig. 8.11. To maximize efficiency at high power densities and mitigate the current ripple losses, this design utilizes standard thin-oxide MOS transistors to implement C_{fly} as well as 32-way interleaving. This level of interleaving was chosen because even at high power densities, the converter's optimal switching frequency is relatively low compared to the intrinsic speed of the transistors.

Since this converter is intended to be co-integrated with the load, measuring the converter's performance requires a careful testing strategy. We will therefore first describe the load structure and its characterization, followed by measured results verifying the design methodology and proposed design techniques.

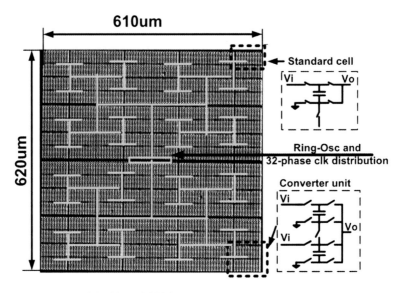

Fig. 8.11 Die photo of the 32 nm SOI SC converter prototype

8.4.1 Test Structure

In order to obtain correct I-V measurements of the on-die loading circuits and thus the efficiency of the converter, the on-die load – which was implemented with a variable-width PMOS device – must be pre-characterized. Four-wire sensing was used to measure the power consumption across the load in order to avoid inaccuracies due to drops in solder bumps, package pins, and PCB traces.

Load characterization was carried out by gating the clock of the converter (disabling it) and then driving the output node Vo of the converter from an off-chip power supply. For each load current (i.e., PMOS transistor width) setting, the voltage supply Vo is swept and the current consumption is measured. Utilizing this data, the power consumed by the load circuit while the converter is in its normal operation can then be extracted simply by measuring Vo.

8.4.2 Measurement Results and Discussion

Figure 8.12 shows the converter's measured efficiency and optimal switching frequencies in the 1/2 mode while supplying the on-die load circuits. For simplicity and in order to obtain optimal efficiency in this demonstration, the switching frequency was adjusted by externally controlling the supply of an on-chip ring oscillator. However, any one of a broad variety of techniques to control switching frequency [18] could be utilized.

8 Fully Integrated Switched-Capacitor DC-DC Conversion

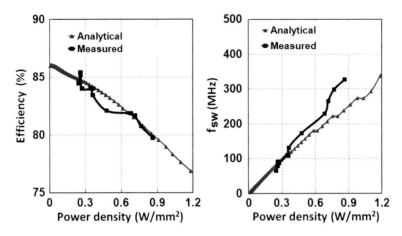

Fig. 8.12 Measured efficiency and optimal switching frequency versus power density in the 1/2 mode with Vi = 2 V and Vo ≈ 0.88 V

The measured converter achieved an efficiency of 79.76 % at 0.86 W/mm^2. The experimental data matches the analytical predictions to within 1.3 % across the range of measured power density (0.24–0.86 W/mm^2). Note that the performance quoted here is better than that reported in [16] due to the availability of new test-chips fabricated in a nearly production version of the process (rather than the developmental process used to obtain the original results).

Figure 8.13 shows the converter's efficiency vs. output voltage in the three operating modes, verifying that the converter functions correctly in all three of the reconfigurable topologies. The measurements in the 2/3 and 1/2 modes match very well with the analytical predictions. The measured efficiency in the 1/3 mode is however much lower. The cause of this discrepancy appears to be un-modeled leakage from Vi and Vo due to over-voltage-stress (~1.4 V) on switches M1, M2, M4, M6, M7 and M9 powered off of the Vi-Vo rails. Therefore, a practical implementation that uses the 1/3 conversion ratio would likely require a lower input voltage (~1.8 V), higher voltage devices, and/or switch cascoding. Despite this issue with the 1/3 conversion ratio, the two reconfigurable topologies enable the converter to maintain an efficiency of over 70 % for most of the output voltage range from 0.7 to ~1.15 V.

The converter's performance is summarized and compared with other work in Table 8.1. This prototype experimentally verifies that by following the design methodology and techniques proposed in this work, both boundaries in efficiency and power density of the previous works in [6] and [7] can be achieved with an implementation in a current commercial process. At 79.76 % efficiency and 0.86 W/mm^2, the proposed design could potentially be integrated into the same space as that already required for decoupling capacitors (as well as serve the same function) in a processor or SoC targeting mobile applications where the load operates at ~100 mW/mm^2. In fact, our recent work [20] demonstrated the efficient

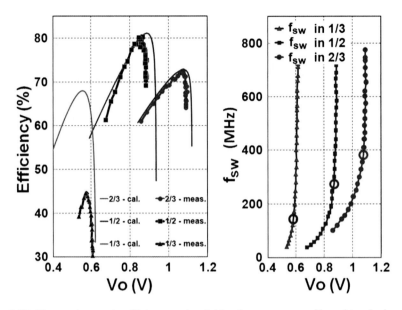

Fig. 8.13 Measured converter efficiency and switching frequency across Vo and topologies with Vi = 2 V and the load circuits set to $R_L \approx 0.9\ \Omega$ at Vo = 0.88 V

Table 8.1 Comparison of recently published fully integrated SC converters

Work	[6]	[7]	[19]	This work
Technology	130 nm bulk	32 nm bulk	45 nm SOI	32 nm SOI
Topology	2/1 step-up	2/1 step-up	1/2 step-down	2/3, 1/2, 1/3 step-down
Capacitor technology	MIM	Metal finger	Deep trench	CMOS oxide
Interleaved phases	16	32	1	32
C_{out}	400 pF (= C_{fly})	0	Yes	0
Converter area	2.25 mm^2	6,678 µm^2	1,200 µm^2	0.378 mm^2 (1.4 % used for load)
Quoted efficiency (η)	82 %	60 %	90 %	79.76 % (in 1/2 step-down)
Power density @ η	0.67 mW/mm^2	1.123 W/mm^2	2.185 W/mm^2	0.86 W/mm^2

integration of a converter interfacing directly to a Li-ion battery and generating a ~1 V output in a standard 65 nm CMOS process. This further opens up the opportunity for mobile SoC's to eliminate the need for external DC-DC converters completely.

In order to expand the applicability of SC converters to even higher performance processors operating at ~1 W/mm^2, the work reported in [19] utilizes ~200 fF/µm^2 deep trench capacitors and achieves 90 % efficiency at a power density of 2.185 W/mm^2. This further experimentally verifies the benefit of high density capacitors in increasing efficiency and power density – as also predicted in Eq. 8.16 and Fig. 8.7. In fact, the analysis from Sect. 8.2 predicts that with 200 fF/μm^2 deep trench capacitors and modern CMOS switches, an optimized SC design may achieve over 88 % efficiency for power densities up to 10 W/mm^2. Thus, the application of the techniques outlined in this chapter along with existing high-density capacitor technologies appears promising in enabling the broad adoption of fully integrated SC converters for on-die power distribution and management.

8.5 Conclusions

As parallelism increases the number of cores integrated onto a chip, there is a clear need for fully integrated DC-DC converters to enable efficient on-die power management. With the availability of high density and high quality capacitors in existing CMOS processes, switched-capacitor DC-DC converters have gained significant interest as a cost-effective means of enabling such power management functionality.

The area required by a fully integrated SC DC-DC converter to deliver a certain amount of power to the load has direct implications on both cost and efficiency. This chapter therefore describes a methodology to predict and minimize the losses of such a converter operating at a given power density. The chapter further introduced a circuit and physical design strategy to enable topology reconfiguration and hence efficient generation of a wider range of output voltages.

Measured results from a 32 nm SOI prototype confirm the methodology's predictions of ~80 % efficiency at a power density of ~0.5–1 W/mm^2 for a 2:1 step-down converter operating from a 2 V input and utilizing only MOS capacitors. Topology reconfiguration enables the converter to maintain >70 % efficiency for most of the output voltage range from 0.7 to ~1.15 V. Given that this performance was achieved in a standard CMOS process with no modifications or additions, these results illustrate that fully integrated switched-capacitor converters are indeed a promising candidate for low-cost but efficient power management on a per-core or per-functional unit basis.

References

1. Q. Li, Y. Dong, F.C. Lee, High density low profile coupled inductor design for integrated point-of-load converter, in *IEEE Applied Power Electronics Conference (APEC)*, 2010, pp. 79–85
2. G. Schrom, et al., A 100MHz eight-phase buck converter delivering 12A in 25mm2 using air-core inductors, in *IEEE Applied Power Electronics Conference (APEC)*. 2007, pp. 727–730
3. J. Wibben, R. Harjanai, A high-efficiency DC–DC converter using 2 nH integrated inductors. IEEE J. Solid-State Circuit **43**(4), 844–854 (2008)

4. D.S. Gardner, G. Schrom, P. Hazucha, F. Paillet, T. Karnik, S. Borkar, Integrated on-chip inductors with magnetic film. IEEE Trans. Magn. **43**(6) (2007)
5. J. Lee, G. Hatcher, L. Vandenberghe, C.K. Yang, Evaluation of fully integrated switching regulators for CMOS process technologies. IEEE Trans. VLSI **15**, 1017–1117 (2007)
6. T. Van Breussegem, M. Steyaert, A 82% efficiency 0.5% ripple 16-phase fully integrated capacitive voltage doubler, in *IEEE Symposium on VLSI Circuit*, June 2009, pp. 198–199
7. D. Somasekhar, B. Srinivasan, G. Pandya, F. Hamzaoglu, M. Khellah, T. Karnik, K. Zhang, Multiphase 1GHz voltage doubler charge-pump in 32nm logic process. IEEE J. Solid-State Circuit **45**(4), 751–758 (2010)
8. K.D.T. Ngo, R. Webster, Steady-state analysis and design of a switched-capacitor DC-DC. IEEE Power Electron. Spec. Conf. **1**, 378–385 (1992)
9. B.R. Gregoire, A compact switched-capacitor regulated charge pump power supply. IEEE J. Solid-State Circuit **41**(8), 1944–1953 (2006)
10. M.D. Seeman, S.R. Sanders, Analysis and optimization of switched-capacitor DC–DC converters. IEEE Trans. Power Electron. 841–851 (2008)
11. D. Maksimovic, S. Dhar, Switched-capacitor DC-DC converters for low-power on-chip applications. IEEE PESC **1**, 54–59 (1999)
12. E. Alon, M. Horowitz, Integrated regulation for energy-efficient digital circuits. IEEE J. of Solid-State Circuit **43**, 1795–1807 (2008)
13. D. Ma, F. Luo, Robust multiple-phase switched-capacitor DC-DC power converter with digital interleaving regulation. IEEE Trans. VLSI Syst. **16**(6), 611–619 (2008)
14. G. Patounakis, Y. Li, K.L. Shepard, A fully integrated on-chip DC-DC conversion and power management system. IEEE J. Solid-State Circuit **39**(3), 443–451 (2004)
15. Y.K. Ramadass, A.P. Chandrakasan, Voltage scalable switched capacitor DC-DC converter for ultra-low-power on-chip applications, in *IEEE Power Electronics Specialists Conference (PESC)*, 2007, pp. 2353–2359
16. H-P Le, M.D. Seeman, S.R. Sanders, V. Sathe, S. Naffziger, E. Alon, A 32nm fully integrated reconfigurable switched-capacitor DC-DC converter delivering 0.55W/mm^2 at 81% efficiency. IEEE ISSCC Dig. Tech. Pap. 210–211 (2010)
17. Y. Ramadass, A. Fayed, B. Haroun, A. Chandrakasan, A 0.16mm^2 completely on-chip switched-capacitor DC-DC converter using digital capacitance modulation for LDO replacement in 45nm CMOS. IEEE ISSCC Dig. Tech. Pap. 208–209 (2010)
18. S.K. Enam, A.A. Abidi, A 300-MHz CMOS voltage-controlled ring oscillator. IEEE J. Solid-State Circuit **25**(1), 312–315 (1990)
19. L. Chang, R. Montoye, B. Ji, A. Weger, K. Stawiasz, R. Dennard, A fully integrated switched-capacitor 2:1 voltage converter with regulation capability and 90% efficiency at 2.3A/mm^2, in *IEEE Symposium VLSI Circuit*. 2010
20. H.-P. Le, J. Crossley, S.R. Sanders, E. Alon, A Sub-ns response fully integrated battery-connected switched-capacitor voltage regulator delivering 0.19W/mm^2 at 72% efficiency. IEEE ISSCC Dig. Tech. Pap. 372–373 (2013)
21. M. Seeman, S.R. Sanders, Analysis and optimization of switched-capacitor DC-DC converters, in *10th IEEE Workshop on Computers in Power Electronics (COMPEL)*, 2006, pp. 216–224.
22. M. Seeman, *A Design Methodology for Switched-Capacitor DC-DC Converters*, University of California, Berkeley, Technical Report No. UCB/EECS-2009–78, 2009

Chapter 9
Battery Management in Mobile Devices

Francesco Rezzi, Luca Collamati, Maurizio Costagliola, and Massimo Cutrupi

Abstract The demand for higher performance and enriched user experience in mobile devices has steadily increased their power consumption over the past few years, a trend that rapidly outpaced the evolution of the Li-ion battery technology whose energy density simply could not keep up with the ever-increasing power demand. It is not uncommon for intensive users to charge their smartphones at least once a day if not twice. As it will be explain the recharging cycles lead to a deterioration of the battery performance that over time needs to be replaced. Nowadays battery life is among the biggest complaints among smart device users. It is understandable therefore that proper handling of the battery is becoming a hot topic in the engineering community particularly in light of the fact that more and more devices moved or are moving to non replaceable batteries. The paper will review the historical trend of battery technology and address battery and power management techniques aimed to increase battery life and safety with particular focus on smartphone and tablets

9.1 Introduction

At the dawn of the digital communication era, back in the 90s when the GSM digital network became de facto the mainstream communication technology, a typical cellphone would run on the average in the 100 mW range. Sure the GSM transmitter would require large peak current but the functionality of the phone was limited to voice communication and text messages with some basic managing of the internal database. Everyone who had a cellphone in those days may remember pleasantly how he could forget to re-charge it and keep using it for days despite of the fact that the batteries were small and in many cases still used Ni-Cd chemistry a technology

---

F. Rezzi (✉) • L. Collamati • M. Costagliola • M. Cutrupi
Marvell Italia
e-mail: frezzi@marvell.com

A. Baschirotto et al. (eds.), *Frequency References, Power Management for SoC, and Smart Wireless Interfaces: Advances in Analog Circuit Design 2013*, DOI 10.1007/978-3-319-01080-9_9, © Springer International Publishing Switzerland 2014

that is now obsolete. Nowadays smart phones and tablets all include an application processor that transformed a basic communication device in a portable PC with an increasing computational capability that is closing the gap with laptop, yet in a much smaller device. On top of it the communication technology and connectivity expanded from 2G to 2.5G, 3G, LTE, WiFi, BT, NFC not to forget GPS and FM radio. The average power consumption of such a device depends of course on the usage model but it is not uncommon to be in the 1–2 W range a tenfold increase with respect to only few years ago. On the contrary the energy density of the batteries only increased by a factor of two. It is clear then the reason why the battery needs to be recharged more often leading to a faster degradation. In paragraph 2 the paper will review the principle of operation of the most common battery types giving some basic understanding of the physical and chemical processes that underlines the energy management of the battery and how the battery converts it as chemical potential into electrical work (discharge). In paragraph 3 the charging process and charging technique will be illustrated together with the circuit techniques that are more commonly used to control the charging process.

Finally paragraph 4 will explain the battery monitor function with particular focus on the fuel gauging techniques that allow determining the State of Charge (SoC) of the battery and predict its runtime, aging and State of Health (SoH).

9.2 Battery Chemistry

A battery is a device that converts chemical energy into electrical energy. It is composed by two electrodes called anode and cathode separated by an electrolyte. When the electrolyte is different for cathode and anode a membrane needs to separate the two composites. Each electrode with its own electrolyte is called a half cell. All the battery types are characterized by the fact that a positive charge can be permanently stored at the cathode as positively charged ions (the "cations"), while at the anode a negative charge is accumulated either as free electrons or negative ions (called "anions"). This creates a net potential difference between the two electrodes. The process that originates this potential difference varies greatly among the different battery chemistries and it depends both on the chemical and physical properties of the electrodes and the electrolyte and relies on the presence of ions to transport and release charge.

In all cases under static conditions the potential difference that appears in each half cell counteracts any further release or exchange of ions and the battery cell reaches its thermal equilibrium. The net positive voltage difference that appears between cathode and anode is known as Open Circuit Voltage (OCV). Although many times the breaking up of the battery in two half cells may not have any physical real correspondence, the behavior of the battery is always characterized by two separate chemical reactions that happen at the anode and cathode and that are referred as "half-cell reactions" as it will be clear in the following paragraph when discussing about the charge/discharge process.

9 Battery Management in Mobile Devices

Some of the most common type of batteries are the Zinc-Carbon (ZnC), Alkaline, Lead-Acid (commonly used in cars), the Nickel Cadmium (NiCd), the Nitride Manganese (NiMH) and Lithium (Li-Ion). It is beyond the scope of this work to present a full overview of all the batteries types and behavior. In the rest of the paper we will concentrate on the technology that most of all enables all the modern mobile devices, i.e. Li-Ion cells.

9.2.1 Battery Discharge

If we connect a conductor material between the two electrodes the potential difference between them will result in a flow of charge, i.e. a current. The electrons flow between the anode and cathode in the load, and at the same time positive ions (cations) move in the same direction in the electrolyte to neutralize the negative charge. A similar but inverse process happens at the anode where the breakup of the cell thermal equilibrium stimulates the release of more electrons to sustain the current flow.

An example of this type of reaction in the Cobalt-Carbon based Li-Ion batteries is illustrated in the following chemical reaction [1].

| Anode reaction | $LiCoO_2 \leftrightarrows Li_{1-n}CoO_2 + nLi^+ + ne^-$ |
| Cathode reaction | $nLi^+ + ne^- + C \leftrightarrows Li_nC$ |

It can be easily noticed that in the case of Li-Ion batteries the chemical reactions are bi-directional, i.e. reversible. In this case the battery can be recharged by forcing a current in the opposite direction of the discharge (see paragraph 3). During discharge the nLi^+ ions generated at the anode move towards the cathode where they recombine with the electrons and get trapped in the lattice structure on the Carbon electrode (a physical process known as *intercalation*). The opposite happens during charging when the ions move from cathode to anode. The anode and cathode different chemical reactions are what fully describe the behavior of any type of battery [1] (Fig. 9.1).

The effect of the discharge process is the reduction of the overall charge available at the two terminals and therefore the potential difference between positive and negative electrodes diminishes. Eventually when no more charge is available or the potential difference between the two terminals is so low that the battery becomes unusable or, even worst, it may reach chemical instability the battery is said to be completely depleted of equivalently is having a State of Charge (SOC) equal to zero.[1] The concept of SOC will be largely explored in Par. 4 when discussing about the fuel gauging algorithm.

---

[1] It is worth noticing that often times the 0 % SOC is defined based on external conditions as for example the minimum operating voltage of the circuit supplied by the battery although some useful charge may still be available inside the cell.

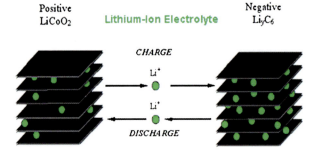

Fig. 9.1 Charge discharge process in Li-Ion batteries

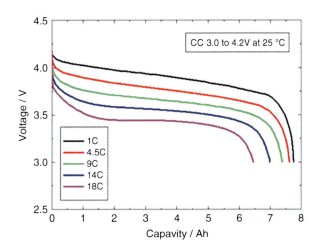

Fig. 9.2 Load discharge curves for Li-Ion battery type

If at any time during the discharge process we remove the load and wait enough time for the cell to regain its equilibrium we can measure the voltage difference between the terminals and plot the OCV *vs* SOC curve. This curve is particularly important when, as we will see in Par. 4, we want to determine the SOC of the battery based on the measurement of its voltage.

Measurement of the voltage can be made also under load conditions at different current rates to produce the so called constant current discharge curves. This maybe more interesting from an application standpoint since they show the behavior of the battery while is delivering power. Figure 9.2 shows such a family of curves for a Li-Ion battery pack with nominal charging voltage of 4.2 V [1].

From these curves it can be seen that for the same SOC (Capacity/Ah is a measure of the charge that can be extracted from the battery) the cell will exhibit different voltage at different load values a phenomenon that can be modeled as a finite output resistance. At the same time under large loads it is impossible to extract all available energy from the battery since the minimum operating voltage is reached before the

SOC is zero. In this figure the term C refers to the nominal unit current of the battery. This current numerically corresponds to the nominal capacity of the current expressed in *Ah* (or *mAh*), that is it is the nominal current that would discharge the current in 1 h. For example for a 1,500 mAh battery the 1 C current is 1.5 A.

9.3 Charging Process

To re-charge a battery it is necessary to pump energy into the cell by reversing the current that would normally flow during discharge, i.e. we need to force a current into the cathode. The circuits or apparatus that control the process are called "battery chargers". Battery chargers need to regulate the so called "*charging profile*". Different types of chemistry have different types of charging profiles that include the maximum charging current, the maximum allowed voltage, the method to terminate the charge, the safety precautions that need to be adopted to avoid chemical instability due to over-charging and the controlling of the battery temperature.

Every cell has a limit to its charge acceptance rate. If we pump charge faster than the chemical process can react to, overheating as well as unwanted chemical reactions could damage the cell. Generally a cell can even accept short and very high current pulses as long as enough time is allowed to transform the reacting material between one pulse and the other.

When all the material has been transformed the cell reaches its maximum capacity. Trying to pump more current after this point may trigger secondary chemical reactions that may cause overheating, venting (i.e. the production of gas material that tend to inflate the cell) and ultimately the cell destruction. A proper termination method must be set in every charging profile.

Basic charging profiles for cell commonly used in mobile device include:

- **Constant Current/Constant Voltage** This method is most commonly used in Li-Ion batteries. A constant current compatible with the maximum charge rate of the battery (less than 1 C) is first applied to the cell until it reaches the maximum allowed charging voltage. After that a constant voltage is maintained on the cell. As the cell gets more charged the current diminishes and the charging process is usually stopped when the current reaches a preset minimum value. A typical charging profile is shown in Fig. 9.3. It should be noticed that if the charge process would stop when the maximum cell voltage is reached (end of constant current phase), then the cell capacity would be limited to 70–80 %.
- **Pulse charge**. Pulsed chargers feed the charge current to the battery in pulses. As previously explained, batteries can accept high current pulses (even much higher than the average acceptance rate) but a suitable time must be allowed between pulses for the chemical reactions to stabilize before the next pulse. When the cell is fully discharge the duty cycle between charge and relaxing time can be kept constant and sufficiently high to minimize the charging time.

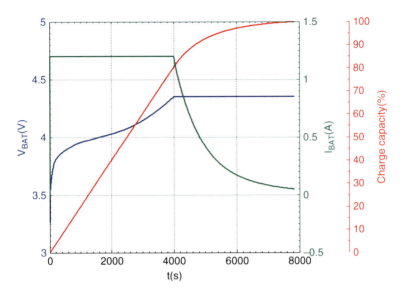

Fig. 9.3 Li-Ion Battery charging profile

However as the cell is reaching its maximum SOC, the high current pulses can temporarily drive the cell over its maximum allowed voltage. In this case a proper amount of time must be allowed to re-gain a proper voltage below its maximum allowed target before applying the next pulse (usually the pulse width tON is fixed). The time between pulses tends to become longer and longer as the cell approaches its full state and the charge can be terminated when the duty cycle falls below a pre-determined value (for example 10 %).

- **Trickle charge**. This term was initially intended for the charging process used to compensate for the self discharge of the battery. Generally this term is now referred to continuous charging of the battery at low rate (below 0.1 C) used particularly when the cell is deeply discharged and its charge acceptance rate is very low. This process is also as pre-charge.
- **Top-off**. This process refers to a very low charge rate (below 0.01 C) that can be applied to the cell for a relatively long time when the charge process is completed. The low rate allows to fill the cell with additional charge without exceeding its maximum allowed voltage.

9.3.1 Charging Circuits

In the following we will mainly review the chargers used for Lithium batteries. In essence these circuits are required to regulate the current that goes into the battery and ultimately limit the maximum voltage that corresponds to its full state of charge.

9 Battery Management in Mobile Devices

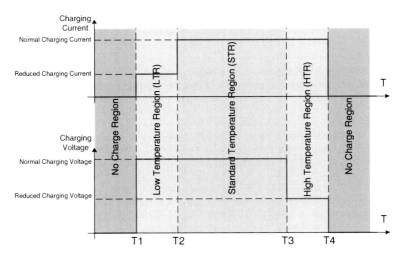

Fig. 9.4 Example of JEITA compliant charging profile

It is then obvious that the circuits will have to include a current regulation loop and a voltage regulation loop. On top of that there are additional circuits that are used to monitor the state of the battery and provide fault protection mechanisms. These circuits include:

(a) Thermal monitoring of the battery

Lithium batteries cannot be charge at any temperature. Below a certain temperature the cell acceptance rate drops significantly and it is no longer safe to charge them. The same applies if the temperature is too high. In general the charge acceptance rate of the battery (as measured by its output impedance) is still good at high temperatures but secondary chemical reactions gets activated that may lead to the generation of gas (venting) that may inflate the cell and ultimately damage it. The Japanese standard JEITA recommends applying different charging profiles outside a first temperature range and differentiating the profiles for high and low temperatures. A second and larger temperature range is then defined outside which the charge must be stop (Fig. 9.4). The temperature cell is generally measured using a thermistor placed either inside the battery pack or, if external, placed very close to the battery in order to maximize its thermal coupling.

(b) Thermal monitoring of the silicon charger

Battery chargers are usually high power devices. The thermal management of the silicon charger and, more generally of the mobile device is very important in order to prevent mis-functioning of both the charger and nearby devices. It is important to prevent the temperature of the charger from increasing too much. Most commercially available chargers include a temperature sensor and will stop the charge if the junction temperature gets too high. More sophisticated chargers also include a "Thermal loop" that limit the current (and therefore reduce power dissipation) in order to keep the maximum temperature constant without stopping the charge process

(c) Current detection and measurement

The charger not only needs to regulate the current during the constant current phase but also needs to monitor it in order to stop the charge when the current falls below a preset value. It is generally pretty simple to detect the current delivered by the charger and several topologies of current sensing circuits can be implemented. However, most of the time during the charging period the system supplied by the battery is active. In these conditions, depending also on the charger architecture, some (or even all) of the current delivered by the charger may end up in the Power Management Unit (PMU) instead of the battery. In these cases the current level in the battery may decrease despite the fact that the cell is not fully charged or even worst the battery may end up supplementing current if the power requested by the system exceeds the power capability of the charger. So monitoring the current delivered by the charger or flowing into the battery it is not always the optimal choice to terminate the charge and some kind of awareness of the system status is needed before terminating the charge. For example mobile systems go often in low power states to preserve battery life (called also sleep states) and the current level in the battery can be detected during these periods of rest or the observation of the current behavior can be prolonged over a longer period of time in order to average out current peak demand from the PMU.

(d) Voltage detection and measurements

Chargers must always include a way to monitor the battery voltage. In its simplest way this can be implemented with a series of comparators with bandgap referred thresholds that would trigger some action on the charger's hardware. More sophisticated chargers may also integrate an ADC that may report to the host the real battery voltage. Reporting may include averaging, min and max and also set programmable alert thresholds through interrupts.

(e) Over/Under voltage protection

(f) Two of the most important thresholds in the battery chargers are the over and under voltage protections. Over voltage protection is meant as a secondary protection should the voltage limiting circuit fail. The under voltage protection is meant to prevent the battery from over discharging. Usually mobile systems are turned off well before the battery gets over discharged sacrificing part of its capacity in favor of preserving the cell and prolonging its life. However most manufacturer require a double fault protection mechanism for the cell and an additional safety measure must be implemented in the charger on top of what already exists in the battery pack. In order to completely prevent the battery to over discharge there should be a way to completely disconnect the system from the battery in order to stop any leakage. This can be done with a switch placed in series with the battery. However this switch must present very low impedance (40 mΩ or less depending on the current requirement) in order to avoid large voltage drops and loss in efficiency during normal operation

9 Battery Management in Mobile Devices

(g) Safety timers

Safety timers give an additional level of protection should any of the algorithms aimed to stop the charging process fail. Safety times are programmed from a few to several hours depending on the expected charging time. Charging time, however, is predictable only in a standalone setup (charger + battery) or when the system is in standby. If we operate the mobile device during charging, it can become un-predictable since the amount of power delivered to the battery diminishes. If the charging time becomes too long the charge could be terminated before reaching the end. Also in this case some sort of system awareness can help. For example the timers can be stopped during high system activity and resumed when the activity is low.

(h) Battery ID system

(i) Usually the charger is programmed to charge a particular type of battery. If incidentally the battery is swapped it may create safety concern because for example the new cell may have lower capability and therefore it should be charged with a different current level. In order to prevent that, phone manufacturers use batteries with a unique mechanical design that prevents insertion of a battery not suited for that. A second level of authentication is the so call Battery ID resistor that is usually inserted inside the battery pack connected between an extra pin and ground. The battery manufacturer associates to a resistor value a particular charging profile. If the charger detects a wrong resistor value then the charging will not take place. Smarter system can actually measure the BID resistor value and adjust via SW the charger profile. Smart batteries then provide a communication link between the charging system and the battery pack. This link (that is regulated by the BIF a 1-wire MIPI interface specification []) is used to exchange information between the battery pack and the host system that regulates the charging process. Smart batteries include sophisticated electronic HW that may report information as temperature, SOC, current, voltage, capacity, State of Health and impedance but it may also include an encryption that tightly couple the battery pack with its charging system preventing the use of a wrong battery.

(j) Battery detection system

Last but not least the charger must provide a way to indicate if the battery is present or not. If the battery is removable it may either be removed on purpose or it may fall off accidentally after dropping the phone for example. In either case the sudden loss of the system power may cause problem to some critical component in the system starting from the non-volatile memories that can get corrupted if a proper power down sequence is not applied. A fast indication of the battery loss may help mitigating the problem if we have enough time to protect the memory (few ms before the system completely looses power). The better way to fast detect the battery removal is through the BID resistor or the thermistor inserted in the battery pack. Another way would be to detect the battery current going to zero or the sudden voltage drop. The methods based on the monitoring of electrical variables are however generally too slow and may get tricked by other user scenarios.

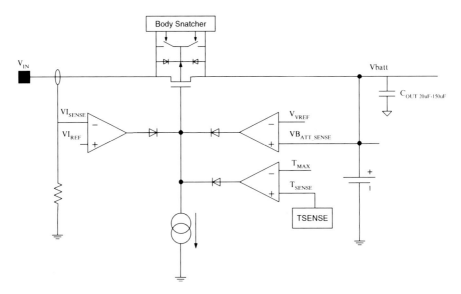

Fig. 9.5 General scheme for linear charger

The type of regulation circuits that regulate the charger operation can be divided in three main categories

- Linear chargers
  Linear chargers are essentially linear regulators with a precise current limit. Figure 9.5 shows a typical configuration for a linear charger which includes two regulation loops for voltage and current that are usually single quadrant, i.e. they do not act if the electrical variable is below the preset value, but they only serve as a limiter or a clamp. In this way when the battery voltage is low only the current loop is active, while when the voltage tends to increase the voltage loop will intervene and as the current diminishes its controlling loop will give away its control. Some chargers make a hard switch decision between the two loops while others pass the control more gently and progressively between the two.

  Linear chargers are relatively simple and inexpensive but may suffer of severe thermal limitation. The power dissipation of the charger is $(V_{IN}-V_{BATT}) \times I$. For a discharged battery at the beginning of the charging cycle when both the current and the voltage drop are the highest, the power dissipation may account for several Watts. So their use is limited to low current use cases and nowadays are progressively being abandoned as the battery size and current demand increases. In any case introducing a thermal loop or a proper thermal management is a must for linear chargers.

  Another essential characteristic of every charger is the *inverse leakage protection*. When disconnecting the input voltage we need to avoid any leakage path from the battery towards the input. In Fig. 9.5 this is achieved by the body snatcher circuit that will reverse the diode polarity of the PMOS bulk when

9 Battery Management in Mobile Devices

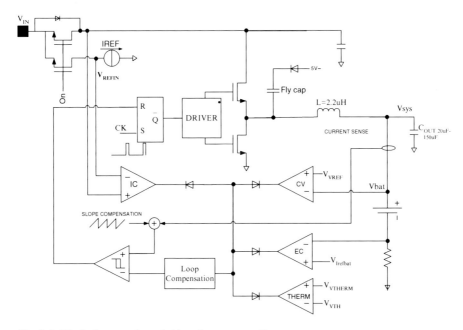

Fig. 9.6 Block diagram of a switching charger controller

VIN < VBATT. Other methods include inserting an isolation diode (or better MOSFET) in series with VIN [2].

- Switching Chargers
  Switched mode battery chargers use switching regulators to control both current and voltage. They achieve high efficiency therefore limiting the power dissipation in the chargers. They also can optimize the power transfer from the source to the battery reducing the charging time. Figure 9.6 shows the typical configuration of a switching battery charger.

  As opposed to classical voltage regulators where the duty cycle of the output bridge is controlled exclusively by a voltage loop, in the battery charger other loops can take control. Among these we have:

- An inner current loop controls the battery flowing into the battery limiting its value to a preset level. In this case accurate current sensing relies on sense resistor placed in series with the battery. This sense resistor must be in the order of tens of mΩ to limit voltage drop and the efficiency losses in the battery. Ultra low offset comparator or amplifiers are needed for accurate current sensing. Example of such circuits can be found in references [3], [4] and [5].
- An external current loop limits the maximum current that can be sunk from the source. Due to the efficient power transfer input current differs from the output current that can be higher. Limiting the maximum input current allows protecting the input source for being overloaded. This has two advantages.

Fig. 9.7 V-I profile of a typical wall adapter AC/DC converter

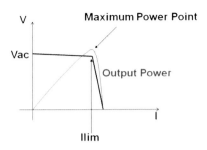

In first place if the source is a USB host usually after enumeration the current is limited to 500 mA and cannot be exceeded without violating the USB specification. Before enumeration there is a limit even lower of 100 mA. Secondly any power source (whether it is a USB host, a dedicated USB charger or any AC/DC wall adapter converter) has a maximum current capability. If the current is exceeded the voltage suddenly drops. Figure 9.7 shows a typical V-I curve of a power source.

If the input voltage drops then the switching regulator enters its dropout region where Vout ~ Vin and will deliver a current that is equal to the current limited by the input source. This behavior however is undesirable since it limits the maximum power the input source can deliver. As it can be seen from Fig. 9.7 the maximum power is available at the knee of the V-I characteristics and this would be the sweet spot where the system should operate in order to maximize the power transfer. So the charging system should adopt some sort of algorithm to detect the maximum current capability of the source. These algorithms are generally defined as Maximum Power Point Tracking since they tend to put the system into its maximum power transfer state. For example the charger could recognize the drop in the input voltage and back off the input current limit until no more drooping occurs.

- A thermal regulation loop similar to what was already explained for linear charger should be present. Although the internal power dissipation is limited in switching chargers depending on their efficiency, for high power chargers it could still reach values that may rise significantly the internal silicon temperature. In the case of switching chargers the system should act the input current to reduce the overall power intake and limit the power dissipation

- Pulse chargers
Pulse chargers have been adopted successfully for years although nowadays are being gradually abandoned. The pulse charge algorithm has already been explained at the beginning of this session. From a hardware standpoint a pulse charger is essentially a switch controlled by a state machine that determines the turn on and turn off times based on the battery voltage value. When the switch is turned on the input power source is short circuited with the battery and will enter

9 Battery Management in Mobile Devices

its current limit zone (see Fig. 9.7). Pulse chargers rely exclusively on the current limiting capability of the source wall adapter that needs to be custom designed. For this reason it cannot be used with USB port and this, in addition to the safety concerns related to the overdriving the battery has contributed to its decline. We will only remark that pulse chargers are cheap, simple and extremely thermal efficient for the mobile device. In fact, most of the power loss happens in the power source that continuously runs in its current limit region.

9.4 Fuel Gauge

Essential part of battery management in a mobile device is the monitoring of the state of charge of the battery. All the algorithms that perform this task go usually under the name of "Fuel Gauge" algorithms. This section describes the three main algorithms used for a battery monitor system for cellphone applications. The main definitions are given in the first subsection. In the second subsection the OCV (Open Circuit Voltage) dependence on SOC is investigated. The fuel gauge algorithms are reviewed in subsection three. Finally some experimental results are presented in subsection four.

9.4.1 Definitions

- Maximum battery capacity
  $Q_{T,max}$ is the maximum charge that can be stored in the battery.
- Remaining battery capacity (battery charge)
  $Q(t)$ is the charge stored by the battery at time t.
- Nominal battery capacity
  $Q_{T,nom}$ is the nominal value for Q_T provided by the battery manufacturer.
- Battery capacity
  In order to increase battery lifetime the battery is not charged at its maximum value. Q_T is the charge stored in the battery at the end of charge.
- Coulomb counting
  From the measurements of the battery current $I_L(t)$ it is possible to estimate the charge variation ΔQ. If we assume that at the initial time t_i the battery capacity is Q_i and that the final capacity at time t_f is Q_f then we have

$$\Delta Q = Q_f - Q_i = - \int_{t_i}^{t_f} I_L(t)\, dt$$

The coulomb counter is the equipment that measures the load current and integrates it.

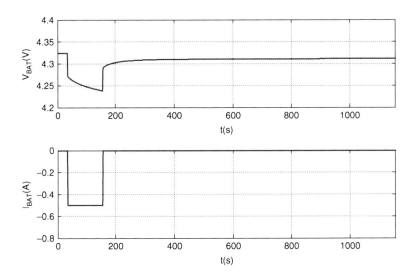

Fig. 9.8 Battery transient step response

- Relaxation time and Open Circuit Voltage
  If we apply a current step (Fig. 9.8) we can observe how the voltage response evolves to a steady state condition [6]. The time required to reach steady state is the relaxation time t_{rel}. We can notice that there are two dominant time constants and the slowest one is in the order of 100 s.
- Open circuit voltage (OCV)
  If the battery terminals are floating or the load current is zero then the battery reaches a steady state condition where the difference of potential between the terminals is constant. This voltage is called Open Circuit Voltage (OCV).
  For example the OCV measured in Fig. 9.8 is 4.31 V.
- End of charge
  The charger voltage is set to a value suggested by the manufacturer. Typical values for Li-ion batteries are in the range from 4.2 to 4.35 V [7] although nowadays we start seeing battery with a charge voltage up to 4.5 V. When the battery charge is close to the maximum battery capacity then the charger current starts to decrease (Fig. 9.3). The end of charge is set to a current that depends on the battery and on the required battery capacity. Typical values for Li-ion batteries are in the range from 0.01 C to 0.1 C. Those values guarantee a battery capacity Q_T above 95 % of maximum battery capacity $Q_{T,max}$. If the battery reaches the end of charge condition then we say that the battery is fully charged or it has a 100 % of SOC.
- End of discharge
  Undervoltage supervisors are usually inserted in the battery pack and in the equipment in order to avoid battery deep discharge. Over discharging decreases battery lifetime and can also permanently damage the battery (Fig. 9.9) [7]. For Li-ion battery the undervoltage is usually in the range from 2.4 to 2.7 V.

Fig. 9.9 Operating range of Li-ion battery [7]

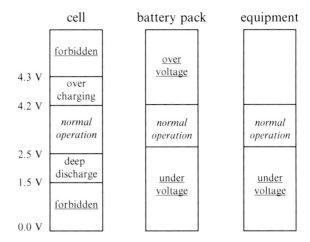

The equipment under-voltage is slightly higher in the range from 2.6 to 3.0 V. Therefore both battery pack and equipment prevent the battery to over-discharge (double fault protection). The end of discharge is set to a value OCV_{EOD} that is above the minimum discharge limit and it is a tradeoff between the need to avoid deep discharge and the need to use all the available charge. We call Q_{EOD} the remaining battery capacity. The battery can be charged up to Q_T and discharged down to Q_{EOD}. So the *battery available capacity* is

$$Q_T' = Q_T - Q_{EOD}$$

The SOC(t) at any time t is defined as

$$SOC(t) = \frac{Q(t) - Q_{EOD}}{Q_T - Q_{EOD}} = \frac{Q(t) - Q_{EOD}}{Q_T'}$$

- OCV versus SOC
  One of the most important characteristics of a battery is the OCV versus SOC. An example of such a curve is plotted in Fig. 9.4.

The OCV *vs* SOC dependence is of paramount importance in the analysis of the state of charge of the batteries. It is related to the chemistry and physics of the battery and the OCV can be easily measured with good accuracy by means of multimeter. Batteries of the same chemistry have the same OCV *vs* SOC within an error margin less than few mV. Aging and temperature have very little impact on the OCV dependency on SOC. For Li-ion battery also the hysteresis between charge and discharge of OCV(SOC) curve is negligible. Therefore only one curve is required to determine the SOC from OCV with accuracy in the order of 1 %. The problem in using this curve may come from the need of measuring the battery in its relaxed state. In fact in real systems that continuously operate under the battery the exact moment when to measure the battery in this state may be difficult to establish.

9.4.2 Fuel Gauge Algorithms

Fuel gauge algorithms can classified in three categories, coulomb counter, coulomb counter plus OCV correction and voltage only.

(a) Coulomb counter

Coulomb counter is based on the measurement of the load current $I_L(t)$. From the definition of SOC we obtain

$$SOC(t) = \frac{Q_T - \int_{t_i}^{t} I_L(t') \, dt' - Q_{EOD}}{Q_T - Q_{EOD}} = 1 - \frac{1}{Q_T'} \int_{t_i}^{t} I_L(t') \, dt'$$

Note that an error in the measurement of Q_T' affects the SOC as a gain error. Moreover every offset error in the $I_L(t)$ measurement is integrated so that its influence on the SOC increases with integration time [8].

Just as a reference in order to have an error of 1 % over a Q_T' of 1 Ahr it takes about 10 h if the measured current offset is 1 mA.

Coulomb counter needs an accurate current sensing. A sense amplifier multiplies the voltage across a sense resistor [3] and provides the input voltage to a converting ADC. Offset canceling techniques like choppering must be used and also provide rejection for 1/f noise. Signal is amplified to a level readable by the converting ADC. Effect of ADC offset is usually negligible due to the amplifier gain in from of it. Since choppering usually stabilize gain and residual offset against temperature variation any residual error can be calibrated and compensated in the digital domain. This architecture is usually more power hungry and it is suggested when an ADC is already present and active in the system. A 10–12 bit ADC usually suffices to provide fast and punctual reading of the current values that can also be useful for the system power management but in general fuel gauge algorithms need a higher dynamic range (15–16 bits) that can be achieved through averaging.

Another common fuel gauging systems make use of a 1-bit $\Sigma\Delta$ converter [9]. A simple Comb filter (accumulate and dump) may be used as decimator. To reach the desired current N bit resolution it takes the accumulation of 2^N 1-bit samples. Also in this case offset compensating techniques like CDS or choppering can be used to suppress offset and 1/f noise. This is a very low power circuit that can use as low as 20–30 uA current but can only provide average current value and not punctual reading of the current. For a typical 32 kHz sampling rate, in fact, to reach a 15-bit resolution the integration time is 1 s. It is usually an always ON circuit when the battery is in a valid state.

When dealing with ADC based integrated current sensors we need to account for the effect of sampling. Not only offset is important but we need to account for aliasing of frequency spectrum components at multiple of the sampling frequency that may fold back at DC. In this perspective a continuous

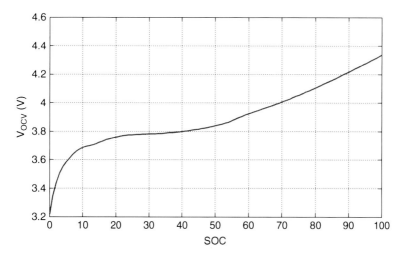

Fig. 9.10 Example of OCV(SOC) plot for a Li-ion battery

time ΣΔ converter offers the interesting property of a signal transfer function STF(f) that has zeros at multiples of the sampling frequency offering a natural anti-aliasing effect without the need of external filtering component.

(b) Coulomb counter plus OCV

The main issue with the coulomb counter algorithm is the integration of the offset error. This error can be periodically corrected by the SOC estimate based on the open circuit voltage. Usually battery powered equipments like cell phones periodically enter a very low power consumption mode to save battery life. If a counter is available to measure this time interval, after a delay longer than the battery relaxation time the battery voltage V_{BAT} can be measured. By storing the OCV vs SOC curve, the measured V_{BAT} can be used to extract the SOC value and compared with the coulomb counter projected value eventually correcting for large error superior to a certain threshold.

Therefore the integrated error is not accumulated over the time but may be reset periodically. However errors larger than 10 mV in measuring the battery voltage can cause large errors in the SOC estimate particularly in the flat region of the OCV(SOC) curve (Fig. 9.10).

(c) Voltage only

The voltage only is the more challenging algorithm. It is based on an accurate electrical model of the battery. This model must take into account of the nonlinear OCV(SOC) characteristic and the dynamic response of the battery. Moreover the model of the transient response must include the instantaneous response and the memory effects. One of the more popular models accounts for the memory effects with two time constants, a fast one in the order of 100 msec–1 s and a slow one in the order of 10 up to 100 s. This is a good tradeoff between the model accuracy and the algorithm computational cost. The overall battery electrical circuit model is shown in Fig. 9.11 [10]. C_{SOC} models the battery capacitance,

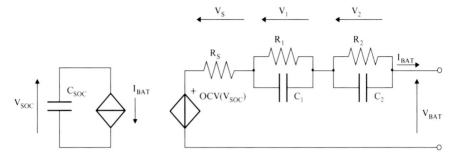

Fig. 9.11 Electrical model of the battery

he nonlinear voltage controlled voltage source models the OCV(SOC) dependence, the linear current controlled current source I_{BAT} models the discharge/charge of the battery due to the battery output current, the resistance R_S models the instantaneous response, while R_1C_1 and R_2C_2 model the short and long time constants. In some cases also a self discharge resistance in parallel to C_{SOC} is added. However for Li-ion battery this effect is negligible for practical cases.

Knowing the parameters of the battery impedance and measuring the voltage at the battery terminals, the I_{BAT} can be calculated solving the differential equations that regulate the electrical circuit. Since the voltage is sampled at fixed time intervals the differential equation can be solved in the discrete time domain assuming the OCV does not change significantly between a sampling time and the next.

A fix current I_{BAT0} can be calculated as

$$I_{BAT0} = (OCV - V_{BAT})/(R_S + R_1 + R_2) = (OCV - V_{BAT})/R_{tot}$$

Therefore an accurate characterization of battery total resistance R_{tot} is required [11]. R_{tot} depends both on battery chemistry and the electrical characteristics of the battery. The chemical resistance is due to the limited mobility of the ions within the electrolyte whereas the conduction electrical resistance depends on the electrodes conductivity as well as the contact resistance and board traces. Therefore a proper voltage sensing as close as possible to the battery pack is required. The chemical resistance has strong dependence on the temperature, the SOC, and also on the discharging/charging current (Fig. 9.12). All these factors must be taken into account in the battery model.

Battery parameters depend also on aging. This variability could be taken into account using by adaptive algorithms that are able to track battery impedance.

9.4.3 Experimental Results

Some results for a voltage only algorithm are provided in the following figures. A comparison between the algorithm implemented in the Marvell chip PM812 (red curves) and the value measured by accurate lab equipment (black curves)

9 Battery Management in Mobile Devices

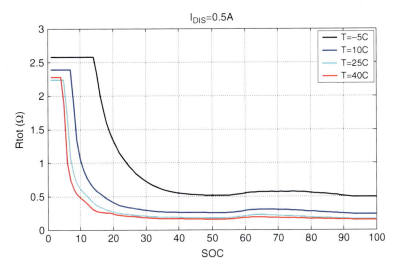

Fig. 9.12 R_{tot} versus SOC at discharging current of 0.5 A and temperature form -5 C to 40 C

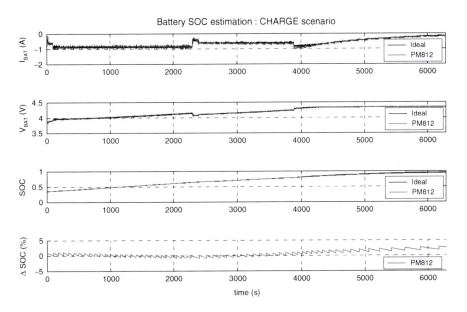

Fig. 9.13 Voltage only algorithm. Scenario 1

is highlighted. Last row of each figure represents the error between the SOC extracted by the algorithm and the reference SOC.

Figure 9.13 shows a charge scenario. It's possible to distinguish three phases by observing the I_{BAT} values. In the first phase the $I_{charge} = -900$ mA (there is no activity on the phone cell and the display is turned off) while in the second the

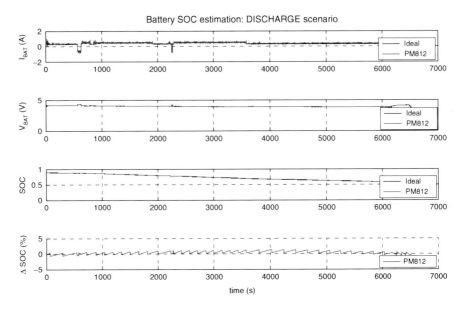

Fig. 9.14 Voltage only algorithm: scenario 2

absolute value of the current is lower because some applications were activated on the phone. Last phase is characterized by a reduction of the absolute value of the current due to the approaching of end of charge.

Figure 9.14 reports a discharge scenario. The load is related to random activity of the phone. Note that the accuracy is pretty good over a broad range of battery discharge.

Finally, a mixed scenario (charge – discharge) is highlighted in Fig. 9.15. Note that the error accumulated at high current level is nullified by a long period at low current. When IBAT is almost zero than VBAT is very close to OCV and SOC value can be extracted by OCV(SOC) characteristic.

9.5 Conclusions

The paper presented an overview of the main issues encountered in battery management in modern mobile devices. The main purpose of battery management system is to preserve battery life while providing good user experience in term of battery extended use and fast charging time. This task will be more and more important as many devices like tablets and high-end smartphone make use of non-replaceable batteries that offer the advantage of being narrower with respect to the replaceable one. Monitoring of the battery status not only helps in extending its longevity but can also provide to the system information on its capacity and

9 Battery Management in Mobile Devices

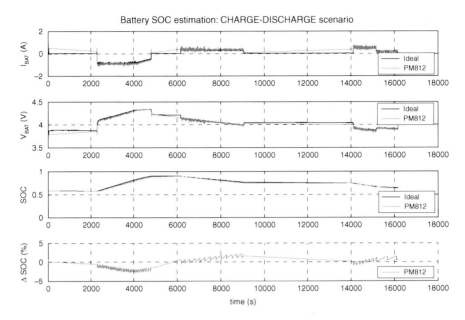

Fig. 9.15 Voltage only algorithm: scenario 3

program the maximum load based on its SOC status. Future battery technologies will provide more capacity per volume and extended voltage range but also will pose more challenges for the electronics that needs to supervise and monitor the battery behavior as the power demand of the mobile device will further increase.

References

1. http://www.mpoweruk.com/
2. Bq24040, 800mA, Single-Input, Single Cell Li-Ion Battery Charger with Auto Start, datasheet from www.ti.com
3. J.F. Witte, J.H. Huijsing, K.A.A. Makinwa, *A current-feedback instrumentation amplifier with 5μV offset for bidirectional high-side current-sensing*, ISSCC (Delft, Delft University of Technology, 2008)
4. Linear Technology Corp., LTC6102 data sheet, http://cds.linear.com/docs/en/datasheet/6102fd.pdf
5. Texas Instruments, Small Size, Low-Power, Unidirectional, CURRENT SHUNT MONITOR Zerø-Drift Series, http://www.ti.com/lit/ds/symlink/ina216a1.pdf
6. L. Gao, S. Liu, R.A. Dougal, Dynamic lithium-ion battery model for system simulation. Compon. Packag. Technol. IEEE Trans. **25**(3), 495–505 (2002)
7. Sony, Lithium IonRechargeable Batteries – Technical Handbook, available at http://www.sony.com.cn/products/ed/battery/download.pdf. Accessed 31 Jan 2013

8. Maxim Application Note 485, Evaluating accuracy of coulomb-counting fuel gauging systems, Dec 2000
9. http://www.maximintegrated.com/datasheet/index.mvp/id/4560/t/al DS2780 Datasheet
10. M. Chen, G.A. Rincon-Mora, Accurate electrical battery model capable of predicting runtime and I-V performance. Energy Convers. IEEE Trans. **21**(2), 504–511 (2006)
11. S. Abu-Sharkh, D. Doerffel, Rapid test and non-linear model characterisation of solid-state lithium-ion batteries. J. Power. Sources **130**, 266–274 (2004)
12. http://mipi.org/specifications/battery-interface

Chapter 10
Is Digital SMPS Ready to Eliminate Analog Regulators for Portable Applications Power Management?

S. Cliquennois and A. Nagari

Abstract This chapter reviews the challenges which integrated voltage regulators have and will have to tackle for power management of portable applications while focusing on how the Digital Switched-Mode Power Supplies (SMPS) technology, already widely used for medium and high power systems, is able (or not) to challenge the classical analog loops. The study will explore mainly step-down architecture, and analyses the challenges in many aspects of integrated regulators design: efficiency, area, speed, flexibility, current estimation, low-power modes, multi-phases and control sharing, voltage scaling and EMI.

10.1 Introduction

Integrated Switched-Mode Power Supplies have become ubiquitous regulator architectures in portable applications, and have superseded the classical Low-Dropout (LDO) linear regulators for every power hungry (i.e. more than half a watt) supply needed on portable, battery-operated devices. Although LDO provide low-cost, fast and low-noise regulated output voltages, their typically very low efficiency – less than 28 % for a typical output voltage of 1 V supplied by a regular 3.6 V battery – is disqualifying them for the very demanding core of portable applications (processors, modems, memories, I/Os...). More than 85 % peak efficiency can be expected from a noisy and expensive – because using an external coil – well-sized SMPS [1].

The main and specific challenges which SMPS designers have to face when working for portable applications will be reviewed, and discuss if so-called digital SMPS can bring added value with respect to their analog counterparts, a debate open for many years [2].

S. Cliquennois (✉) • A. Nagari
ST Microelectronics, Grenoble, France
e-mail: sebastien.cliquennois@st.com

A. Baschirotto et al. (eds.), *Frequency References, Power Management for SoC, and Smart Wireless Interfaces: Advances in Analog Circuit Design 2013*, DOI 10.1007/978-3-319-01080-9_10, © Springer International Publishing Switzerland 2014

While efficiency at high load currents is a critical factor of merit of integrated SMPS, it is not the only one: in order to optimize speed of processors, which can dynamically adapt their speed and power supplies depending on their operating mode, regulators need to be very precise and fast – a trend that is becoming more stringent as the processors supply voltage lowers, and the maximum current capability requirement grows [3]. Reaching this transient speed specification often requires using coil-current sensing techniques which are essential in modern, fast SMPS. The different control techniques will also be discussed.

An often over-looked feature in literature is the capability for switched regulators to keep a very good efficiency figure of merit on their full output load range, which requires dedicated and crafted low-power modes, and on-the-fly efficiency optimization tricks [4].

While the trend in the previous decade was to aim at higher and higher switching frequencies to reduce external component footprint – mainly coil – the recent surge of tablet computers is actually orienting integrated SMPS in another direction: even if integration remains important, the need for higher current capabilities at reasonable efficiency is nowadays driving development of integrated multi-phase DC-DC converters, where digital control architectures can bring some benefits.

10.2 Generic Analog and Digital SMPS Architectures

Figures 10.1 and 10.2 illustrate the core components of "analog" and "digital" buck (i.e. step-down) converters. The power stage is driven by logical signals, hence actually possesses an intrinsic digital control, although design of integrated power MOSFETs is a purely analog task, where robustness, channel on-resistance, efficiency of driving buffers are the main aspects to be controlled. This important part of switched-mode power supplies is exactly the same whatever the type of control chosen.

An essential differentiation lies on the control loop. 'Pure Analog' SMPS rely on a number of electrical signals to perform feedback for regulation. Output voltage is generally sensed, and most of fast control structures, may they be current-control [1] or sliding-mode controllers [5] use either directly a current sensing resistor, or better, a lossless current sensing circuitry in order to re-construct the current information. Note that the intermediate LX node can also be sensed, as it provides very relevant information (on current direction in coil for example).

On the other hand, digital SMPS are all characterized by presence of one of several ADC converters which at least convert the error voltage ($Vout$-$Vref$) into a digital signal (common architectures are windowed flash, delay-lines or SAR ADC), and generally also input voltage (battery voltage in portable applications), an information which is essential to many compensation or current estimator schemes. It must be noted that the coil current bandwidth is much more important that the switching frequency, hence digital estimation of current has to rely on

10 Is Digital SMPS Ready to Eliminate Analog Regulators for Portable... 171

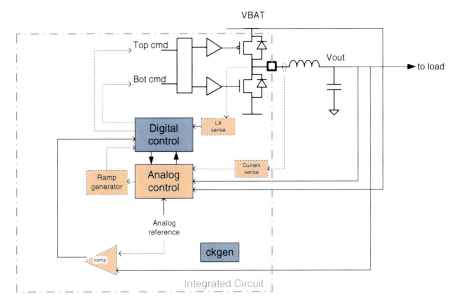

Fig. 10.1 Generic "analog" integrated SMPS block diagram

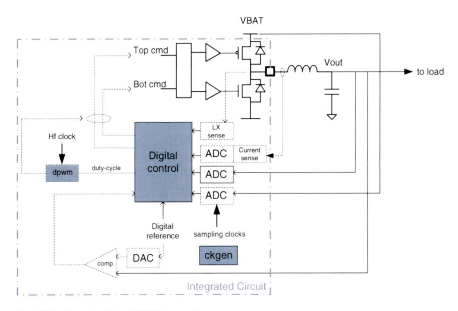

Fig. 10.2 Generic digital SMPS block diagram

sampled information of coil current, one among the challenges that digital control has to face.

The different control strategies will be discussed more in detail, but essential part of control is to generate a two-leveled (i.e. digital) signal which will control power switches. An historically important family of controller generates Pulse-Width Modulated signal, may it be analogically thanks to a clock-synchronized ramp generator or digitally by the means of a Digital PWM, for which many different implementation details have been proposed [6].

Another important family does not rely on PWM, but directly generates the control signals using for example a sliding mode approach [7]. This approach, which leads to very fast response, has been implemented both in analog and digital world.

10.3 Efficiency

Efficiency is a key parameter of any power management system. It will have a direct impact on system thermal dissipation, as well as on battery life in portable systems.

While theoretical efficiency of SMPS is 100 % [1], losses which affect this figure can be split into three main categories: Ohmic losses, which main components are power FETs on-resistance, coil series resistances and parasitic resistance. Switching losses, which account both for losses due to buffers needed to charge and discharge power mosfets gates, as well has losses due to switching parts of control system. Quiescent losses, which are static DC currents needed in control parts.

Ohmic losses and switching losses are essentially identical in analog and digital SMPS, but they can be minimized by adapting the size of power mosfets to output currents: at low currents, ohmic losses are less important, so reducing the size of power stage has a very beneficial impact on switching losses. An adaptive size selection can be done digitally. Note that discussion is only concerned with synchronous rectification, asynchronous rectification when a simple diode replaces the bottom switch being much less efficient and generally avoided in integrated designs.

Unlike analog parts which in general require to be biased during operation, digitally-controlled SMPS can generally operate at zero-DC bias.

However, because losses are dependent on output load, the efficiency curve has to be optimized in the four areas where buck converters operate:

- The high current area, where ohmic losses are dominant, and proper sizing and choice of MOSFETs (e.g. use dual-Nmos architectures instead of Nmos-Pmos, which reduces gate capacitance for same on-resistance)
- The medium current area, where switching and ohmic losses are dominant. The two previous areas correspond to SMPS supplies circuits in full to medium activity.
- The low current area, where if switching is kept, switching losses become way to dominant, so some pulse-skipping designs, or pulse-frequency modulation scheme have to be devised. This area corresponds generally to load circuits with little activity.

- The no-load area, which is an essential part, and often overlooked feature: SMPS need to supplies circuits which are not working (in retention state) and only compensate for leakage of supplied parts, with no transient load involved.

10.4 Scalability, Flexibility and Partitioning

Integrated SMPS for portable application have to face the tough challenge of area optimization, which has a direct impact on cost of solution. Whereas digitally controlled-SMPS comprises a control part which scale well with process, the area of power mosfets, which does not scale very well with technology – is dominant in most of integrated implementations. Actually, the problem is even getting worse with more advanced technology nodes: below 65 nm, it is rare that native mixed technologies support devices that can stand directly battery voltage (up to 4.35 V) as drain-source voltage, making usage of cascoded power stage compulsory – which added to more area and losses dues to more complex gate drivers.

On the other hand, a very scalable architecture for digital SMPS is shown on Fig. 10.3: the System-In-Package (SiP) approach allows assembling different circuits with different technologies in a single package. Typically, a product like ST-Ericsson's M7350 comprises of an "analog" die dedicated to power management, and a "digital" die – which here includes a modem circuit. Classical SMPS are fully embedded in analog die, but digital SMPS can have their (digital) control

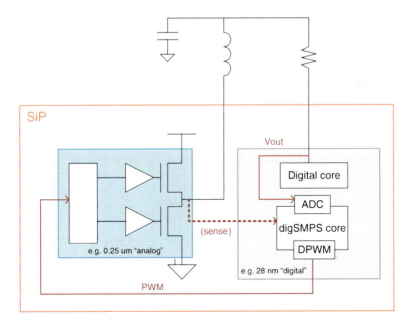

Fig. 10.3 Proposed system-in-package partitioning

part, ADC and DPWM embedded in digital die, while power part remains in analog die. Not only this approach allows a better scalability of SMPS, but is also give more flexibility and control to the digital die, which can directly control its supply internally. Remote sensing, which allows a better voltage stack optimization can be directly done inside digital chip. This approach requires an additional LDO regulator that allows for startup and can be turn-off when SMPS auto-supplies its digital part. The main drawback of this approach is that package thermal characteristics should be able to handle both power FETs and load (i.e. digital IC), which is not the trend for mobile processors.

10.5 Current Estimation Challenges

While voltage-mode (VM) loops have been the first ones to be proposed and are still widely used, the advantages of current-mode (CM) or sliding-mode control in terms of load transient speed, maximum current limitation, and multi-mode optimizations such as automatic power-stage sizing or automatic mode switching are tremendous [1]. Many CM analog implementations are using a discrete sense resistor, which is a solution to be avoided in portable applications, both because of sense resistor cost and additional loss in efficiency.

Several analog solutions without external sense resistor have been proposed:

- Internal current sensing: principle schematic is shown on Fig. 10.4. This solution generally shows a poor precision due to inherent poor matching -between very big power MOS and integrated sense-FETs. Bigger sense-FETs would imply un-acceptable efficiency losses.
- External R-C sense [8]. This type of sensing is lossless and relies on the fact that sensed voltage is equal to

$$V_{sense}(s) = I_L(s) \cdot R_L \cdot \frac{1 + s\frac{L}{R_L}}{1 + sR_fC_f}$$

i.e. $V_{sense}(s) \sim I_L(s)$ if $L/R_L = R_f C_f$. The spread in R,C and R_L are such that this solution requires calibration.

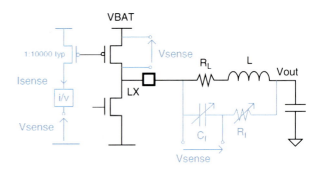

Fig. 10.4 Different current sensing strategies

- Structures based on LX node sensing (Fig. 10.4). This approach generally relies on sensing and amplifying *VBAT-LX* voltage, which is proportional to output current on the top mosfet conducting phase.

On the control digital side, a first approach consists in implementing the above analog solutions, and using an ADC to convert sensed current/voltage. Unlike ADC used for output voltage sensing which is generally windowed around reference, the current sensing ADC should be full range.

Another scheme consists in directly estimating coil current in digital world: a first digital-only first order current estimator has been proposed [9], and relies on the following approximation:

$$V_{OUT} \approx DV_{BAT} - I_L(DR_{ONP} + (1 - D)R_{ONN} + R_L)$$

It allows calculating I_L, but requires a precise measurement of *Ronp*, *Ronn* and R_L, requiring extra (analog and digital) circuitry for calibration. What's more, this equation is only valid in CCM.

Another digital-only solution is to implement a digital current estimator [10], which can be based on a state-space representation of system. It provides a robust solution, but comes at a cost of implementing the estimator, requiring real-time complex operations such as matrix multiplications, hence limiting the switching frequency of system.

10.6 Low-Power Modes

While literature concentrates generally on CCM, where the transient and efficiency performance of SMPS are critical, commercial SMPS must keep an acceptable efficiency even on low output current, and particularly when load is such that current in coil in below critical conduction current, hence naturally inverting current in coil if system stays in CCM [22].

10.6.1 CCM Detection

Detecting this threshold is fundamental so that system can switch to a mode where conduction is such that coil current is not inverting anymore, and consequently, the system now passes some time in hi-impedance mode, which can be done for example by skipping pulses when voltage is above a determined thresholds, or changing control scheme to Pulse-Frequency modulation (PFM) where a fixed current pulse is output at a variable frequency.

If a very precise current sensor is implemented, it can be used for this threshold detection, however, it is preferable for precision to use a dedicated circuitry. It must

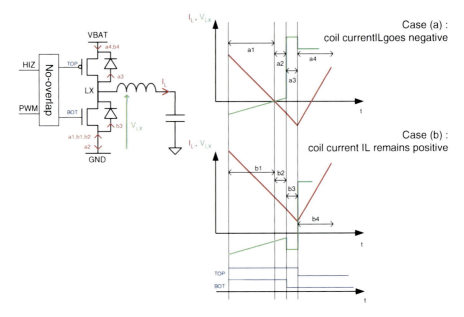

Fig. 10.5 LX node sensing for current inversion detection

be noted that if CCM detection is done with a current actually above the actual critical conduction current, system will oscillate between the full power (CCM) and whatever the low power mode chosen, creating an important ripple on output voltage.

Most CCM detection schemes rely on analog principles, but they output a single-bit digital signal which can be used in digitally controlled SMPS. A common principle for bucks consists in relying on body-diode conduction at the end of conduction period (cf. Fig. 10.5): when current in inverted in coil at end of conduction period, during the dead time, current is evacuated through top mosfet body diode, instead of bottom mosfet body diode when coil current is above critical conduction threshold. This conduction on top MOS can be either detected by a fast sampled comparator on LX node [11], or detecting phase delta on LX node.

10.6.2 *PFM Digital Implementation*

PFM mode is generally used for ultra-low current modes, where higher ripple is acceptable. Switching losses are reduced to a minimum, because the system is no clocked anymore and only generating constant *Ton* pulse (in classical implementations). Using such a scheme for "digital" SMPS requires to be able to calculate ideal $Ton = Vref/Vbat$, and to generate the needed PWM pulse.

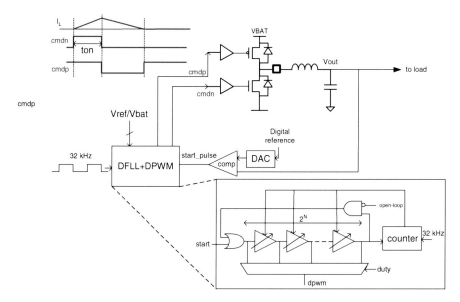

Fig. 10.6 Digital PFM loop with combined DFLL + DPWM (simplified)

However, a proposed digital implementation consists in using an open-loop DFLL/DPWM which internal clock, digitally compensated, is only woken up when output comparator states that system need to be compensated (shown on Fig. 10.6). The DFLL being open-loop can drift if supply or temperature is evolving – and these changes are compensated by running DFLL in closed-loop based on a very low-frequency clock, with a negligible impact on power consumption.

In order not to invert current in coil – which is necessarily the case if duty cycle is taken to be the ideal *Vref/Vbat*, and adaptive duty-cycle compensation scheme, using the previously described CCM detectors can be used.

It however appears that in the area of very low power, digital SMPS show little improvement over analog ones, because digital requires clock presence, which implies power consumption well above acceptable thresholds for ultra-low power modes where analog quiescent current can be as low as 15 µA.

10.7 Precision and Voltage Stack

By essence, regulators are designed to maintain output voltage as constant as possible, whatever the load and input voltage variations. Load and line regulation describe the DC variation of output voltage to output current and input voltage on their full range, while load and line transient are concerned with transient response to sudden variation in output current or input voltage.

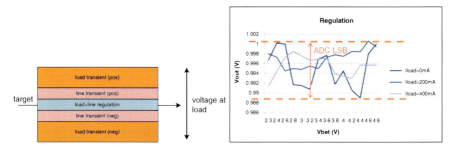

Fig. 10.7 Simplified voltage stack and typical regulation of digital SMPS

10.7.1 Load Transient and Voltage Stack

Load transient performances are of uttermost importance when designing (integrated) power supplies: processors – which supply voltage trend is to lower, while they current consumption is growing because of supplemental cores for example – tend to dynamically update they voltage request in order to set it at the minimum value to be able to work flawlessly at a given frequency.

But actually, the "minimum" value the processor shall require has to take into account the fact that the voltage which effectively reaches the core will have to account for the line and load regulation of SMPS, as well as its line and load transient in the region of operation. All this summed-up constitute the voltage stack (Fig. 10.7) which had to be minimized, and for which load transient represents an important challenge.

Load transient performance is a direct outcome of regulation loop performance, so in this matter, analog or digital implementation show significant differences.

Classical compensation analog compensation loops are voltage-loop PID controllers [1], but there are superseded by many regulation scheme using current sensing (and regulation loop) such as current-programmed control as far as linear controller are concerned.

Digital controllers have first started to implement digital PID, which performance where poorer than their analog counterparts and are now exploring a wide-range of non-linear techniques, ranging from non-linear PID to model-predictive controllers.

All controllers are actually trying to approach the optimal response to load transient, while keeping good line transient and regulation properties.

For a given LC output filter and a given load transient, the optimal transient response is known (Fig. 10.8), and it means, for a positive load transient, to turn on top mosfet on for a time T_{on} than off for time T_{off}, these two times being calculated using output capacitor charge balance approach [13]. This approach requires complex calculations that can only be handled by digital implementation and most certainly offline e.g. using look-up tables (LUT) for fast switching circuitries.

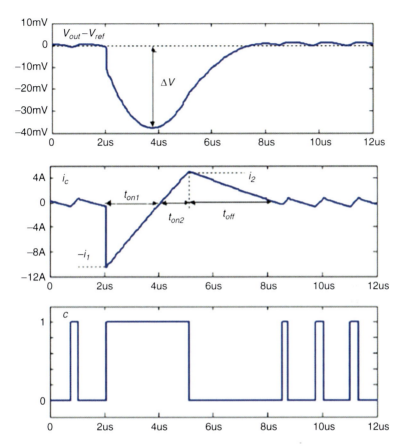

Fig. 10.8 Proximate time-optimal digital control [12]

Several proposed implementation are combining a non-linear implementation for transient response and linear controller for steady-state control.

Digital control can theoretically calculate the ideal or near-ideal response [12], but it will be limited by several factors:

- Lag due to calculation time, which can be minimized using fast processes and fast computing units, or LUT techniques.
- Lag due to ADC conversion times, and A/DC throughput rate, particularly on coil current measurements.
- Imprecision due to ADC resolution (quantization effects)

On the other hand, analog system such as sliding mode controllers, when implementing very precise current sensors, shows a transient response which (at least in theory) can nearly match optimal response [7].

Another solution for digital controllers is to implement digital sliding mode, but the three previous limitation factors will also applied, making transient response worse than its analog counterpart.

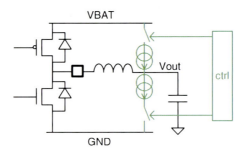

Fig. 10.9 Augmented buck

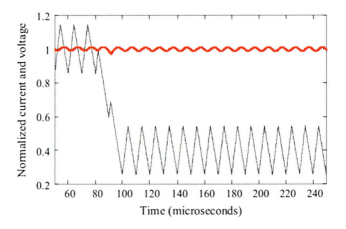

Fig. 10.10 Simulated ideal augmented control [14]

10.7.2 Augmented Systems

Nevertheless, Digital SMPS can show better load transient performance than optimal analog controllers, if they implement augmented systems [14, 20] as showed on Fig. 10.9. When a load transient is detected (on output voltage) digital turns on an extra current source which help to sink or source current in the load, hence reducing the over/undervoltage.

The extra current source is then turned off when output voltage has recovered. Figure 10.10 illustrates the ideal case when output current need is integrally compensated by current source, which is never the case. However, actual (unpublished) silicon implementations have shown that load transient can be reduced by a factor of 2 (this figure depends on size of current source). The silicon area cost depends on the size of the extra current sources which are chosen, but these types of techniques which need to be further studied are good contenders to optimal analog controllers, with an extra area cost.

10.7.3 Digital SMPS Precision and Ripple

A major drawback of most digital SMPS structures is that they use an ADC which LSB limits the resolution of system. A fundamental stability equation [15, 23] states that in order to avoid limit cycles, resolution of DPWM must be high enough to ensure that output voltage can be set to lie into ADC zero-bin. Because of limited feasible resolution of DPWM, this condition greatly limits the precision of digital SMPS, compared to analog SMPS where precision can be only limited by loop gain. A typical precision of reported digital SMPS lies between 5 and 10 mV, whereas analog loop can achieve regulation in the range of less than 1 mV.

What's more, the needed reported DPWM precision required is typically around or more than 10 bits, making direct DPWM implementation impossible, because of required time resolution of such a system. It is hence necessary to use dithering or delta-sigma modulation to achieve an equivalent average resolution with less "physical" bits in DPWM. This technique generates low-frequency spectral content on output voltage, visible as a low frequency ripple.

10.7.4 Line Transient and Feed-Forward

Most analog Bucks use a supply voltage feed-forward, which not only makes the loop gain independent of supply voltage, but also allows to almost reduce line transient (i.e. supply voltage variations) to zero. The classical implementation [1] consists in using a ramp generator with a gain proportional to inverse of input voltage.

Implementation of feed-forward in digital systems imposes more constraints: not only digital image of supply voltage is required, generally through an ADC, but the compensation loop gain has to be multiplied by the invert of this signal. A proposed solution [11] uses a look-up table to implement division as a multiplication, but more efficient techniques have probably to be proposed to improve feed-forward. On top of this problem, quantization of input voltage creates a less than ideal non-linear response to line transient which adds to voltage stack.

10.8 Auto-Tuning

Integrated IC suffer from large process variation effects on both passive and active devices, which make integrated analog control difficult to tune in all process corners, and gives an important advantage to digital implementation of filters (e.g. typically in a PID controller), which are not subject to process variations.

Another important advantage of embedded digital processing capabilities and ADC, which are required for digital control, it that it eases the auto-measurement and auto-tuning methods, which have been a popular research subject, may it be for regulator performance [16, 17], or current sensor self-tuning. These techniques are somewhat adaptable to analog controllers, but in this case require dedicated logic and converters.

10.9 Multi-phases and Control Sharing

Maximum current loads requirements continue to rise in portable application, and a practical solution to keep with this current while keeping acceptable load transient and ripple without requiring a huge output capacitor consists in using multi-phase SMPS, where several coils in parallel are providing current to a single load (Fig. 10.11) [23]. In order to minimize ripple, the different phases are spread, and this allow for Digital SMPS which are using a unique time-multiplexed ADC to perform all conversions. If control calculation allow for it, the processing unit can also be shared between the phases.

The same ADC sharing technique can be used to share ADC and digital controller among several independent DC-DC converters, as long as their phases are spread, as shown on Fig. 10.12. The sharing can be extended to use the same ADC to sample output voltage and battery voltage [11], hence allowing further area savings.

Sharing techniques are unique to digital controlled SMPS and can provided consistent saving both in area of control part, as well as in power consumption.

An additional point in multi-phases SMPS is that a current-sharing external loop is necessary to balance current in all coils. This requires sensing and processing of very coil current, which can also be done in an analog or a digital way.

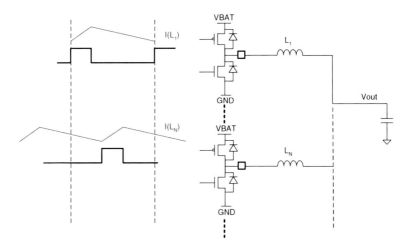

Fig. 10.11 Multi-phase power stage

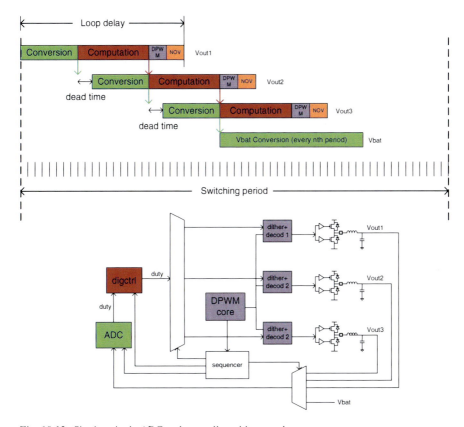

Fig. 10.12 Sharing single ADC and controller with several power stages

10.10 Voltage and Frequency Scaling

A common feature to regulators is ability to allow for Dynamic Voltage Scaling, allowing changing output voltage on the fly. This technique requires a filtering on reference in order to prevent destructive current in-rushes if reference voltage is changed abruptly. Including this as a digital filter inside digital control makes its somehow more flexible than adding an additional control part before a static DAC, which is the main option in analog structures.

Another technique consists in changing on the fly the switching frequency of SMPS at low currents in order to reduce the switching losses which become dominant in this area – at expense of slightly bigger ripple.

However, DPWM is not very flexible concerning frequency control: its resolution is generally frequency related – most architectures are not designed to support multiple switching frequencies. Yet, a digital DCM architecture proposed in [18] shows an efficient way of reducing the switching frequency, while keeping then *Ton* duration between *Vref/Vbat* and *Vref/Vbat/sqrt(2)*.

10.11 EMI Mitigation

Portable applications typically include cellphones where conducted and radiated noise due to switching power supplies can have a dramatic effect on RF part and should be minimized. On top of using expensive ferrites to isolate parts, some important EMI mitigation techniques are used both in analog and digital SMPS:

- Slope on power mosfet controls can be reduced so that current drawn on battery contains less high harmonics, at the expense of efficiency loss
- System clock can use some dithering so that some spectrum spreading occurs. But this technique should be used with care in digital SMPS, because a minimum clock period is generally required to sample and calculate next period duty-cycle.

Yet, the most efficient and programmable spectrum-spreading technique, i.e. random wrapped-around pulse-position modulation (RWAPPM) [19] can only be implemented easily on digital SMPS: while in analog systems duty-cycle value is unknown at the beginning of a conduction period, most DPWM system actually require that is this duty cycle is calculated before conduction period begins. This allows to simply implementing a RWAPPM scheme by randomly position the start of conduction pulse (Figs. 10.13 and 10.14).

10.12 A Conclusion on Analog Versus Digitally-Enabled Versus Digital SMPS

Table 10.1 summarizes the main differences between analog and digital SMPS (assuming that analog SMPS are developed in low-cost, older process, while digital SMPS take advantage of cutting edge digital process – no partitioning involved here). While Digital SMPS is a clear winner for flexibility – which however comes at cost of area and power consumption – for EMI reduction complex schemes and advanced sharing for complex structures, analog structures still keep an advance when pure transient performance is required, because the advances in control which have been reported for discrete medium-to-high power SMPS (up to kW range) are difficult to transpose to integrated high-speed, low-area SMPS.

One could wonder if, with such a picture, "digital SMPS" will ever become an option in integrated SMPS for portable applications.

Yet, digital SMPS is already there, even if not for main loop control: the many different modes, and controls, and calibration today require many, many more gates in an 'analog' SMPS. An example from a ST-Ericsson commercial analog SMPS is that the (analog) control part uses six times as many "digital" transistors than analog transistors! So integrated SMPS is already truly a mixed-signal system, which should be designed and architecture as such, and digital control loop will in the long run probably becoming an "option" for this complex system, when, for a given design requirement, its strong assets in configurability and flexibility will justify to sacrifice some of the transient performance.

10 Is Digital SMPS Ready to Eliminate Analog Regulators for Portable...

Modulation Scheme		Peak Spectral Power (dBFS)	Ripple Noise (mV$_{rms}$)[#]
PWM		0.4	0.9
RPPM		−1.0	1.6
RPWM		−2.0	115.1
RCFM-FD (δ = 15 %)		−16.1	0.9
RCFM-VD (δ = 15 %)		−16.2	39.3
PFM	(δ = 15 %, β = 20)	−12.9	2.1
	(δ = 15 %, β = 10)	−10.2	1.3
CFM	(δ = 5 %)	−16.4	1.0
	(δ = 15 %)	−21.1	1.1
	(δ = 30 %)	−24.2	1.1
Proposed RWAPPM		−26.6	2.0

Fig. 10.13 From [19], analytical peak spectral density and ripple noise of various modulation schemes

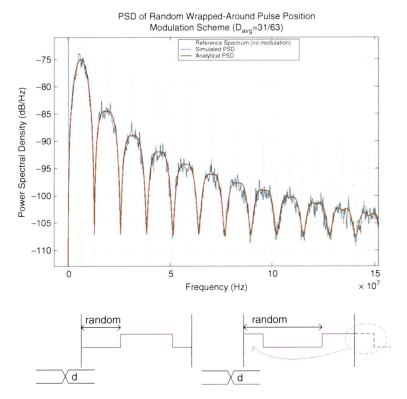

Fig. 10.14 Implemented RWAPPM scheme vs. simulation and theory

Table 10.1 Summary of analog versus digital SMPS

	(Fully) Analog SMPS	(Fully) Digital SMPS
Input voltage range	Low cost 5 V compliant process	High cost process. (40, 28 nm)
	Low digital integration capability	5 V capability through expensive options
Low profile components	Almost independent of the control type	Almost independent of the control type
Efficiency	Very competitive RDSon/Cg/Area	Cascoding increases design complexity. Transistor area almost equivalent. Requires integrated capacitors for decoupling
	20 uA Iq achievable with simple PFM control	
	35 uA Iq achievable with Pulse Skipping control	Very low quiescent possible
Fast transient	Current mode sensing can be done in pure passive way. Very simple and cheap	Voltage mode requires at least 1 ADC for feedback. Current mode requires additional sensitive system
Clock synchronization capability	Easy to medium (PLL for hysteretic control)	Easy
EMI	Basic	Easy implementation of complex modulation schemes
Flexibility	Possible (with digital): configurable compensation network, information exchange with digital requires extra ADC/DAC	Possible: parameterized algorithm. Requires additional hardware
		Possible to take benefit from processor activity status
Control sharing	Impossible	Possible (depending on controller bandwidth)
DVS and reference management	DAC for reference	No conversion required

References

1. R.W. Erickson, D. Maksimovic, *Fundamentals of Power Electronics*, 2nd edn. (Springer, 2001)
2. F. Carabolante, Digital power: From marketing buzzword to market relevance, in *2006 I.E. COMPEL Workshop*, Troy, NY (2006)
3. J. Rabaey, *Low Power Design Essentials*, (Springer, 2009)
4. J. Xiao, A. Peterchev, J. Zhang, S. Sanders, A 4-uA quiescent current dual-mode digitally controlled buck converter IC for cellular phone applications. IEEE J. Solid-State Circuit, 39(12), 2342–2348 (2004)
5. P. Mattavelli, L. Rossetto, G. Spiazza, P. Tenti, General-purpose sliding-mode controller for DC/DC converters applications, in *Power Electronics Specialists Conference, PESC'93*, Seattle, USA, (1993)
6. A. Syed, E. Ahmed, E. Alarcon, D. Maksimovic, Digital pulse width modulators architectures, in *IEEE Power Electronics Specialists Conference (PESC)*, Aachen, Germany, (2004)
7. B. Labbe, B. Allard, X. Lin-Shi, D. Chesneau, An integrated sliding-mode buck converter with switching frequency control for battery-powered applications, IEEE Trans. Power Electron. **28**(9), 4318–4326 (2013)

10   Is Digital SMPS Ready to Eliminate Analog Regulators for Portable...   187

8. S. Saggini, D. Zambotti, E. Bertelli, M. Ghinoni, Digital autotuning system for inductor current sensing in voltage regulation module applications. IEEE Trans. Power Electron **23**(5), 2500–2506 (2008)
9. A. Prodic, D. Maksimovic, Digital PWM controller and current estimator for a low-power switching converter, in *Computers in Power Electronics, COMPEL 2000. The 7th Workshop on*, pp. 123–128, Blacksburg, VA (2000)
10. A.G. Beccuti, S. Mariethoz, S. Cliquennois, S. Wang, M. Morari, Explicit model predictive control of DC–DC switched-mode power supplies with extended Kalman filtering. Ind. Electron. IEEE Trans. **56**(6), 1864–1874 (2009)
11. S. Cliquennois, A. Donida, P. Malcovati, A. Baschirotto, A. Nagari, A 65-nm, 1-A buck converter with multi-function SAR-ADC-based CCM/PSK digital control loop. Solid-State Circuit IEEE J. **47**(7), 1546–1556 (2012)
12. V. Yousefzadeh et al., Proximate time-optimal digital control for synchronous buck DC-DC converters. IEEE Trans. Power Electron. **23**(4), 2018–2026 (2008)
13. G. Feng, E. Meyer, Y.F. Liu, A new digital control algorithm to achieve optimal dynamic performance in dc-to-dc converters. IEEE Trans. Power Electron **22**(4), 1489–1498 (2007)
14. P. Krein, Feasibility of geometric digital control and augmentation for ultra-fast DC-DC converter response, in *IEEE COMPEL Workshop*, Troy, NY (2006)
15. A. Peterchev, S. Sanders, Quantization resolution and limit cycling in digitally controlled PWM converters. IEEE Trans. Power Electron. **18**(1), 301–308 (2003)
16. Z. Lukic, S.M. Ahasanuzzaman, Z. Zhao, A. Prodic, Self-tuning sensorless digital current-mode controller with accurate current sharing for multi-phase DC–DC converters, in *Proceeding IEEE Applied Power Electronics Conference (APEC)*, pp. 264–268, Washington, DC (2009)
17. M. Shirazi, R. Zane, D. Maksimovic, L. Corradini, P. Mattavelli, Autotuning techniques for digitally-controlled point-of-load converters with wide range of capacitive loads, in *Proceeding IEEE Appied. Power Electronics Conference (APEC)*, pp. 14–20, Anaheim, CA (2007)
18. J. Chen et al., DPWM time resolution requirements for digitally controlled DC-DC converters. in *IEEE APEC*, Dallas, TX (2006)
19. V. Adrian, J.S. Chang, B.H. Gwee, A randomized wrapped-around pulse position modulation scheme for DC–DC converters. Circuit Syst. I Regul. Pap. IEEE Trans. **57**(9), 2320–2333 (2010)
20. L. Amoroso et al., Single Shot Transient Suppressor (SSTS), in *Applied Power Electronics Conference and Exposition, APEC '99*, Dallas, TX (1999)
21. N. Raman, A. Paranyadeh, K. Wang, A. Prodic, Multimode digital SMPS controller IC for low power management, in *IEEE International Symposium on Circuits and Systems, (ISCAS 2006)*, (2006)
22. H. Peng, A. Prodic, E. Alarcon, D. Maksimovic, Modeling of quantization effects of digitally controlled DC–DC converters. IEEE Trans. Power Electron **22**(1), 208–215 (2007)
23. T. Carosa, R. Zane, D. Maksimovic, Implementation of a 16 phase digital modulator in a 0.35 µm Process, in *2006 I.E. COMPEL Workshop*, Troy, NY (2006)

Chapter 11
A 2.2A, 4 MHz Switch-Mode Battery Charger for a Cellular Power Management Unit

Jay Ackerman, Mike Baker, Ryan Desrosiers, Vipul Katyal, Marc Keppler, John McNitt, Russ Radke, Mark Rutherford, Scott Savage, and Kerry Thompson

Abstract Battery chargers are a necessary part of any mobile electronic device. Because a mobile phone battery is a high-energy storage unit, care must be taken in the design of the entire charging system. This paper discusses the overall charging system for a mobile phone as well as some key factors that make battery chargers unique. The design of a switch-mode battery charger is presented, including subsystem circuit architecture, stability analysis, and sequencing logic, plus the key performance parameters of the design are summarized.

11.1 Introduction

A battery charger is an essential component of any mobile device because mobile devices require some energy storage element to enable use without a wired power source. Although a battery charger may appear as just another step-down regulator, several external conditions make the design of battery chargers particularly challenging.

The organization of this paper is as follows: Sect. 11.2 describes an overview of battery charging in a cell phone. Section 11.3 presents the design of the battery charger with partial, lossless current sensing [1]. Section 11.4 presents the experimental results of battery-charging operations in different operating modes. Section 11.5 presents the conclusions reached during the course of this study.

J. Ackerman (✉) • M. Baker • R. Desrosiers • V. Katyal • M. Keppler
J. McNitt • R. Radke • M. Rutherford • S. Savage • K. Thompson
Broadcom Corporation, Fort Collins, CO, USA
e-mail: jaya@broadcom.com

A. Baschirotto et al. (eds.), *Frequency References, Power Management for SoC, and Smart Wireless Interfaces: Advances in Analog Circuit Design 2013*,
DOI 10.1007/978-3-319-01080-9_11, © Springer International Publishing Switzerland 2014

11.2 Battery Charging Overview

A block diagram of a typical charging subsystem is shown in Fig. 11.1. Although the energy source shown in Fig. 11.1 is typically a wall adapter, it can also be a Universal Serial Bus (USB) supply or another energy source such as a wireless power receiver, non-rechargeable battery, or even a manually operated electric generator.

For most cell phones, the battery is either a lithium-ion or lithium-polymer cell. The cell is packaged in a battery pack that also includes safety circuits that limit the source and sink current from the battery as well as circuitry that prevents excessive discharge of the battery cell itself. Excessive discharge can damage the battery cell and reduce its storage capacity.

The battery charger shown in Fig. 11.1 is the interface between the energy source and the battery pack and is responsible for regulating both the current and voltage to the battery pack during a charging cycle. The rest of the electronics for the phone is labeled as the System.

11.2.1 Types of Battery Chargers

The most common battery chargers are linear chargers and switch-mode chargers. A linear battery charger is often seen in low-end (i.e., limited-feature) phones. This type of charger is inexpensive and offers good output current and output voltage regulation at the battery terminals. Other than the usual decoupling capacitors on the battery terminals, no additional components are required to implement the linear charger. The primary disadvantage of the linear charger is that power dissipation is usually quite high and is dependent on input (adapter) voltage, output (battery) voltage, and the charging current. In some cases, excessive power dissipation can limit the amount of other activity that can occur with the phone. For a low-end phone, where cost is critical, and the typical charging scenario occurs during periods of non-use (e.g., overnight), this type of charger may be the preferred charger.

For best efficiency and highest performance, a switch-mode charger is usually employed. Due to the addition of an external inductor and a more complicated on-chip control scheme, this type of charger is more expensive. Along with the

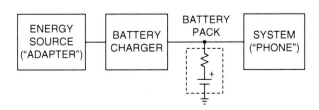

Fig. 11.1 Cell phone charging subsystem

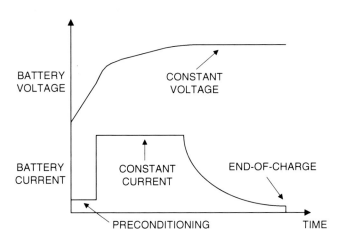

Fig. 11.2 Battery charging cycle

increased cost, however, there is an increase in power-transfer efficiency, which results in lower power dissipation and a reduced probability of limiting the phone usage cases.

The overall goal of the charging system is to charge the battery as quickly as possible without compromising safety. A typical charging cycle [2] is shown in Fig. 11.2.

11.2.2 Description of a CC-CV Battery Charging Cycle

If the battery is fully depleted, the charger must first output a small current. The application of this small current is the preconditioning phase of charging. This low-level current closes the safety switch inside the battery pack and brings the battery terminal voltage to a level where the charge current can be increased. Because the current value is low (typically 1/10 of the maximum charging current), the preconditioning phase can be done with a linear current source.

The next phase in the charging process is a transition to a higher output current. Depending on the battery type and capacity, the charging current may go directly to its maximum value. Because the current is relatively constant during this phase of charging, it is called the constant current (CC) phase of the charging cycle. Typically, the maximum charging current for a battery is 0.8 C (where C is the capacity of the battery in amp-hours or AHr). For a 1.5 AHr battery, the maximum charge current is approximately 1.2A.

As the battery voltage rises to the target (or 'float') value, the charger transitions from CC to constant voltage (CV) mode. During the CV phase of charging, the battery voltage is relatively constant, whereas the current exponentially decays from its maximum value toward zero. It is important to note that the actual output current from the battery charger may include both the current used to charge the battery and any other system current that the phone requires.

Depending on the system loading (i.e., the current required by the phone), the charger may transition randomly between CC and CV modes. The charging circuit must be designed to handle these transitions gracefully without compromising the safety of the charging subsystem or the rest of the phone platform.

The final phase occurs as the output current drops below the end-of-charge (EOC) value. At this time, the charging process terminates until the battery voltage drops below some predetermined 'maintenance' value. Once the battery voltage is below this maintenance value, the charging circuits reengage.

If the battery is not fully depleted when the adapter is initially connected to the phone, the charging cycle will usually start in CC mode and then transition to CV mode as the battery voltage increases.

11.2.3 Unique Requirements on Battery Chargers

Conceptually, the battery charger can be thought of as a current-limited buck switching regulator. There are several external factors that must be considered for the charger, however, that do not need to be considered for a typical buck converter. Circuitry must be added within the charger to gracefully deal with these factors.

The first factor to consider is safety. Lithium-ion batteries can be a potential hazard, particularly if overcharged. Safety standards such as IEEE-1725 [3] and JEITA [4] were specifically created to ensure the safety of the charging subsystem. These specifications address maximum charge current, maximum battery voltage, and the temperature range over which battery charging is allowed.

The second factor to consider is the variability of the energy source. The voltage/current characteristics for various AC-to-DC (wall) adapters are shown in Fig. 11.3. Based on the data in Fig. 11.3, it is obvious that the maximum current available from the adapter source is not guaranteed. Although standards such as USB Power Delivery have attempted to reduce the variability of the input source, aftermarket adapters are readily available from multiple sources and are often preferred to original manufacturer adapters due to their low cost. The battery charger must be able to safely handle conditions where the energy source (i.e., the adapter) cannot deliver the power expected by the charging system.

Finally, the power dissipation of the charger can very likely limit the overall power dissipation of the phone platform. If excessive power dissipation occurs, the charging current will usually be reduced until the cell phone (or power management unit IC) temperature is within acceptable limits. Additional circuitry must be added to manage this thermally limited usage case.

11 A 2.2A, 4 MHz Switch-Mode Battery Charger for a Cellular Power Management Unit   193

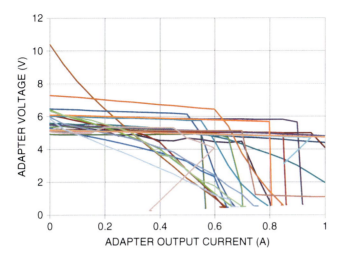

Fig. 11.3 V-I characteristics for various battery charging adapters

11.3 Description of the Switch-Mode Battery Charger

An ideal CC-CV battery charging system consisting of a CC-CV charger and battery is shown in Fig. 11.4. The charging circuit consists of an ideal voltage source at V_{FLOAT} in series with circuitry that limits the output current to the I_{CC}. The ideal battery consists of a very large capacitor (in excess of 1 kF) with some equivalent series resistance (ESR). When a discharged battery is connected to this ideal charger, current will flow into the battery at an increasing rate until the output current value equals the current limit of the charging source.

At this point, the charger source is in the CC phase and looks like an ideal current source charging the battery at I_{CC}. As the battery voltage increases, the current begins to drop due to the battery ESR, and charging enters the CV phase. As the battery terminal voltage approaches V_{FLOAT}, the current into the battery drops to zero and charging is complete. The goal in developing any battery charger circuit for a cell phone is to emulate this ideal charging model as efficiently as possible.

A synchronous, step-down switch-mode power supply (buck regulator) with a precision current limit is an excellent candidate for a battery charger due to the high efficiency achievable with this circuit topology. The switch-mode topology, unlike that of a linear regulator, is ideally lossless, allowing for maximum power transfer from the input source to the output load. A switch-mode charger operating in current limit exhibits a "transformer ratio" in which the current going into the battery is ideally scaled by the ratio of the adapter voltage to battery voltage.

$$I_{BATTERY} = I_{ADAPTER} {}^{*} V_{ADAPTER} / V_{BATTERY} \qquad (11.1)$$

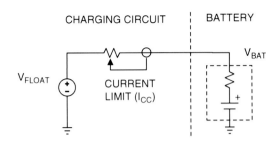

Fig. 11.4 Ideal CC-CV battery charger

In Eq. 11.1, $I_{BATTERY}$ is the average battery charging current, $I_{ADAPTER}$ is the average current from the adapter, $V_{ADAPTER}$ and $V_{BATTERY}$ are the adapter and battery voltages, respectively. In addition to being highly efficient, the switch-mode charger significantly reduces charge time. For a charging source limited to 5 V at 1.5A (7.5 W), the charge current in the CC phase charging a 3.5 V battery can be as high as 2.14A. This higher current value represents a 40 % increase in CC current compared to a linear regulator design limited to 1.5A. Thus, the switch-mode charger offers lower power dissipation (due to high efficiency) and shorter charge time (due to the transformer ratio) compared to a linear design. Synchronous switch-mode chargers designed with low resistance MOS output devices for the high- and low-side switches offer efficiencies in excess of 90 %. In the example above, 90 % efficiency for the switch-mode charger would result in a charge current of 1.9A, which is still significantly above the linear regulator limit of 1.5A. As adapter voltages increase, the benefit from the switch-mode topology increases even more.

As described earlier, the switch-mode charger is essentially a current-limited step-down (buck) regulator. For buck regulators, there are many different control topologies to choose from, including voltage control, current control, and hysteretic control [5–8]. For the switch-mode charger circuit described here, a constant-frequency, peak-current-mode control scheme was chosen for several reasons. First, in peak-current-mode control, the duty cycle is controlled by the inductor current. Because the inductor current is equal to battery current, the design has a built-in mechanism to measure and regulate the battery current. By measuring peak inductor current, the circuit also has the ability to measure and limit cycle-by-cycle inductor current, which is very important for battery safety [3, 4]. Second, in peak-current-mode control, a lossless current-sense scheme [1] using only the high-side (PMOS) device can be used. Sensing the PMOS device current eliminates the need for a low-value sense resistor in series with the inductor to measure inductor current. The lossless sensing scheme reduces the external component count (and associated cost) as well as the power loss from the sense resistor. Third, the peak-current-mode control scheme eliminates the double pole due to the LC output filter making for a more area-efficient compensation scheme in the IC [9]. Finally, the constant frequency aspect of the control scheme makes potential EMI issues more manageable. Switching frequency is well-defined, and the spurious energy can be spread out via a spread-spectrum PLL, if needed, to reduce potential switching noise spurs below the required EMI levels.

Fig. 11.5 CC-CV switch-mode battery charger diagram

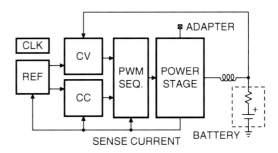

A high-level block diagram of the switch-mode charger presented in this paper is shown in Fig. 11.5. The circuit consists of a clocking system for generating the 4 MHz switching frequency reference clock, a voltage reference circuit for generating the CC/CV control point reference voltages, a CC control loop, a CV control loop, a PWM sequencer, and a power stage. The power stage consists of the output devices, predrivers, a nonoverlap generator, the zero-current detector, and the PMOS (high-side) current sense circuitry. In this implementation, the CC and CV loops operate in parallel. The loop (CC or CV) that requires the lowest duty cycle (i.e., the lowest current) controls the on-time of the high-side device. When the battery voltage is below the CV setting, the CC loop is in control, limiting the output current with the CV loop railed. The CC and CV control loops will be described separately, after which the combined behavior will be discussed.

11.3.1 *Constant-Current (CC) Control Circuitry*

The circuit diagram for the CC loop is shown in Fig. 11.6. The CC loop consists of a high-frequency inner loop that controls the sensed inductor peak current and an outer loop that controls the average of the sensed inductor current. The rising edge of the reference clock (CLK) begins the cycle causing the high-side PMOS to turn on. With the high-side device on, the current through the PMOS device and the external inductor increases. This inductor current is sensed and converted to a voltage through a resistor R_I. The sense voltage across R_I is then added to a ramp voltage V_{ART} (voltage across C_I) to generate the voltage V_{RAMP}. The voltage V_{RAMP} is then compared to the integrator node V_{CI}. When V_{RAMP} exceeds V_{CI}, the high-side PMOS device is turned off, and the low-side NMOS is turned on, which ramps down the inductor current for the remainder of the reference clock cycle. The waveforms associated with the circuit in Fig. 11.6. are shown in Fig. 11.7.

The artificial ramp (V_{ART}), described above, is generated by a constant current source I_{ART} into capacitor C_I. As described in [10], the ratio of the voltage ramp slope generated by the current sense and the artificial voltage ramp slope can impact inner loop stability. Ensuring stability over the full range of expected input and

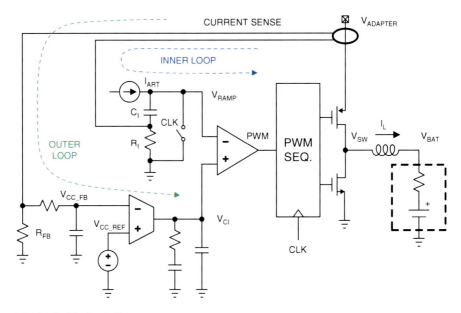

Fig. 11.6 CC circuit diagram

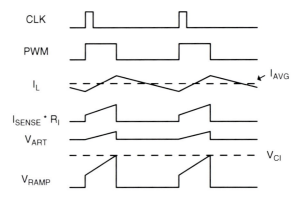

Fig. 11.7 CC loop waveforms

output voltage requires careful design of the ramp voltage. For a switch-mode charger, duty cycles can range from 20 % to 100 %, adding to the challenge of keeping the inner loop stable.

The average current going into the high-side (PMOS) device is controlled by the outer loop. A replica of the sensed current is applied to resistance R_{FB} and is heavily filtered to generate a voltage that is an accurate representation of the average input current. The error voltage (difference between V_{CC_REF} and V_{CC_FB}) is then integrated by a G_m-C filter stage to generate the V_{CI} control signal in Fig. 11.6. The loop is locked when the average sensed voltage (V_{CC_FB}) equals the reference voltage (V_{CC_REF}).

11 A 2.2A, 4 MHz Switch-Mode Battery Charger for a Cellular Power Management Unit

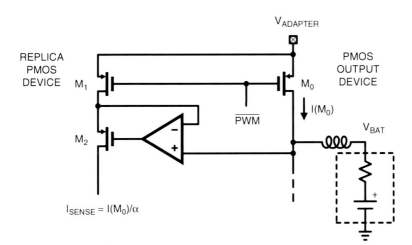

Fig. 11.8 CC replica sense circuit diagram

The current sense circuit for the high-side PMOS switch is critical for an accurate battery charger because this circuit sets the overall CC current accuracy of the charger. The current sense must maintain high accuracy and high speed for a 4 MHz switching frequency. To achieve acceptable accuracy in the sensed inductor current, the current sense typically must start tracking the inductor current within the first 10 % of the switching cycle.

A functional schematic of the circuit used to sense the high-side current is shown in Fig. 11.8. In this circuit, M_0 is the high-side PMOS switch in the charger output stage, and M_1 is a ratio-metric device sharing the gate and source connections. A high-speed operational amplifier drives the drain voltage of the replica device M_1 to the same potential as the drain voltage of the high-side (PMOS) device M_0 by controlling the gate of control device M_2. When M_0 turns on (via gate signal PWMb going low), the op amp circuit drives the V_{DS} of the two devices so that the current through M_1 is ratio-metric with the current through M_0 (i.e., $I_{DS}(M_1) = I_{DS}(M_0)/\alpha$). The current through M_2 is replicated, as needed, for use in both the inner and outer control loops.

The voltage generated across R_{FB} in Fig. 11.6 represents the instantaneous input current flowing through M_1. When the sensed inductor current (I_{DS} of M_1) is applied to resistor R_{FB} and averaged by a RC low-pass filter, the resulting V_{CC_FB} voltage becomes an accurate representation of the average adapter input current.

$$V_{CC_{FB}} = (I_{ADAPTER}/\alpha)^* R_{FB} \qquad (11.2)$$

To control the average input current in the system, the voltage V_{CC_FB} is then compared against the reference voltage V_{CC_REF} in the outer loop.

The circuit used to generate the reference voltage V_{CC_REF} (in Fig. 11.6) is shown in Fig. 11.9. This circuit operates in two modes. In input current-control mode, the switch S_0 is held open to control average input (adapter) current.

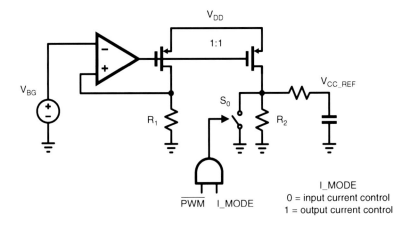

Fig. 11.9 CC reference circuit diagram with PWM modulation

In output current-control mode, the PWM signal modulates the switch S_0 to control output current.

When the circuit is operating in input current-control mode, the signal I_MODE is held low; therefore, switch S_0 is always open. In this mode, voltage V_{CC_REF} is equal to $V_{BG}{}^*(R_2/R_1)$. Thus, when the loop is in lock, the input current is set by Eq. 11.3.

$$V_{CC_{REF}} = V_{CC_{FB}} \rightarrow V_{BG}{}^*(R_2/R_1) = (I_{ADAPTER}/\alpha)^* R_{FB} \qquad (11.3)$$

$$I_{ADAPTER} = V_{BG}{}^*\alpha^* R_2/(R_1{}^* R_{FB}) \qquad (11.4)$$

In Eq. 11.3, α sets the maximum desired sense quiescent current, and R_{FB}/α is set to maximize the sensed voltage, given the available supply voltage headroom and maximum sensed input current. With α and R_{FB} set, R_1 and R_2 are then adjusted to control the input current. By making R_1 or R_2 adjustable, the circuit can accommodate a range of input current-control levels.

For regulating the output current, the signal I_MODE in Fig. 11.9 is high. In this mode, the PWM signal is applied to switch S_0 such that the on-time of S_0 is proportional to (1-D), where D is the duty cycle. Therefore, for a 100 % duty cycle (D = 1), the switch is never closed, whereas for a 50 % duty cycle, the switch is on half the time. The switched voltage generated on resistor R_2 is averaged by a low-pass RC filter. As a result, the voltage at V_{CC_REF} is given by:

$$V_{CC_{REF}} = D^* V_{BG}{}^*(R_2/R_1) \qquad (11.5)$$

Thus, when the loop is in lock

$$V_{CC_{REF}} = V_{CC_{FB}} \rightarrow D^* V_{BG}{}^*(R_2/R_1) = (I_{ADAPTER}/\alpha)^* R_{FB} \qquad (11.6)$$

11 A 2.2A, 4 MHz Switch-Mode Battery Charger for a Cellular Power Management Unit

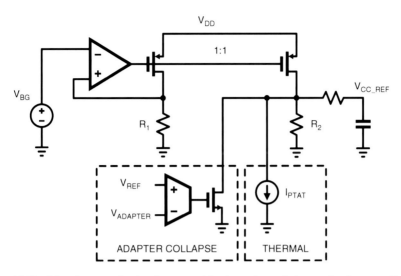

Fig. 11.10 CC reference circuit diagram with thermal regulation and adapter collapse modifications

$$I_{ADAPTER} = D^*V_{BG}{}^*\alpha^*R_2/(R_1{}^*R_{FB}) \quad (11.7)$$

Because, for an ideal step-down regulator, output power is equal to input power, we can rewrite Eq. 11.7 as

$$I_{ADAPTER} = I_{OUT}{}^*D = D^*V_{BG}{}^*\alpha^*R_2/(R_1{}^*R_{FB}) \quad (11.8)$$

$$I_{OUT} = V_{BG}{}^*\alpha^*R_2/(R_1{}^*R_{FB}) \quad (11.9)$$

Thus, the V_{CC_REF} voltage, generated when I_MODE is high, can be used to accurately regulate the charger output current. Note that the equations above apply when the regulator is operating at constant frequency in continuous conduction mode (CCM). Complications with this circuit arise when the regulator is skipping pulses in discontinuous conduction mode (DCM) due to the presence of a third active state (i.e., high-impedance output state). The circuits in Figs. 11.6 and 11.9 require the charger to operate in CCM when the charger is in the CC phase.

A slight modification can be made to the reference circuit to correctly handle either an adapter 'collapse' (the requested current exceeds the capacity of the adapter) or a reduction of current due to excessive on-chip temperature. This modification (without the PWM modulation circuit) is shown in Fig. 11.10. Note that V_{CC_REF} is heavily filtered to keep these additional loops stable.

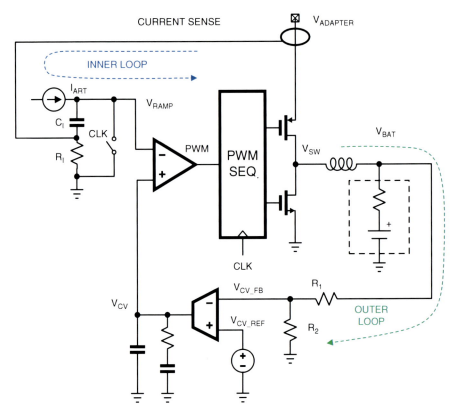

Fig. 11.11 CV circuit diagram

11.3.2 *Constant-Voltage (CV) Control Circuitry*

The circuit for the CV loop is shown in Fig. 11.11. This circuit is very similar to that of the CC loop in Fig. 11.6 described above. The only difference is that the outer control loop is set by comparing a divided-down battery voltage VBAT to the CV reference voltage, V_{CV_REF}. The inner current loop is identical to that of Fig. 11.6. In CV mode, the circuit is operating as a conventional peak-current controlled buck regulator described in [10]. When the outer voltage loop is active, the charger is in the CV phase, and the current is gradually reduced as the battery approaches the CV voltage set by V_{CV_REF}.

The CV loop accuracy is critical in a battery-charging system. To maximize the energy in the battery, the actual CV voltage must be as close to the desired CV voltage as possible. The CV voltage, however, should not exceed the desired float voltage because excessive voltage at the battery terminals can degrade battery safety. In this design, the higher-bandwidth loop was assigned to the CV loop to ensure an accurate float voltage in the presence of system load current modulating the terminal voltage through the battery ESR.

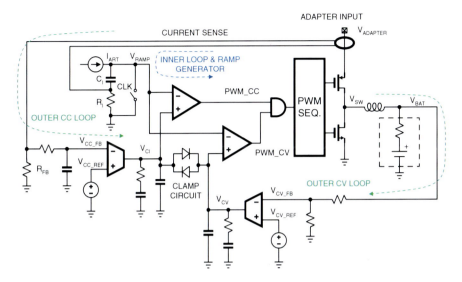

Fig. 11.12 CV-CC circuits combined

11.3.3 Combining CC and CV Control Loops

The combined CV-CC control circuit is shown in Fig. 11.12. In the implementation of this design, the inner current control loop and artificial ramp generator are shared between the two outer loops. The CC and CV loops have independent Gm-C integrators that are compared to the sensed current plus the artificial ramp voltage. The CC and CV loops generate PWM signals that are combined by an AND gate and sent to the PWM sequencer. The loop that ends the PWM cycle (i.e., PMOS on-time) first turns off the high-side (PMOS) device and then turns on the low-side (NMOS) device until the next 4 MHz clock period starts. Thus, the loop that requires the lowest duty cycle (i.e., lowest current) is in control.

The loop that is not in control will integrate error in the Gm-C stage causing the integrator node (V_{CV} or V_{CI}) to drift to the supply. This situation can create a problem with the transient response of the charger. If the charger suddenly transitions from CC to CV due to a large system current load, the previously inactive loop must slew to get back into control. The delay due to slewing can cause a large overshoot in the controlled current or voltage for the charger. To prevent excessive overshoot, a clamp circuit (shown conceptually as back-to-back diodes in Fig. 11.12) is implemented, which keeps the two integrator nodes within a few hundred millivolts of each other. Keeping both integrator node voltages close together significantly reduces the recovery time when transitioning between modes, allowing for a faster transient response.

A linear model for the system in Fig. 11.12 was developed to analyze loop behavior and design loop parameters using the methodology developed in [10–12].

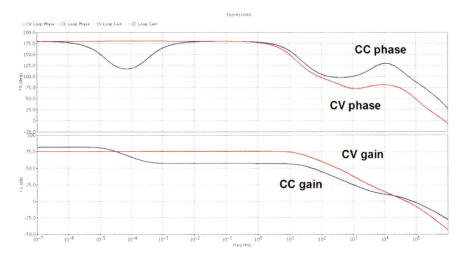

Fig. 11.13 Stability simulations for CC and CV feedback loops

The simulation results for the linear model of the CC and CV loops are shown below in Fig. 11.13. The linearization of the loop also allowed fast transient simulation of the system to analyze the handoff between CC and CV loops.

11.3.4 PWM Sequencing

The sequencing of the power switches is an extremely important and often overlooked aspect of any switching regulator, but it is key to clean, glitch-free operation. Battery charging adds further requirements to this part of the system.

The PWM sequencing block and associated waveforms are shown in Fig. 11.14. PWM operation is controlled by a single flip-flop. When PWM is high, the high-side PMOS is on, and when PWM is low, the high-side PMOS is off, whereas the low-side NMOS is on.

Initially, assume that all PWM_SKIP_{1-M} signals are low. The switching cycle starts when the rising edge of the clock sets the output of the PWM flip-flop. The switching cycle ends when one of the PWM_END_{1-N} signals goes high. Included in the PWM_END signals are the CC and CV comparators and the battery over-voltage and inductor overcurrent indicators. When one of the PWM_END_{1-N} signals goes high while BLANKb is high, CLR goes high, which clears the flip-flop and forces the PWM signal low until the next clock cycle.

During CC operation, the loop is controlled by the CC comparator. When the ramp signal (V_{RAMP}) crosses the threshold set by the CC integrator (V_{CI}), the CC comparator clears the flip-flop (see Fig. 11.7 for V_{RAMP} and V_{CI}). In CV operation, the flip-flop is cleared by the CV comparator. CC and CV modes are distinguished

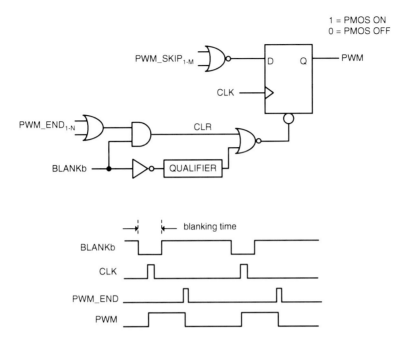

Fig. 11.14 PWM sequencing circuit and waveforms

solely by which comparator clears the flip-flop first, which is determined by which integrator output voltage (V_{CI} vs. V_{CV}) is lower.

There are a number of other PWM_END signals that can reset the flip-flop. For example, safety-related events such as an overvoltage condition on the battery or an overcurrent condition in the inductor can end the PWM high time (i.e., PMOS on-time).

The BLANKb signal controls the minimum and maximum PWM duty cycles. If the CC or CV comparator were to go high too early, a very narrow high-going pulse on the PWM could result, which may cause unpredictable behavior from the output device block. To set the minimum high pulse width, BLANKb gates the PWM_END signals for a short time at the beginning of each cycle.

Similarly, if the battery charger is programmed to draw more current from the adapter than it can supply, the adapter may go into a current limit mode, and the input voltage of the charger will collapse. When the adapter voltage collapses, the CC loop duty cycle will approach 100 % with narrow, low-going pulses on PWM signal during the PMOS-off phase. To filter these narrow pulses, BLANKb gates the PWM_END signals for a short time at the end of each cycle. Instead of pulsing low for a short time, the PWM signal will stay high through the rest of the cycle.

If, for some reason, the master clock stops (e.g., reference clock failure occurs) while the BLANKb signal is low, PWM will be stuck high, and the PMOS will remain on indefinitely. Having the PMOS device on for a long time results in an uncontrolled charging current, which compromises the safety of the system. To avoid having the PMOS stuck on in the event of a clock failure, a time qualifier circuit (QUALIFIER in Fig. 11.14) is triggered when BLANKb goes low. If BLANKb fails to go high before the qualification time expires, the flip-flop is cleared, and the PMOS is turned off, thereby eliminating the possibility of the PMOS device being turned on indefinitely.

During CC operation, the output current is well-known because the charger is operating in CCM, and the CC loop is regulating the output current to the desired value. During CV operation, the charging current can get low enough to have the charger in DCM. To prevent unnecessary switching in DCM, the CV integrator voltage is monitored. When the integrator voltage drops below a certain value, one of the PWM_SKIP signals goes high, preventing the PWM flip-flop from going high on the next rising edge of the clock (and preventing the next switching cycle from starting). Thus, at the rising edge of each clock cycle, the CV integrator output voltage is evaluated, and a switching cycle starts only if the integrator output voltage is above a fixed threshold.

11.4 Experimental Results

Experimental results are shown below, including transient behavior, as well as key performance criteria over a wide population of devices.

11.4.1 Charging Cycle Operation

The battery charger waveforms are shown in Fig. 11.15. Figure 11.15a shows the battery current (I_{BAT}), switching (output) node (V_{SW}), adapter ($V_{ADAPTER}$), and battery (V_{BAT}) voltages. The waveforms show predictable behavior for an input

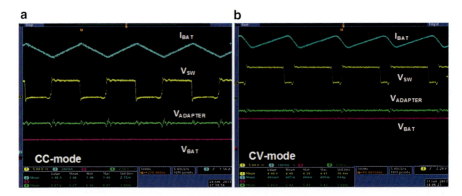

Fig. 11.15 Waveforms in CC (**a**) and CV (**b**) modes of operation

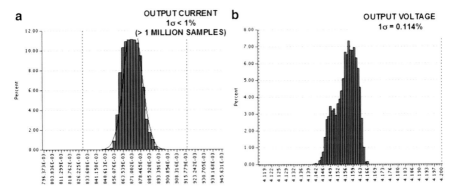

Fig. 11.16 CC (**a**) and CV (**b**) accuracy

voltage of 6.1 V, a battery voltage of 2.6 V, and an output current of 1.5A in CC mode. The plot in Fig. 11.15b also shows predictable behavior with the charger operating in CV mode. Other than by measuring the battery voltage, it can be difficult to determine which mode of operation the charging is in by observing the terminal waveforms only, especially at high current levels. If the entire charging cycle is observed, the CV current will eventually drop to a level where the charger is in DCM, with the associated high-impedance state being observed on the switching node.

11.4.2 CC and CV Accuracy

The accuracy in CC mode is primarily dependent on the accuracy of the current-sensing circuitry, the feedback resistor (R_{FB} in Fig. 11.12), and the input referred offset of the CC integrator. Because the variation of resistor sheet ρ can be significant, a trim circuit is added to R_{FB} to trim the output (or input) current for each part. The trim is done at one value of adapter input current only, typically to meet the input requirement for a USB adapter (500 mA maximum input current limit). Variations across the range of output current scale well as long as the CC integrator and current sense accuracy exhibit low drift across temperature. The measured accuracy of the charger is shown in Fig. 11.16. The 1-sigma variation is slightly less than 1 %, giving excellent yield for a sample size of 1.28 million parts.

Accuracy in CV mode is dependent on the on-chip band gap reference, the CV voltage reference generator, the battery resistive feedback divider, and the CV integrator (see Fig. 11.11). To achieve accuracy better than 1 % (the approximate band gap variation over process, voltage, and temperature), the reference circuit is also trimmed on a part-by-part basis. The data in the right plot of Fig. 11.16 shows a 1-sigma variation of approximately 4.7 mV (0.114 %), achieving a ±0.5 % accuracy specification for the float voltage assuming ±4−σ limits.

11.5 Conclusions

A dual-loop, CC-CV battery charger has been presented. The battery charger can regulate either input current or output current using internal, lossless current sensing. The performance of the complete charger is summarized in Table 11.1. A plot of the circuit layout is shown in Fig. 11.17. The total area for the charger including ESD protection is 3.6 mm^2.

Results from production silicon show excellent correlation with simulations in both CC and CV modes. In addition, the output current accuracy for a large sample of parts demonstrates the viability of the lossless current sensing approach. Accuracy in CV mode is better than ±1 %, allowing for maximum energy storage in the cell phone battery.

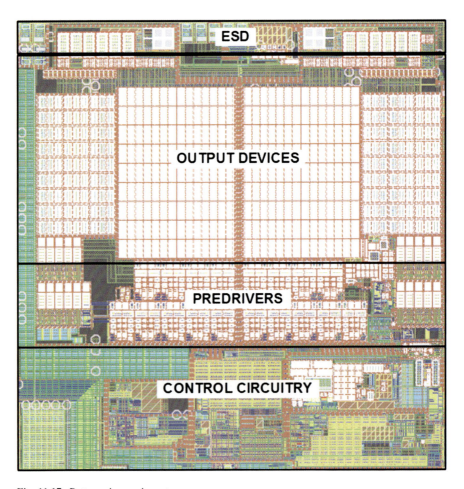

Fig. 11.17 Battery charger layout

Table 11.1 Battery charger performance summary

Specification	Value
Technology	0.18 μm HV
Input voltage (charging)	4–10 V
Input voltage tolerance (non-charging)	0–20 V
Output voltage range	3.6–4.4 V
Output current range	0.1–2.2 A
Switching frequency	2–4 MHz
CC accuracy	±5 %
CV accuracy	±0.5 %
Charger area	3.6 mm^2

References

1. R. Pagano, M. Baker, R. Radke, A 0.18-um monolithic Li-Ion battery charger for wireless devices based on partial current sensing and adaptive voltage reference. IEEE. JSSC **47**(6), 1355–1368 (2012)
2. A.A. Hussein, I. Batarseh, A review of charging algorithms for Nickel and Lithium battery chargers. IEEE Trans. Veh. Technol. **60**(3), 830–838 (2011)
3. IEEE Standards Website. [Online]. Available: http://standards.ieee.org/findstds/standard/1725-2011.html
4. Japan Electronics and Information Technology Industries Association Website. [Online]. Available: http://standards.ieee.org/findstds/standard/1725-2011.html
5. F.-C. Yang, C.-C. Chen, J.-J. Chen, Y.-S. Hwang and W.-T. Lee, Hysteresis-current-controlled buck converter suitable for Li-Ion battery charger,. in *2006 International Conference on Communications, Circuits and Systems Proceedings*, vol 4, 2006, pp. 2723–2726
6. Y.-S. Hwang, S.-C. Wang, F.-C. Yang, J.-J. Chen, New compact CMOS Li-Ion battery charger using charge-pump technique for portable applications. IEEE Trans. Circuit Syst. I Regul. Pap. **54**(4), 705–712 (2007)
7. Y.-H. Liu, J.-H. Teng, Design and implementation of a fully-digital Lithium-Ion battery charger, in *TENCON 2006. 2006 I.E. Region 10 Conference*, vol. no, pp. 1–4, 14–17 November 2006
8. Y. Sun, X. Wu, M. Zhao, Li-Ion battery charger with smooth-switch-over four-stage control, in *Proceedings of the 2009 12th International Symposium on Integrated Circuits*, pp. 49–52, December 2009
9. R.W. Erickson, D. Maksimović, *Fundamentals of Power Electronics*, 2nd edn. (Kluwer, Dordrecht, 2001)
10. R. Ridley, A new, continuous-time model for current-mode control. IEEE Trans. Power Electron. **6**(2), 271–280 (1991)
11. V. Vorperian, Simplified analysis of PWM converters using model of PWM switch part I: continuous conduction mode. IEEE Trans. Aerosp. Electron. Syst. **26**(3), 490–496 (1990)
12. V. Vorperian, Simplified analysis of PWM converters using model of PWM switch part II: discontinuous conduction mode. IEEE Trans. Aerosp. Electron. Syst. **26**(3), 497–505 (1990)

Chapter 12
Power Gating and State Retention Applied to SOC Standby Power Management

David Flynn

Abstract Power Gating, PG, is a well-established technique for mitigating leakage power when a subsystem in a SoC is in some form of standby power state with clocks stopped. Register contents are lost in basic PG, requiring a reset on re-powering. State Retention Power Gating, SRPG, trades off a little more power when in standby to retain some or all of the register state values in a circuit in exchange for a more efficient and responsive wake-up and continued execution with known state. This paper addresses promising approaches to enhance PG and SRPG, which are appropriate to digital designers without the need to resort to full-custom design techniques. The aim is also to increase designer understanding of how the essentially analog circuit challenges can be abstracted for a richer set of standby power management schemes. Example implementation experience and results are described for silicon technology demonstrators developed as part of the work

12.1 Motivation

Power management is of increasing concern and challenge to SOC and product designers [1]. Power Gating, PG, is now well understood as a technique for reducing static leakage power when circuits are idle [2, 3]. State Retention enhancements in hardware [4] can address fast wake-up latency and transparency to system software but have area, performance and robustness/reliability impacts that require minimizing [5].

Current EDA tools support for Power Gating is tuned around "logic-level" drive of power gates. The new techniques that are described and contrasted build on the

---

D. Flynn (✉)
ARM Ltd, Cambridge, UK

ECS, University of Southampton, Southampton, UK
e-mail: david.flynn@arm.com

A. Baschirotto et al. (eds.), *Frequency References, Power Management for SoC,*
and Smart Wireless Interfaces: Advances in Analog Circuit Design 2013,
DOI 10.1007/978-3-319-01080-9_12, © Springer International Publishing Switzerland 2014

multi-voltage aware tools and formats to add enhanced power gate performance as well as addressing state retention without the traditional area and timing penalties.

The work described in this paper is at an applied research phase and has been undertaken in collaboration with researchers in the Electronics and Computer Science faculty of the University of Southampton in the UK; the technology demonstrator implemented in Silicon (on a 65 nm Low Leakage process) was co-developed and fabricated using the EUROPRACTICE "mini@sic" Multi-Project Wafer service [9] with TSMC Inc as the semiconductor foundry.

12.2 Industry State-of-the-Art in Power Gating

Outside of companies using custom layout techniques, design flows and tools, mainstream SoC development companies using synthesis, place & route and layout flows are now able to exploit on-chip Power Gating (PG) effectively when working with sub-90 nm process technologies where leakage power has to be addressed.

The EDA industry has standardized around "power-intent" enhancements that allow annotation of power gating domains and controls, either using Unified Power Format,[1] UPF, or Common Power Format,[2] CPF. Both UPF and CPF provide similar functionality for define voltage domains, power-gating inference, isolation or clamping inference on signals at domain boundaries, plus the notion of "power state tables" that define power states and the transition arcs between states.

Power gates or switches, isolation clamps and "always-on" buffers are now well understood as additional elements to standard cell libraries that have additional library multi-voltage support attributes that the EDA tools understand and can infer and use in PG implementation flows. Similarly multi-voltage simulation and verification enhancements are now provided in the mainstream design and sign-off tools.

Designers understand that on-chip power gates are non-ideal: they suffer from non-zero on resistance that always results in an additional voltage drop across the switch when active current is drawn. This typically always costs a small de-rating of maximum operating frequency compared with a non-power-gated circuit. Similarly the off resistance of the switches is non-infinite and the off-leakage current is non-zero, especially at elevated temperatures. The primary trade-off at design time is balancing the total number of switches (in parallel) to keep active IR-voltage drop below a chosen threshold – while realizing the off-leakage power function scales in proportion.

Both 'Header' PMOS and 'Footer' NMOS power gates are of interest, power gating the standard-cell VDD or VSS rails respectively; the I_{on}/I_{off} ratios are a function of the mobility of the PMOS and NMOS transistors which varies with

---

[1] UPF is a power-intent standard developed by Synopsys Inc., Mentor Graphics Inc. and Magma Inc. [7].

[2] CPF is a power-intent standard developed primarily by Cadence Inc. [8].

12 Power Gating and State Retention Applied to SOC Standby Power Management   211

semiconductor process generation and "strain" engineering. In the examples discussed in this chapter PMOS header power switches are used as the example and all circuits share a common VSS ground rail. Clamp gates are required to ensure outputs from power gated regions do not float to non-logic values and cause crowbar currents to flow in downstream standard-cell logic.

The main challenges for designers verifying PG functionality are those of physical power planning and electrical characteristics with respect to inrush currents associated with turn-on. Significant work can go into the control network sequencing to ensure ground (or power) rail "bounce" is kept within limits to avoid voltage drop-related timing impact on adjacent circuit blocks. Power grid design and worst-case IR-drop analysis is always a challenge and PG adds yet another variable impedance into the voltage analysis problem – and the dynamic vector test sets that are used for power rail integrity sign-off.

12.3 Advanced Power-Gating Options

Power-gating today uses "always-on" control buffer networks to drive the gate terminal of the Header or Footer transistor power-switches. This section introduces approaches to improving on this baseline technology and EDA tools support.

The traditional academic name for such logic-level drive power gating is Multi-Threshold CMOS, or "MT-CMOS"; a high threshold voltage power switch (a high off-resistance characteristic) is inserted in series with the standard cells to be made power-gated, which are typically built of lower threshold-voltage transistors (higher-performance but leakier characteristics).[3] Figure 12.1 depicts the basic power gating structure with a high Vth Header power switch in series with the high-current VDD supply rail, controlling a gated virtual rail (VVDD) shared across low or mixed Vth circuitry.

Multi-Voltage-aware EDA tools understand the requirement to handle the (non-power-gated) control buffer networks that will be hooked up in the UPF/CPF description to the appropriate state machine control signals.

Multi-Voltage CMOS, "MV-CMOS" [10] enhances the power gating control by utilizing "Gate-Bias"; the power gate still remains between the high-current primary rail and the switched virtual rail, but the gate terminal of the power switch transistor is over-driven beyond the supply rail (positive relative to VDD for header switches, negative below VSS for footers).

The effect is to 'pinch-off' the off leakage current by an appreciable (process dependent) factor, controlled by the level of over-drive, typically in the range of 50–150 mV extra gate voltage above the main VDD rail.

---

[3] Academic papers traditionally proposed *both* Header and Footer series transistors but in industrial usage, where on-resistance is typically the key parameter to avoid degrading performance unduly, only one power-gate device can usually be tolerated.

Fig. 12.1 Logic-drive "MTCMOS" header control

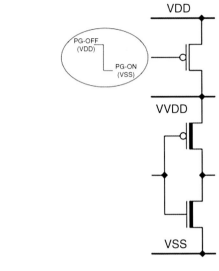

Fig. 12.2 Gate-bias drive MTCMOS (**a**) LVth or (**b**) HVth power gates)

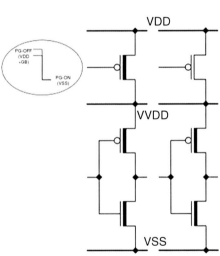

Figure 12.2a shows the simple extension to MT-CMOS control; the power switch gate terminal is overdriven by controlled amount from low-current gate-bias supply. MV-CMOS was academically applied to low-Vth power gates (reduced PG area cost by using higher on-current switches) but to drive the leaky power gates hard off for an acceptable off current (close to MT-CMOS high-Vth off current). Alternatively in Fig. 12.2b the approach can be applied to high Vth switches where the off leakage current can be reduced over and above logic-voltage PG drive.

Such techniques are well understood in full-custom design flows, and more expert designers have been able to apply such functionality in "ring" structures around a specific voltage domain, but it has not been obvious how best to support MV-CMOS

Fig. 12.3 Power gate with integrated SCCMOS gate bias drive

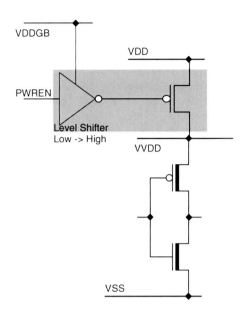

techniques in UPF/CPF PG inference flows. Simplistically adding a buffer network on an additional gate bias supply rail results in the tools seeking to infer level shifters and clamps between the voltage domains – which is not what is required.

Figure 12.3 shows a prototype MVCMOS power- gate with internal buffering.

The enhanced switch cell takes a standard logic-level control input signal, and provides one extra VDDGB supply port that requires low-current hook-up to a gate bias control voltage grid that the implementer has to provide. Internally the control signal is level shifted to drive the power gate into Super-Cut-off mode, known as SC-CMOS.

Buffered switch variants that also provide a daisy-chained output control signal use standard VDD-rail buffering such that the MV tools simply see a standard "always-on" logic interface at the periphery and have no knowledge of the multi-voltage cell internals.

12.4 Industry State-of-the-Art in State Retention Power Gating

Conventional PG approaches typically power down all standard cell logic, combinational and sequential. After turning the power back on all register state is unknown and effectively has to be re-initialized similarly to the reset required to such circuits at external power-on. Explicit hardware or software schemes can then be built at the system level to manage check-point saving and restoring of state for example where the illusion of state persistence is beneficial to the system.

State Retention with Power Gating (SRPG) is a hardware enhancement to this that allows the state of some, or all, register state to be retained or held during "sleep-mode" power gating, such that the combinational logic can fully power gated off. On re-powering the sequential state is preserved and the combinational logic simply re-evaluates the logic terms for next state clocking.

SRPG is richly supported in UPF and CPF power intent standards but is limited in use in industry. For ARM microprocessor-based designs this is typically in the low-power embedded control application area (Cortex-M® class family CPUs) where low idle power must be balanced with minimal wake-up and service response latencies; hardware supported state retention for small-scale subsystems works well here.

For higher performance mainstream consumer-product designs SRPG is not favored industrially. EDA compatibility assumes that state-retention registers encapsulate the retention latch structure with isolation (both functional and electrical), have an additional retention supply voltage port, and also one or two extra control signals that are required to manage state saving, state holding and state restoration functionality; all in addition to the baseline clocked flip-flop or Master–slave register operational functionality. To get the best SRPG power benefits, high-Vth structures, or non-minimal gate-length transistors are used and the rest of the register internal circuitry and output buffering can all be power-gated with the logic supply rails. However this impacts negatively on both the maximum performance and the area of the retention registers.

In performance critical design and implementation such retention registers are too costly, especially where there are a large number of registers.

Retrofitting an existing design with only partial or selective state retention unless it has been designed and verified with exactly this state retention sub-set is fraught with problems and it can be nearly impossible to fully validate and re-verify safely. (The state space explosion caused by the product of retained and non-retained/re-initialized states and how this impacts on cones of logic that are factored into clock gating circuitry etc. is orders of magnitude greater than conventional validation from power-on-reset conditions).

Given the area and performance costs, hardware-based SRPG is currently applied industrially to niche areas only. And such a SRPG scheme is only transparent and reliable if the state retention functionality is 100 % robust and safe. State-bit errors may cause malfunctions or control system deadlock so the power-gated design overall must guarantee not to cause noise injection that affects weaker retention latch structures in retention domains.

12.5 Advanced State Retention Power-Gating Options

In increasingly leaky semiconductor technologies, the leakage power of inter-connect and subsystems whenever logic is stalled or quiescent for numbers of cycles matters, no matter how efficient the clock gating may be. Light-weight

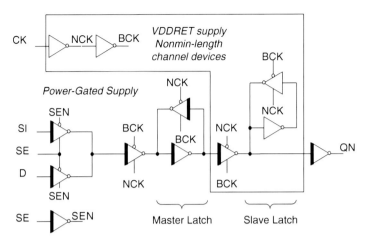

Fig. 12.4 Split-supply-rail DFF for SRPG

retention mechanisms in hardware will increasingly become important for short term static power reduction in addition to the explicit software-based "architectural" state save and restore functionality that is valuable for longer periods of sleep or power-down.

Rather than bearing the cost of retention controls on every retention register, and by approaching the performance problem by building the Master–slave flip-flops as full performance designs but with split supply rails, it becomes feasible to address both the area and performance problems. Figure 12.4 illustrates an example register design that maintains the clock-to-output performance by keeping the High Vth (or longer channel devices) off the critical path. As a retention device, the retention power saving is not as optimal as that of traditional retention registers, so energy break-even point analysis needs to be factored into the sleep mode cost-functions.

However such a split-rail register is not internally isolated from floating power-gated clock (and reset) buffer trees. Clock trees in particular are tuned for performance and latency balancing and can exhibit significant leakage power, which must be addressed by power gating. A system level approach is adopted whereby the final clamping of the clock to the split-rail registers is subsumed into a retention-control clock gating cell, such that, in the case of rising-edge clocked registers the clock to the flop is forced low from the clock gate at the final leaf of the clock distribution; the high buffer strength clock trees up-stream can all be power-gated as normal. A design flow requirement is the ability to add dummy clock gates to any remaining clocked flip-flops; this is present thanks to the requirement to address this for clock mesh or grid implementation schemes, and can be invoked when using more conventional clock tree synthesis.

Figure 12.5 depicts the scheme where the absolute minimum of the sequential circuitry is kept alive, shown by the colored marking of the slave latches and retention clock clamping that will be powered by an independent retention supply.

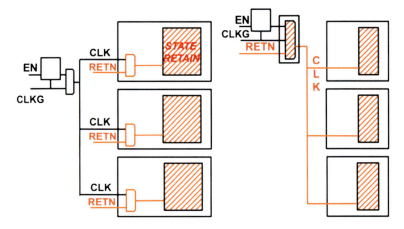

Fig. 12.5 Retention control re-partitioning, conventional (*left*) new (*right*)

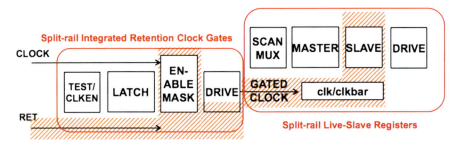

Fig. 12.6 'Split-rail' retention register deployment

Figure 12.6 illustrates in more detail the split-rail voltage domain view where the hashed areas marked indicate power from a retention voltage supply, and the rest of the circuitry in the integrated clock gate and the master latch/output drive of the flip-flop are powered from the same power gates as the combinatorial logic. The retention voltage rail would typically be implemented as an independently power-gated rail in order to support a deep power-down state when the entire sequential state can be turned off as well as the logic for minimum leakage power. Supporting more than one power-gated supply adds some implementation complexity but the total switched current demands are the same and the retention supply grid carries considerably less current than the main logic power gated rail.

Such a retention approach has the potential to address the area and performance shortcomings, but there are a few caveats:

- Retention control has to be architected into the system and state-machine controller: e.g. retention relies on controlling the clock phase – rather than relying on clock-independent retention register designs to hide this from the designer.

- Both clocks and asynchronous reset/set networks must be made controlled. Initialization signaling must be asserted in (retention) power on cases, but must be held de-asserted during retention sleep states to avoid losing current state.
- There is no way of inferring "reset-clamps" in the same way as dummy integrated-clock-gates so more typically an always-on reset control network has to be implemented (some extra leakage overhead cost but much less of a concern than clock buffer tree which conventionally has to be power-gated).
- UPF and CPF implementation tools were not designed to expect retention registers that have "no" control, and certainly do not understand the concept of applying selective retention to a cluster of registers associated with specific clock gates; full-state retention within a subsystem is currently the most preferable approach.
- The need to provide a weak VDDR retention supply grid, also switched.

12.6 State Retention Integrity

State retention can be a highly desirable low-power functional mode when leakage power is significant on a technology implementation node. However, state retention must be guaranteed to be 100 % robust; if any state bits get corrupted then a subsystem may malfunction or potentially become deadlocked in the system. Retention state is not only subject to substrate and retention supply noise while holding state but also reliant on the inrush current management in the power-gating sequencing of the logic virtual rails when entering and exiting retention mode.

Some specific structures have been designed to analyze and characterize state retention integrity and which provide a programmable level of control over both on-chip noise generation and real-time parallel monitoring of register state.

Figure 12.7 illustrates the approach. Two identical banks of 8,192 registers are designed for synthesis and layout on a shared ground rail and which can support both standard power gating and virtual rail retention voltage scaling. Either bank can be configured as a large 32-bit wide barrel-shifter that can rotate programmable patterns to provide state toggle patterns from zero to 8,192 state toggles per clock cycle. The other bank can have arbitrary programmed state patterns written in, two-dimensional parity is then generated for the entire (128×64) array, held in the HPAR and VPAR registers outside the array, and the array is then placed into voltage-scaled retention mode.

The 2-D parity scheme supports real-time monitoring of the entire register array state when observing for the first bit-corruption. The Vertical and Horizontal chains are routed through level-shifters to support low-voltage visibility and the X- and Y-coordinate pinpoint the logical register that shows first failure.

Subsequent work on using scan chains for both error detection and correction is continuing [11].

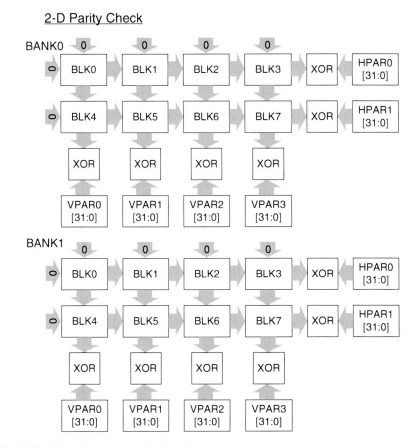

Fig. 12.7 2-D parity state retention integrity test structures

12.7 Experimental "Drowsy-Retention" SCCMOS Power-Gates

Figure 12.8 shows an enhancement to the Super-Cutoff power gate introduced earlier. If the advanced SRPG scheme has higher state retention leakage power, then judicious retention voltage scaling is an attractive option. However, as already been stated, state integrity is crucial and with diminished supply voltage headroom on many deep sub-micron process technologies the traditional simple approach of a full "Vth" voltage drop (equivalent to a diode forward voltage drop) in series with the nominal supply voltage is typically unsafe. Latch feedback structures are more noise sensitive at scaled-down voltages due to the local transistor variability and mismatch.

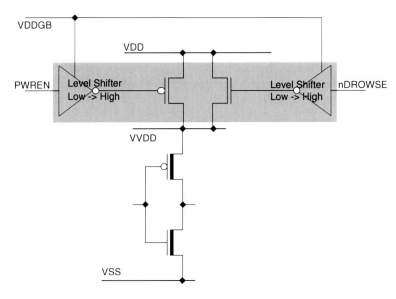

Fig. 12.8 Experimental SCCMOS power gate with additional drowsy SRPG

Therefore use is made of the raised potential of the same gate bias control voltage used for super-cut-off of the power gates, to additionally provide an optional "boosted-gate" drive to a parallel NMOS voltage-drop device supplying the virtual rail when the main power gate is not turned on. This device provides a weak bleed current to the virtual rail to maintain it a Vth drop below the VDDGB rail, typically 100 mV safer than dropping this from the VDD rail.

The Gate-Bias supply rail may be made adaptive to tune the behaviour to match temperature and safe "drowsy" voltage retention level to a "VDDR" retention virtual rail that this would typically be applied to.

12.8 Experimental "Sub-Clock Cycle" SRPG

In the case of very low power embedded controllers that operate with highly constrained energy budgets, as is the case with energy-harvesting applications, it is not possible to be able to minimize leakage power by running fast and then power gating. The available supply may be at reduced voltage or current and the CPU or system may have to throttle clock rate to make forward progress without exceeding a restricted instantaneous power budget. Running at reduced frequency results in a higher active leakage energy cost per cycle.

In this field of application area, for certain process technologies it becomes attractive to power gate the logic between clock cycles in the case where the leakage energy saved outweighs the cost of re-powering back the logic and re-evaluating the combinational state ready for the next active clock edge [12].

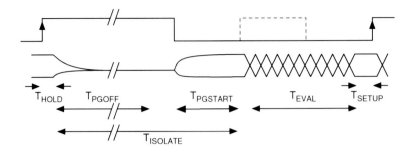

Fig. 12.9 Sub-clock-cycle SRPG

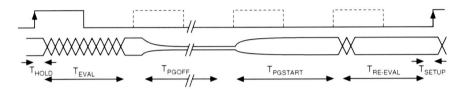

Fig. 12.10 Sub-clock SRPG – with sub-threshold drowsy voltage scaling

Figure 12.9 shows the time-domain behaviour conceptually where the high-phase of the clock is extended, and the low-phase of the clock used to power gate back on the logic and includes controlled power-up as well as standard static timing analysis for worst-case logic function evaluation.

The glitch energy associated with having to evaluate from scratch every cycle is state-dependent and simulation with full power gating in place requires fast SPICE simulator tools compared to the transparency of applying clock-gating to a design. For full speed operation a mode is provided to override the power gating to allow standard Fmax timing closure and operation.

An alternative to fully power gating the logic between active clock edges is to provide a simple form of switchable voltage scaling of the virtual rail(s). For deep long-term sleep the voltage scaling would be turned off, but for short-term sleep this drowsy logic voltage scaling mode has the advantage that the cones of logic can be scaled to run sub-threshold, avoiding the power-gating glitch energy cost, albeit at a slightly higher standby current.

Figure 12.10 shows the conceptual waveform behaviour when both Header and Footer power gates are employed as the primary power switches, and NMOS and PMOS switchable "diode = drop" sources are used for sub-threshold voltage scaling of the logic once the cone of logic between register stages has evaluated (T_{EVAL}). The uses of both header and footer power gates has the advantage that the virtual rail collapse towards mid-rail rather than primary VDD or VSS, which results in symmetric reverse well bias for the N-channel and P-channel transistors. When power gated back on ($T_{PGSTART}$) most of the logic cones are restored to super-threshold voltage, but for safety an additional clock period is provided to ensure any sub-threshold logic terms that may have been disturbed by noise are allowed time to re-evaluate ($T_{RE-EVAL}$) to ensure robust functional behaviour. The additional (T_{PGOFF}) leakage when the power-gated logic is scaled to drowsy sub-threshold level compared to the full power gated (T_{PGOFF}) of Fig. 12.9 improves the energy break-even point compared to the glitch energy cost of Fig. 12.10 ($T_{PGSTART}$) + (T_{EVAL})

12.9 The "Tokachi" Reference System Design

To evaluate and characterize the techniques described in this paper, and to determine the design flow steps required to prove that standard EDA power intent flows could be exploited or coerced in practice given the theory discussed, an R&D demonstrator test-design was put together in 2011. A multi-project-wafer was targeted, in this case limited to a tiny 2×2 mm die-size. To get best use of the limited area an ARM Cortex®-M0™ CPU was used as reference subsystem for power management strategy evaluation, and a total of 14 CPUs were integrated onto the technology demonstrator.

The process technology chosen was 65 nm LP, Low Leakage, platform from TSMC Inc. Standard ARM Artisan™ libraries, memories and power-management kits were used plus additional R&D prototype PG and SRPG cells to support the advanced power gating, advanced state retention and sub-threshold, sub-clock-cycle SRPG experimental deployment. Synopsys Inc implementation and verification EDA tools were used, building on a UPF-based multi-voltage design flow.

Figure 12.11 shows an annotated plot of the final layout with the overlay annotating the experimental layouts and implementations in the test vehicle and 12 of the Cortex-M0 CPUs used are highlighted.

- Reference CPU: standard single-voltage Cortex-M0 implementation, 333 MHz Worst Case (WC) sign-off
- Reference PG CPU: standard Power Gated (non-retention) implementation of Cortex-M0, 333 MHz WC sign-off
- Reference SRPG CPU: standard traditional State Retention Power Gating implementation of Cortex-M0, 333 MHz WC sign-off
- Advanced-SRPG CPU: 'split-rail' Live-Slave SRPG (with independent PG retention voltage) Cortex-M0, 333 MHz WC sign-off
- Drowsy/SCCMOS CPU: Advanced gate-bias PG, 'split-rail' Live-Slave SRPG (with independent PG retention voltage) Cortex-M0, 333 MHz WC sign-off
- SCPG CPU: Sub-clock-cycle SRPG CPU, with independent optional sub-threshold drowsy logic virtual-rail voltage scaling, 66 MHz WC sign off
- State Integrity blocks: dual 8,192-register based noise generators/drowsy retention integrity analysis structures

Fig. 12.11 Tokachi-1 R&D test chip layout (TSMC65LP)

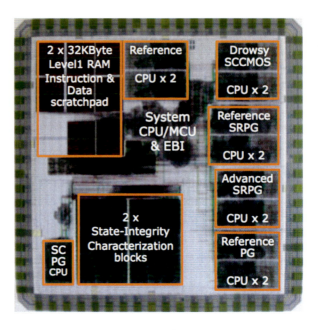

Table 12.1 Tokachi-1 technology demonstrator statistics

Process node	TSMC 65 nm LP
Die area	3.5 mm^2
Unique macros	9
CPU count	14 × Cortex-M0™ processors
Transistor count	5,126,848
Power supplies	3 VDD supplies, 1 gate bias supply
Power domains	13
Memory	32 KB SRAM, 2 KB flop-memory
Libraries	Artisan® 12 & 8 track RVT & PMK 12 track R&D prototypes ARM 'POP' fast cache memory instances

The system-on-a-chip is controlled by a further Cortex-M0 CPU that has basic microcontroller peripherals and an External Bus Interface (EBI) controller to support banks of off-chip Flash memory and pseudo-static SRAM, plus external USB host interface and an OLED display panel controller for diagnostics.

The technology demonstrator was fabricated and packaged through the EuroPractice "mini@sic" program and bonded into an 84-pin JLCC package for evaluation and characterization.

Table 12.1 summarizes the details of the test structures implemented for this technology demonstrator. Three core VDD supply rails (nominal 1.2 V for this TSMC 65LP process technology) and 13 power domains were implemented using UPF power intent to manage the power-gating, isolation, state retention and inter-domain voltage level shifting – with some careful workarounds for the

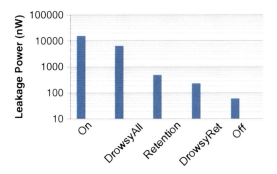

Fig. 12.12 Measured sleep-state leakage power (TT silicon, 22C)

advanced retention and sub-clock SRPG which were not straightforward to verify with the standard EDA flows.

The packaged silicon was delivered July 2011, fully functional for characterization and analysis and has been valuable for understanding the value and potential for some of the techniques described in this chapter.

Figure 12.12 shows the leakage power measurements, at room temperature (22C), plotted on a log-scale on the Y-axis, illustrating the additional standby and leakage mitigation modes over and above basic clock gating (left-most column) and fully power gated (right-most column).

The other labelled bars are explained below:

- **Retention**: this is measured for the Live-Slave split-rail retention, with the clock stopped (low) and the logic power gated off and the power-gated retention rail held on at full voltage to maintain the slave latch state.
- **DrowsyAll**: the static leakage power when the clock is stopped, the outputs clamped and the logic (and register) power gated rails voltage scaled using the Drowsy retention boosted-gate structures described in Sect. 12.7.
- **DrowsyRet**: the static leakage power when the clock is stopped, the outputs clamped, the logic fully power gated off and the retention latch virtual rail voltage scaled using the Drowsy retention boosted-gate structures.
- **Off**: the fully off power-gated case for deep sleep that has the lowest leakage power but where state is lost and a reset condition must be driven at power up to reinitialise (re-boot) the subsystem.

12.10 Conclusions and Future Work

Given the baseline support for power gating and state retention that can be inferred with industry standardized power intent formats, a number of more advanced techniques now look promising for SoC designers without having to resort to full custom circuit design techniques.

- **Enhanced power gating** where gate-bias techniques are applied to the gate of power switch devices is shown to be clean to implement in power-gating design flows providing the control and voltage level-shifting can be abstracted within the power gate structure. Providing gate-bias (overdrive) voltage as a weak-grid enables distributed power switch structuring.
- **Enhanced state retention with power gating** with split-rail master–slave registers requires rather more implementation complexity to amortize the clock-clamping overheads (and reset clamping of registers that have asynchronous set/reset initialization). However the goal of near-zero impact on area and performance compared the conventional SRPG registers currently expected/preferred by EDA tools is demonstrated to be effective, especially where total state retention of a subsystem is required.
- **State retention integrity analysis** has been important to demonstrate and provide characterization vehicle support that is not something an end-customer would expect to have to build and analyse but is proving useful to provide guidance for designers in terms of the impact and effects of poorly managed power-gating inrush currents.
- **Drowsy-voltage scaling applied to power gated rails** also has promise, especially when reusing a low-current gate-bias supply rail to support voltage scaling for both logic and/or retention state, and where a full threshold-voltage drop from the primary supply would not leave sufficient safe margin for holding state reliably.
- **State Retention Power Gating applied within the clock cycle** appropriate for low-clock frequency subsystems that are powered, for example, by energy harvesting supply sources has also been demonstrated to be effective and implementable with multi-voltage implementation tools. Enhancements that add both header and footer switched rails, and add switchable sub-threshold voltage scaling modulation of the virtual rails, appear to show effective techniques to suppress the glitch-energy from full-swing on-off basic power gating, adding new intermediate standby SRPG modes to a standard PG design flow.

Future work is focussed on migrating the techniques developed on basic microcontroller Cortex-M0 CPUs to apply these to more complex Applications processors that support internal level-1 caches and typically are deployed in small multi-core clusters.

Targeting the same small 2 × 2 mm MPW mini-ASIC die size as Tokachi-1, a dual-core Cortex-A5™ design is chosen for the design-flow proving and SCCMOS power gating and live-slave drowsy-voltage retention. This is provided for CPU register state, the 8Kbyte level-1 Instruction and Data caches as well as SCCMOS fully-off, non-retained deep power savings modes.

Figure 12.13 shows a plot of a follow-on technology demonstrator, codenamed Tokachi-4A, that implements the dual-core Cortex-A5 processor with additional level2 RAM, all supported and power-managed by a Cortex-M0 system control processor and SoC infrastructure. The chip is package footprint compatible with the Tokachi-1 technology described in Sect. 12.9 of this chapter.

Fig. 12.13 Tokachi-4A Dual-Cortex-A5 advanced SRPG demonstrator

A mixed-signal enhancement to this design, taped out late 2012 is to provide adaptive gate-bias charge pumps implemented on chip, one associated with each C-A5 CPU core. The gate bias supplies are adaptive to both temperature and process and are therefore able to adjust the gate-bias voltage applied to the super-cut-off power gates and boosted-gate drowsy retention voltage cells optimally with environmental conditions.

Acknowledgements Thanks are due in particular to:

Jatin Mistry, Sheng Yang, PhD researchers at the University of Southampton, UK, plus staff members Dr Matthew Swabey, Dr Reuben Wilcock and Prof Bashir Al-Hashimi.

James Myers, John Biggs, David Howard, Karthik Shivashankar and Anand Savanth at ARM Ltd, Cambridge UK

Staff at the EUROPRACTICE (E.U. FP7) programme in the IC Service organization for *mini*@sic MPW fabrication [9]

Synopsys Inc for the University of Southampton research sponsorship of the EDA tools and laboratory under the "Charles Babbage Award"

References

1. T. Mudge, Power: A first-class architectural design constraint. Computer **34**(4), 52–58 (2001). doi 10.1109/2.917539 http://dx.doi.org/10.1109/2.917539
2. S. Mutoh et al., A 1v multi-threshold voltage CMOS DSP with an efficient power management technique for mobile phone applications, in *ISSCC*, (1996), pp. 168–169

3. M. Keating, D. Flynn et al., *Low power methodology manual – for system-on-chip design.* (Springer, 2007) ISBN: 978-0-387-71818-7 http://www.lpmm-book.org/
4. D. Flynn, A. Gibbons, Design for state retention: Strategies and case studies. SNUG San Jose 2008, Track TA2
5. D. Flynn, High performance state retention with power gating applied to CPU subsystems – design approaches and silicon evaluation. Poster in *Hot Chips 24* archives (2012). http://www. hotchips.org/wp-content/uploads/hc_archives/hc24/HC24-Posters/HC24.30.p10-State-Reten tion-Gating-Flynn-ARM.pdf
6. D. Flynn, An ARM perspective on addressing low- power energy-efficient SoC designs, in *Proceedings of the 2012 ACM/IEEE International Symposium on Low Power Electronics and Design* (ISLPED '12). (ACM, New York, 2012), pp. 73–78. doi: 10.1145/2333660.2333680 http://doi.acm.org/10.1145/2333660.2333680
7. UPF is IEEE1801 (not 1891) – the official IEEE for 1801–2009 (which includes UPF-1.0) is here: http://standards.ieee.org/findstds/standard/1801-2009.html
8. Si2 Common Power Format, CPF, specification http://www.si2.org/?page=811
9. EUROPRACTICE mini@sic programme: http://www.europractice-ic.com/prototyping_minisic.php
10. M. Stan, Low-threshold CMOS circuits with low standby current, in *Proceedings of the International Symposium on Low-Power Electronics and Design.* (IEEE/ACM, Monterey, 1998), pp. 97–99
11. S. Yang et al., Reliable state retention-based embedded processors through monitoring and recovery. Trans. Comp.-Aided Des. Integ. Cir. Sys. **30**(12), 1773–1785 (2011). doi: http://dx. doi.org/10.1109/TCAD.2011.2166590
12. J. Mistry et al., Sub-clock power-gating technique for minimizing leakage power during active mode. DATE 2011 http://eprints.ecs.soton.ac.uk/21768/

Part III
Smart Wireless Interfaces

Pieter Harpe

The third part of this book is dedicated to "Smart Wireless Interfaces". It discusses recent developments and future trends in wireless communication systems to achieve lower power consumption, higher data rates and to enable new applications. The six chapters show a large diversity of techniques based on new technology, new system architectures and circuit-level innovations.

The thirteenth Chapter discusses three unconventional receiver architectures. The key principle here is to merge building blocks together to achieve lower power consumption, to achieve a higher performance, and to offer highly flexible radios with a simplified architecture.

In the fourteenth Chapter, a non-linear circuit is exploited to improve the interference suppression of a multi-standard radio, thereby alleviating the coexistence problem which becomes more pronounced as more and more radio standards are crammed into small mobile devices.

The fifteenth Chapter is taking advantage of the opportunities that are offered by new technologies. MEMS devices are used here to develop technology-scalable radios with an extremely low power consumption, as needed for wireless sensors and body area networks.

Another direction is taken in the sixteenth Chapter, where radio architectures and circuits are introduced to operate at extremely high frequencies (up to 260 GHz) while using standard CMOS. Such architectures will be needed to accommodate for higher data rates and imaging applications.

Chapter 17 introduces the latest products for ultra wideband radios operating in the 3–10 GHz band. The nature of this technology opens new application directions such as indoor GPS, ranging and localization, and supports high data rates and dense sensor networks at the same time.

Finally, the eighteenth Chapter discusses digital intensive transmitter architectures which are compatible to advanced CMOS processes, support software-defined operation, and allow more complex modulation schemes. Techniques are also introduced to improve the power efficiency.

Chapter 13
Unconventional Receiver Architectures

Rinaldo Castello and Antonio Liscidini

Abstract Three unconventional receiver architectures are presented: a harmonic oscillator with inherent mixer functionalities for an ultra low-power single stage quadrature RF front-end, a resonant mixer which is part of a harmonic rejection architecture that does not requires multiple clock phases and finally a filtering ADC where blocker filtering and analog-to-digital conversion are implemented in a single step leading to a solution particularly suitable for the implementation of a softer/digital defined radio. After a brief overview of the key ideas, several prototypes and measurements are reported.

13.1 LNA-Mixer-VCO Cell

This section describes an ultra compact RF front-end where the harmonic oscillator is used also as a mixer providing a quadrature down-conversion of the RF signal injected through the tail current generator. LNA, mixer and VCO are stacked, sharing the same current and part of the devices, but operate in different frequency domain preserving an independence that makes possible an effective optimization of each block.

13.1.1 LC Oscillator as a Mixer

A traditional LC tank oscillator, as the one shown in Fig. 13.1, intrinsically performs the mixing functionality since any RF signal in the oscillator bias current

---

R. Castello (✉)
Università degli Studi di Pavia, Via Ferrata 1, 27100 Pavia, Italy
e-mail: rinaldo.castello@unipv.it

A. Liscidini
University of Toronto, 10 King's College Road, M5S 3L4 Toronto, Canada
e-mail: antonio.liscidini@utoronto.ca

A. Baschirotto et al. (eds.), *Frequency References, Power Management for SoC, and Smart Wireless Interfaces: Advances in Analog Circuit Design 2013*, DOI 10.1007/978-3-319-01080-9_13, © Springer International Publishing Switzerland 2014

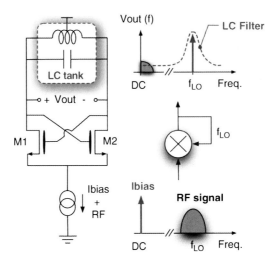

Fig. 13.1 LC oscillator as a mixer

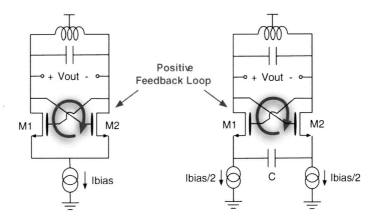

Fig. 13.2 Bias splitting

is down-converted by the switching pair M1–M2. This occurs through the same mechanism by which the DC current of M0 is up-converted to the oscillation frequency. The mixing properties of this structure are generally exploited in the transmitters where the LC tank oscillator is used as up-converter in direct modulation architectures [1]. Nevertheless, when this topology is used as a down-converter, the presence of the inductor prevents any voltage amplification of the signal components around DC (Fig. 13.1).

Unfortunately, any attempt to sense the down-converted signal at the output of the oscillator would degrade the quality factor of the tank increasing the phase noise. Hence, the key idea is to read the down-converted signal at the sources of M1 and M2 where the loop can be opened at low frequency, without perturbing the oscillation. This can be done as reported Fig. 13.2. The sources of M1 and M2 are

13 Unconventional Receiver Architectures

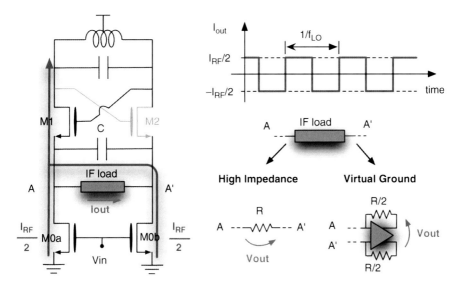

Fig. 13.3 Bias splitted oscillator as a mixer

splitted at low frequency using two bias current generators while a large capacitor closes the loop at the oscillation frequency ω_{LO}. Since the capacitance C degenerates the cross-connected differential pair (M1–M2), which compensates the tank losses, its value must be sufficiently large to guarantee the sustainability of the oscillations (i.e. C>> gm/ω_{LO}).

Having a sufficiently large impedance across the sources of M1 and M2 at IF allows to sense the down-converted signal across an IF load. The configuration shown in Fig. 13.3 demonstrates how the current I_{out}, that flows through the IF load connected between the source of M1 and M2, corresponds to half of the RF signal, injected alternatively by transistor M0a and M0b, multiplied by a square wave at frequency f_{LO} (i.e. mixed down to $\omega_{RF}-\omega_{LO}$).

The down-converted current can be sensed, placing at the output of the cell either a high impedance load (e.g. a resistor) producing a voltage output, or a low impedance (e.g. a virtual ground) producing a current output. As it will be shown in the next section, the second choice is preferred to avoid conversion losses due to parasitic common mode capacitances at the sources of M1 and M2.

Compared to a classical Gilbert cell, the structure proposed in Fig. 13.3 has an inherent loss of 6 dB in the conversion gain since half of the RF current, i.e. the one alternatively injected by M0a and M0b in each half of the clock cycle, is not down-converted but flows directly towards the tank through either M1 or M2. A complete down-conversion of the input current can be obtained inserting an additional switching pair (driven with opposite phases) between the bias current generator and the rest of the oscillator as shown in Fig. 13.4.

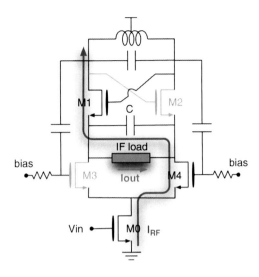

Fig. 13.4 Double switching pair self-oscillating mixer

Notice that the portion of the circuit located above the capacitance C, behaves as a classical cross-coupled oscillator. It follows that the quality factor of the tank can be optimized without having to make any compromise with respect to the design of the down-converter.

13.1.2 *Non-idealities of the Down-Conversion*

The analysis of this section will address the two main loss mechanisms that affect the self-oscillating mixer (SOM). These losses are related to the type of load adopted for both the mixer and the oscillator. In particular it will be shown that a low impedance load at the mixer output (i.e. a virtual ground) and a differential LC tank to be used in the oscillator are the best choice to minimize the conversion losses.

The common mode capacitance at the output of the mixer makes the analyses of the conversion gain particularly tough due to the time-variant nature of the circuit. The problem can be partially overcome applying the superposition principle since to first order approximation this down-converter can be assumed time-variant but linear. Furthermore, to simplify the analysis, transistor M0 will be considered as an ideal transconductor with infinite output impedance. Under this assumption, the current at the drain of M3–M4 is ideally given by the current through M_0 multiplied by a square wave as shown in the equations below:

$$\begin{cases} I_{M3} = \dfrac{I_{M0}}{2}(1 + sign(\cos(\omega_{LO}t))) \\ I_{M4} = \dfrac{I_{M0}}{2}(1 - sign(\cos(\omega_{LO}t))) \end{cases} \quad (13.1)$$

13 Unconventional Receiver Architectures

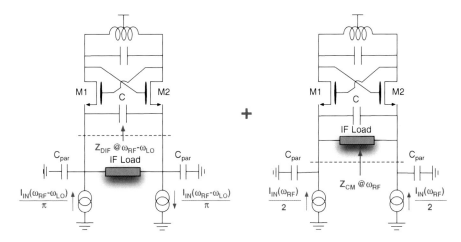

Fig. 13.5 Decomposition of double-switching SOM

These equations show that the total current through M3 and M4 is made up by a differential component that is multiplied by sign(cos(ω_{LO}t)) (i.e. down converted) and by a common mode component not yet down-converted which will be successively mixed by the following switching pair M1–M2. Applying the superposition principle, the double switching pair SOM can now be decomposed in two bias splitted SOMs. For the first one (shown in Fig. 13.5a) the current sources inject a differential signal at $\omega_{RF}-\omega_{LO}$. For the second one (shown in Fig. 13.5b) the current sources inject a common mode signal at ω_{RF} (Fig. 13.5). The optimization of the former circuit will determine the choice of the mixer load, while the optimization of the latter circuit will determine the choice of the oscillator tank.

The signal loss for the circuit of Fig. 13.5a can be computed using a switched-capacitor approach. Assuming the transistors act like switches driven by a clock signal of frequency f_{LO}, M1 and M2 redistribute the charge injected in the capacitances C_{par}, producing an equivalent resistance in parallel with the IF load equal, in the worst case, to $1/2C_{par}f_{LO}$ (assuming a negligible impact of the tank in the charging/discharging mechanism). This parasitic equivalent shunt resistance produces a loss which increases as the impedance level of the IF load is increased. For this reason, a current mode approach, where the IF load is realized with a virtual ground, is chosen. In this case, in fact, the load impedance approaches zero and practically all the output differential current can be collected resulting in a conversion gain for this portion of current equal to $1/\pi$. A more detailed analysis of the SOM circuit can be found in [2]. The quantitative results of such an analysis are shown in Fig. 13.6 that plots the conversion gain of the complete SOM as a function of the parasitic capacitances C_{par}. It can be noticed that a current mode approach, by eliminating the losses associated with the differential current components, leads to a much larger gain especially when the capacitances C_{par} are large. The only remaining losses are associated with the common mode current components.

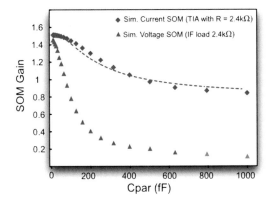

Fig. 13.6 SOM Gain vs. C_{par}: using a current mode (TIA) and a voltage mode approach

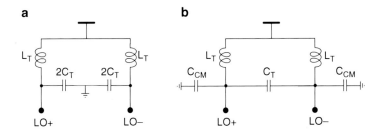

Fig. 13.7 Possible oscillator tank configurations: (**a**) common-mode, (**b**) differential

It turns out that such losses depend on the topology of the oscillator tank as it will be now discussed.

As it was just shown, using an ideal virtual ground as a load, the SOM losses are only associated with the configuration of Fig. 13.5b. i.e. common mode signal components at RF. In Fig. 13.5b only the fraction of RF current that flows into M1–M2 is mixed. For this reason, the losses associated with the common mode RF components derive from the partition between the common mode impedance at the sources of M1 and M2 (Z_{CM}) and the two parasitic capacitors Cpar [2]. The key point to be noticed is that Z_{CM} is a function of the resonant tank used in the oscillator (Fig. 13.7).

If a common mode tank is used in the oscillator (Fig. 13.7a), the common mode impedance at the source of M1–M2 is equal to $\omega_{LO}Q_T L_T/2$, where L_T and Q_T are the inductance and the quality factor of the tank at resonance [2]. In this case there is a trade-off between a small attenuation (when Q_T is small) and a low phase noise (when Q_T is large). This trade off can be broken using a differential tank (Fig. 13.7b) where the resonance occurs just for differential signals. In this case the common mode impedance Z_{CM} is, to first approximation, equal to $\omega_{LO}L_T$, i.e. independent from Q_T [3]. The simulated gain versus Cpar and Q_T for the two cases are reported in Fig. 13.8.

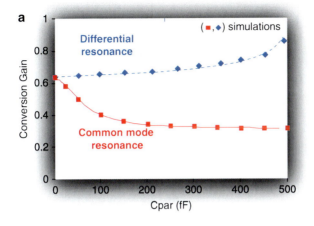

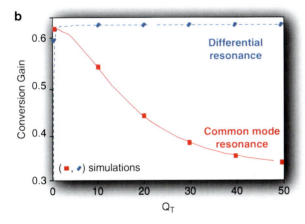

Fig. 13.8 Differential tank resonance vs. common mode tank resonance

13.1.3 Adding LNA Functionality to the SOM

To implement the LNA functionality, no additional active device is needed since a low noise input termination/transconductor can be obtained simply adding an inductive degeneration to the oscillator current generator M0. The result is the (LNA-Mixer-VCO) LMV cell shown in Fig. 13.9 [2]. This structure includes all the main blocks of an RF front-end where both bias current and part of the active devices are shared between the various building blocks, resulting into a compact and very low power architecture.

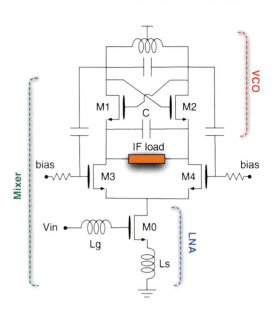

Fig. 13.9 LMV cell

13.1.4 Quadrature Generation

Typically all the RF front-ends adopt a quadrature down-conversion architecture to, among other things, eliminate the unwanted signal sideband that otherwise would be folded on top of the wanted one during the mixing to IF frequency (image rejection). For this reason, inserting the LMV cell in a quadrature receiver is mandatory for an effective use of such a cell in most applications.

The implementation of a quadrature LMV cell can be achieved acting either at the level of the local oscillator (LO) or at the level of the RF signal path. Although the latter approach can appear disadvantageous in terms of signal-to-noise ratio, in ultra-low-power/low-cost applications such as ZigBee it is generally preferred because it is less costly in terms of power consumption and die area [3]. In the following section it will be shown that both solutions are compatible with the LMV cell confirming the versatility of the structure.

(a) Quadrature generation at the local oscillator level

The quadrature LMV cell is shown in detail in Fig. 13.10. Since the top portion of the circuit acts exactly like a traditional LC tank oscillator, the LO quadrature generation can be obtained via the standard cross-coupling of the two VCOs through the two additional differential pairs shown in light grey in the figure [4]. Notice that the additional pairs do not significantly increase the total current consumption while they are capable to guarantee an image rejection adequate for most of applications. However, since the LMV cell (and in particular the resonating tank) needs to be duplicated, the area and the power required by the front-end is doubled.

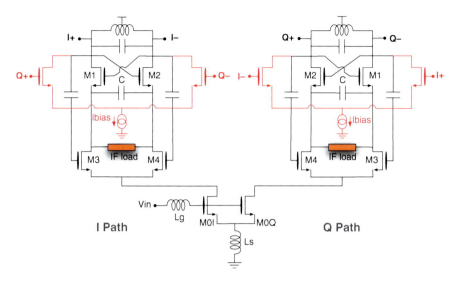

Fig. 13.10 LMV cell: quadrature at oscillator level

(b) Quadrature generation in the RF signal path

When the quadrature is realized in the RF signal path, the tanks of the two LMV cells can be shared as shown in Fig. 13.11. In this way, only a single differential coil is required, thereby significantly reducing the active area. Furthermore, since the bias current of both I and Q paths flow in the same LC load, the total bias current necessary to sustain the oscillation can be reduced (for a given tank Q), thereby resulting in a more power efficient structure [3].

13.1.5 Examples

The two possible LMV based quadrature architectures described above were implemented in a GPS RF front-end and in a ZigBee receiver respectively. In the case of the GPS, the LMV cross-coupled approach was preferred due to the challenging specification in term of sensitivity required by the standard. In the Zig-Bee prototype, thanks to the more relaxed target specification in terms of noise figure and linearity, the quadrature topology of Fig. 13.11 was preferred since, by minimizing the number of integrated coils, it produces a very compact and low cost solution.

(a) GPS Front-end

The micro-photograph of the prototype GPS front-end, fabricated in a 0.13 μm CMOS process is reported below [4] (Fig. 13.12).

The active die area of the RF front end is 1.5 mm^2 and is dominated by the three integrated inductors (one for the source degeneration in the LNA and two

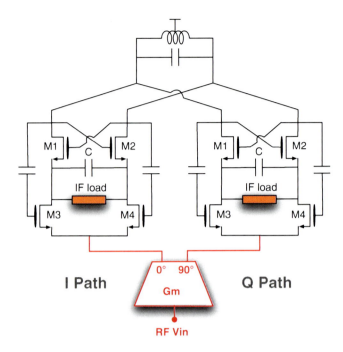

Fig. 13.11 LMV cell: quadrature on RF signal path

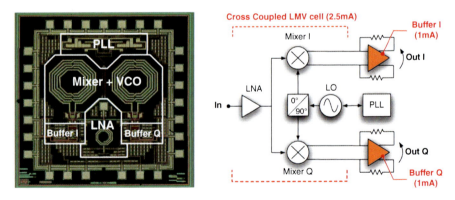

Fig. 13.12 GPS prototype

for the cross coupled LMV cells). To lock the VCO to an external reference crystal oscillator, the quadrature LMV cell is inserted in a phase locked loop. This demonstrates that the oscillator signal can be sensed across the tank, without perturbing the mixer functionality.

A summary of the most relevant measurements results, and a comparison with the state of art of GPS front-ends, are reported in Fig. 13.13. Notice that all the data are referred only to the RF front-end (which however includes the PLL).

	[6]	[7]	[10]	**This Work**
Gain	(25)+60 dB	(50)+80 dB	(33)+60 dB	36dB
NF	4dB	4dB	8.5dB	4.8dB
IIP3	n.a.	−15dBm	n.a.	−19dBm
1dB Comp. Point	−28dBm	n.a.	n.a.	−31dBm
PN @ 1MHz	−95dBc/Hz	−107dBc/Hz	−109dBc/Hz	−104dBc/Hz
LO leak. at input	−66dBm	n.a.	n.a.	−55dBm
Total Power	**35mW**	**27mW**	**19mW**	**11mW***

Fig. 13.13 GPS performance and comparison with the state of the art

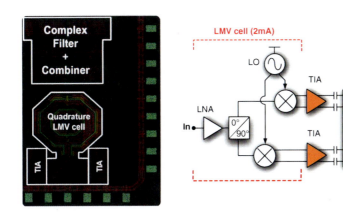

Fig. 13.14 ZigBee receiver

The proposed solution is the less onerous one in terms of power consumption even considering that the PLL was not optimized for the specific application. Notice that the other implementations have attempted to reduce power dissipation either by decreasing the receiver performance [6] or through a partial current reuse [7].

(b) Zig-Bee receiver

The ZigBee receiver has been fabricated in a 90 nm CMOS process. Figure 13.14 shows the micro-photograph of the chip and the implemented architecture. Since the quadrature signal is generated in the RF path, the VCO tanks can be shared which allows the use of only one integrated inductor. This results in an active die area of only 0.35 mm^2 (including the baseband filter), which is less than one fourth the area of the GPS prototype (although at the cost of a much higher noise figure).

A fully differential three-stage variable gain complex gm-C filter, AC coupled at the output of the TIA to suppress DC offset and low frequency noise, performs

	[12]	[13]	This work
NF (dB)	24.7	5.7	9
Sensitivity (dBm)	−82	−101	−94.7
IIP3 (dBm)	−4,5	−16	−12,5
SFDR (dB)	50.3	55.3	53.5
Image Rejection (dB)	---	36	35
PN(3.5MHz) (dBc/Hz)	--	--	−107.8
Vdd (V)	1.8	1.8	1.2
Power dissipation (mW)	15	17	3.6
Number of inductors	6	4	1
Area (mm2)	2,1	0,8	0.35
Technology	0.18	0.18	0.09

Fig. 13.15 ZigBee performance and comparison with the state of the art

the channel selection. In Fig. 13.15, the prototype performance is compared to that of other complete ZigBee receivers present in literature. The noise figure averaged over the band from 1 to 3 MHz is around 9 dB while the IIP3 is −12.5 dBm. The result is a spurious free dynamic range of 55.5 dB with a power consumption about one fifth and an active area less than half compared to the state of art values.

13.2 SAW-Less Harmonic Rejection Receivers

The down-conversion of the signal from RF to either base-band or IF is generally realized by a mixer that multiplies the input signal by a square wave and not by a sinusoid. This occurs even in the presence of a sinusoidal local oscillator (LO) due to the non-linear behavior of the switches. Mixing with a square wave folds the portions of the spectrum close to the odd harmonics of the LO on top of the wanted signal (located around the LO frequency f_{LO}) and this degrades the signal to noise ratio of the receiver as shown in Fig. 13.16.

To limit the amount of spectrum folding, an external surface-acoustic-wave (SAW) filter or a harmonic rejection mixer are generally required. Unfortunately, the use of a SAW filter increases cost and reduces sensitivity while the use of harmonic rejection mixers requires multiple phase clocks which increases the complexity and the power consumption of the receiver chain. A possible solution to alleviate this problem is to use the combination of a filtering low-noise amplifier (LNA) together with a resonant mixer [10].

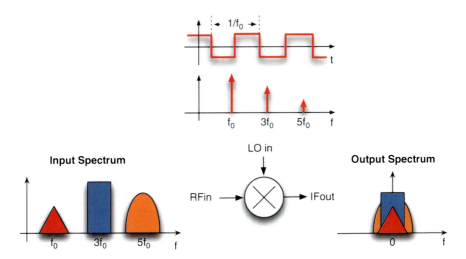

Fig. 13.16 Harmonic mixing

13.2.1 Filtering Single-Ended LNA

In cellular receivers, interferers at multiple of the LO are managed by SAW filters between the antenna and the LNA. For Time Division Duplexing (TDD) the SAW can be removed if the receiver can handle both far way and close in interferers (e.g. in GSM 0dBm 20 MHz away). Once the SAW is removed also the external balun can be eliminated by using a single ended (SE) transceiver further reducing cost and attenuation. For the same sensitivity, a SAW-less SE transceiver can have a noise figure (NF) 2–3 dB higher than a classical one. A transformer-based blocker tolerant filtering SE LNA is shown in Fig. 13.17.

The active portion of the LNA uses a fully differential complementary class A/B common gate configuration. The PMOS and NMOS transistors are coupled to the input by two secondary coils of an integrated transformer that acts like a balun. This gives an immunity from spurious coupling close to that of a differential LNA. The secondary coils have fewer turns than the primary giving 6 dB current gain. To reduce the NF of a classic common gate, the gate-source voltage of the input transistors (M1–M4) is boosted. This is done by a fourth coil with a k of 1. Two feed-forward capacitances (CF) implement a zero at $3f_{LO}$ to attenuate the blockers down-converted in-band through harmonic mixing as shown in below Fig. 13.18.

Using the transformer in normal or inverting mode, the transfer function changes not only in the phase. It turns out that in inverting mode, due to the opposite sign between the signal coupled to the output by the mutual inductance and that provided by the coupling capacitance, a notch is created in the transfer function. By explicitly adding a capacitance between the two coils, the position of the notch can be tuned.

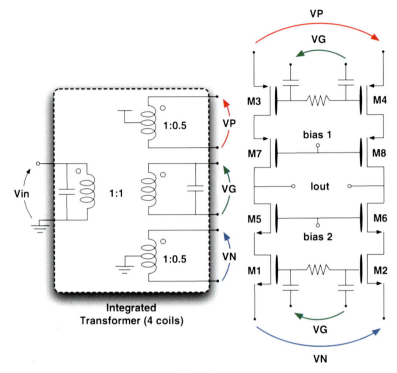

Fig. 13.17 Filtering SE LNA

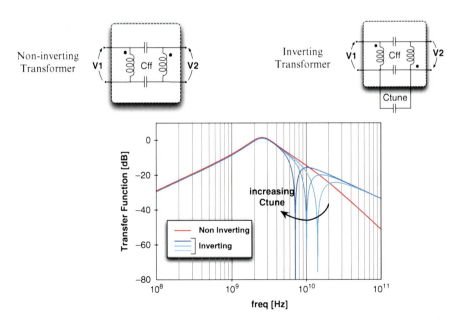

Fig. 13.18 Filtering LNA transformer

13 Unconventional Receiver Architectures

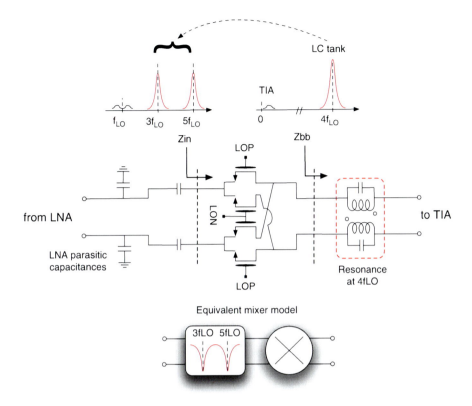

Fig. 13.19 Resonant mixer

13.2.2 Resonant Current-Mode Passive Mixer

The structure of the resonant passive mixer, the impedance level vs. frequency seen at the input and output of the current switches, and the mixer equivalent behavioral model are shown in Fig. 13.19.

The structure is composed of a passive current-mode mixer followed by an LC tank resonating at $4f_{LO}$. At the RF side, due to the bilateral nature of the passive mixer, the resonance frequency of the LC tank appears, to first approximation, up and down converted to $5f_{LO}$ and $3f_{LO}$. This increases the mixer input impedance at these frequencies. Such an effect, combined with the parasitic capacitances at the LNA output, produces two frequency notches in the mixer gain. Contrary to what occurs when using RF filters, this technique does not increase the parasitic capacitance at the output of the LNA which would adversely affect both the mixer and the BB noise. As a consequence, there is no noise/gain penalty. Actually, the frequency notches also reduce in-band noise folding, improving NF.

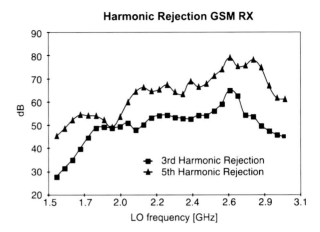

Fig. 13.20 Harmonic rejection

13.2.3 Example

The above resonant mixer was used in a cellular receiver for 2-G/3-G applications [10]. In the chip prototype fabricated in 40 nm CMOS, the measured harmonic rejections at the 3rd and the 5th harmonic reached 60 and 80 dB respectively, due to the combined action of the harmonic rejection mixer and the use of a narrowband (third harmonic notch) LNA in front of the mixer (Fig. 13.20).

13.3 Filtering ADC: Towards a Software/digital Defined Radio

The software-defined radio (SDR) ultimate goal is to substitute all the analog blocks with a more flexible and lower cost digital processor. A first step towards the implementation of an SDR moves the ADC just after the down-conversion mixer. Following this idea, a low-pass continuous-time (CT) $\Sigma\Delta$ ADC that combines interferer filtering, variable-gain amplifier (VGA), and signal digitization is presented [11]. Such a filtering ADC is intended to replace the entire analog BB of a 2G–3G receiver.

The ADC, reported in Fig. 13.21, is based on a Rauch biquad filter in which the feedback resistance is substituted by the cascade of a quantizer and a digital-to-analog converter (DAC). The DAC closes the loop around the forward integrator injecting a current at the input of the filter, which operates in the current domain (the input current I_{in} represents the down-converted received (RX) signal).

13 Unconventional Receiver Architectures

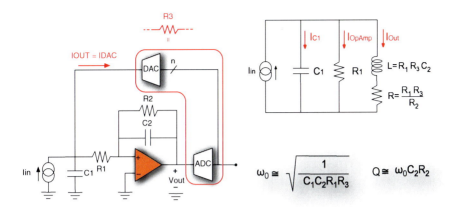

Fig. 13.21 Filtering ADC

13.3.1 Filter Transfer Function

To evaluate the filtering ADC transfer functions, the ADC-DAC cascade can be modeled with a transconductor whose transconductance ($1/R_3$) is given by the ratio between the full-scale current of the DAC and the full-scale reference voltage of the ADC. Under this assumption, the filter embedded in the ADC is equivalent to an LRC shunt network (Fig. 13.21) whose output signal is the current flowing into the inductance. The value of the equivalent inductance is equal to $R_1 R_3 C_2$ and his quality factor is proportional to the damping resistor R_2.

While the DAC current undergoes a second order filtering, the current absorbed by the operation amplifier (OA) undergoes a first order filtering. The presence of the grounded capacitance C_1 ensures that the out-of-band interferers that the DAC and the OA have to manage are only a fraction of the ones present at the input of the base-band. From this point of view the proposed solution differs from existing filtering-ADC architectures where all the input current (coming from the mixer) must be absorbed by the active devices, making the first integrator the most power hungry element of the converter [12].

13.3.2 Filtering ADC Noise

While the amount of filtering sets the maximum out-of-band signal that can be handled by the ADC, the noise floor (i.e., the sum of quantization and analog noise) defines the minimum detectable signal. The main noise sources of the filtering ADC, and its equivalent continuous time model, are shown in Fig. 13.22 (R_s represents the finite driving impedance of the stage preceding the filtering ADC).

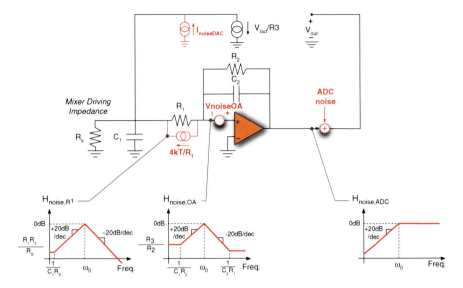

Fig. 13.22 Filtering ADC noise source and transfer functions

The dominant noise contributors are resistor R_1, the OA, the feedback DAC and the ADC. With the exception of the DAC, that injects its noise directly at the input node, all the other noise sources have an in-band zero in their transfer functions. For both digital (quantization) and analog ADC noise, this is the direct consequence of having inserted the ADC in the loop used to synthesize the complex poles. For the other noise sources, the high pass noise shaping is due to the intrinsic mechanism of current filters that occurs because the DAC senses the output signal as a current [13] as opposed to a voltage.

13.3.3 Example

The present filtering ADC was used for the first time in a reconfigurable quadrature DVB-T/ATSC tuner fabricated in a 90 nm CMOS process [11]. A simplified circuit schematic of the receiver chain is reported in Fig. 13.23 (without quadrature path). The ATSC standard presents adjacent channel interferers that for the N + 5 channel can have an average power up to 56 dB higher than the in band signal power. Furthermore, the used low-IF architecture reduces the relative frequency offset of the undesired blockers from the channel edge. This makes the required dynamic range of the ADC extremely high. The advantage in terms of noise of the fabricated prototype with respect to a filter-ADC cascade was 7.5 dB for the integrated quantization noise and 2 dB for the integrated analog noise. This 2 dB reduction allows to use 35 % less capacitance while keeping constant both noise and voltage

13 Unconventional Receiver Architectures

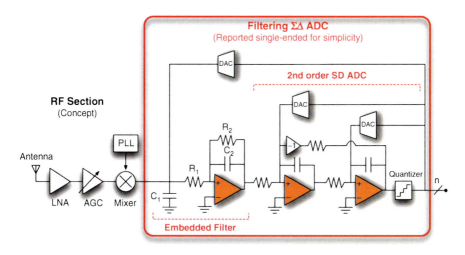

Fig. 13.23 Filtering ADC for DVB-T/ATSC receiver

	BW [MHz]	Fclock [MHz]	Power [mW]	SNDR [dB]	In-band FoM [pJ/step]	Out-of-band FoM [pJ/step]	Area [mm²]	Technology
This	6*	405	54**	74.6	1.03	0.2	0.21	90nm
JSSC 04	1	64	2	59	1.37	0.7	0.14	0.18um
TCAS 09	6.5	96	122.4	70.9	3.28	-	2.15	0.18um
ISSCC 06	8.5	264	375	84	1.7	1.7	2.5	0.13um
ISSCC 08	10	640	100	82	0.49	0.49	0.7	0.18um
JSSC 06	20	640	20	74	0.12	0.12	1.2	0.13um
ISSCC 07	20	340	56	69	0.61	0.61	1.2	0.13um
JSSC 10	25	400	7	52	1.08	1.08	-	90nm
JSSC 10	10	950	40	72	0.61	0.61	0.42	0.13um
ISSCC 09	20	250	10.5	60	0.32	0.32	0.15	65nm

Fig. 13.24 Filtering ADC performances vs. state of the art

swing at the mixer output (as required to preserve linearity for both the mixer and the feedback DAC).

A measurement performance summary is reported in Fig. 13.24. Due to the filtering nature of the proposed ADC, both the signal-to-noise-plus-distortion-ratio (SNDR) and the DR vary with frequency and are different for in-band and out of band interferers. The SNDR is only 1 dB below the DR. The FoM is 1.03 pJ/conv-step for in-band signals and becomes 0.2 pJ/conv-step at 30 MHz. A comparison with the state of art of filtering ADCs is also reported in the table (blue portion). The proposed solution shows the best performance both in-band and out-of-band (the latter evaluated for a frequency four times the signal bandwidth).

13.4 Conclusions

Through the use of unconventional architectures in wireless receivers like the LMV cell, a transformer coupled SE filtering LNA or resonant current mode mixers, either significant area and/or power savings can be achieved. In alternative very difficult functions, like harmonic rejection down conversion, can be implemented at almost no extra cost in terms of area and power consumption (without performance penalty) thereby eliminating the need for external costly and bulky components like SAW filters.

References

1. E. Hegazi, A. Abidi, A 17mW transmitter an frequency synthesizer for 900-MHz GSM fully integrated in 0.35-μm CMOS. IEEE J. Solid-State Circuit **38**, 782–792 (2003)
2. A. Liscidini et al., Single-stage low-power quadrature RF receiver front-end: The LMV cell. IEEE J. Solid-State Circuit **41**(12), 2832–2841 (2006)
3. M. Tedeschi, A. Liscidini, R. Castello, Low-power quadrature receivers for ZigBee (IEEE 802.15.4) applications. IEEE J. Solid-State Circuit **45**(9), 1710–1719 (2010)
4. A. Liscidini et al., A 5.4mW GPS CMOS quadrature front-end based on a single-stage LNA-Mixer-VCO. *IEEE International Solid-State Circuits Conference*, San Francisco, 2006
5. G. Montagna et al., A 35mW 3.6-mm fully integrated 0.18-□mCMOS GPS radio. IEEE J. Solid-State Circuit **38**, 1163–1171 (July 2003)
6. F. Behbahani et al., A 27-mW GPS radio in 0.35□m CMOS. IEEE Int. Solid-State Circuit Conf. Dig. Tech. Pap. **1**, 398–399 (2002)
7. J. Ko et al., A 19-mW 2.6-mm2 L1/L2 dual-band CMOS GPS receiver. IEEE J. Solid-State Circuit **40**, 1414–1425 (2005)
8. W. Kluge, F. Poegel, H. Roller, M. Lange, T. Ferchland, L. Dathe, D. Eggert, A fully integrated 2.4GHz IEEE 802.15.4 compliant transceiver for ZigBee applications. IEEE ISSCC Dig. Tech. Pap. **1**, 1470 (2006)
9. T.K. Nguyen et al., A low-power RF direct-conversion receiver/transmitter for 2.4-GHz-Band IEEE 802.15.4 standard in 0.18-um CMOS technology. IEEE Trans. Microw. Theory Tech. **54**(12), 4062–4071 (2006)
10. I. Fabiano et al., SAW-less analog front-end receivers for TDD and FDD. IEEE ISSCC Dig. Tech. Pap. **1**, 82 (2012)
11. M. Sosio, A. Liscidini, F. De Bernardinis, R. Castello, A complete DVB-T/ATSC tuner analog base-band implemented with a single filtering ADC, in *Proceedings of the IEEE International ESSCIRC*, pp. 391–394 (2011)
12. K. Philips et al., A continuous-time $\Sigma\Delta$ ADC with increased immunity to interferers. IEEE J. Solid State Circuit **39**(12), 2170–2178 (2004)
13. A. Pirola, A. Liscidini, R. Castello, Current–mode, WCDMA channel filter with in-band noise shaping. IEEE J. Solid State Circuit **45**(9), 1770–1780 (2010)

Chapter 14
Smart Self-interference Suppression by Exploiting a Nonlinearity

Erwin Janssen, Hooman Habibi, Dusan Milosevic, Peter Baltus, and Arthur van Roermund

Abstract A 1.8GHz RF amplifier implemented in 0.14um CMOS with frequency-independent blocker suppression is presented. The blocker suppression functionality is obtained by the adaptation of a nonlinear input–output transfer according to the blocker amplitude. Since superposition does not apply to nonlinear transfer functions, the behavior of such a transfer for strong undesired signals is different from the behavior for weak desired signals, which is exploited here. In the presence of a 0 to +11 dBm RF blocker, a voltage gain for weak signals of respectively 7.6–9.4 dB and IIP3 >4 dBm are measured, while the blocker is suppressed by more than 35 dB. In case of no blocker present at the input, the circuit is set to amplifier mode providing 17 dB of voltage gain and an IIP3 of 6.6 dBm while consuming 3 mW. Application areas are coexistence in multi-radio devices and dealing with TX leakage in FDD systems.

14.1 Introduction

Modern handheld devices support a multitude of wireless standards, such as e.g. WLAN, Bluetooth, GSM, UMTS and GPS. In recent years, the number of standards has been increasing steadily. The coexistence of these multiple

---

E. Janssen (✉)
Department of Electrical Engineering, Mixed-signal Microelectronics, Eindhoven University of Technology, Eindhoven, Netherlands

NXP semiconductors, Eindhoven, Netherlands
e-mail: Erwin.Janssen@nxp.com

H. Habibi
Department of Electrical Engineering, Signal Processing Systems, Eindhoven University of Technology, Eindhoven, Netherlands

D. Milosevic • P. Baltus • A. van Roermund
Department of Electrical Engineering, Mixed-signal Microelectronics, Eindhoven University of Technology, Eindhoven, Netherlands

A. Baschirotto et al. (eds.), *Frequency References, Power Management for SoC, and Smart Wireless Interfaces: Advances in Analog Circuit Design 2013*,
DOI 10.1007/978-3-319-01080-9_14, © Springer International Publishing Switzerland 2014

communication standards within a single device becomes therefore an increasingly important issue [1, 2].

Straightforward concepts to achieve reliable coexistence could either use filtering or time-sharing concepts. As filtering is often not sufficient and also not cost effective, present solutions usually apply time sharing. However, the time-shared approach reduces the achievable data throughput and also requires a challenging synchronization between the data packets of the different standards.

Due to the limitations of present coexistence solutions and the increasing number of standards in handheld devices, there is an interest to find alternative solutions to the coexistence problem. In addition, transmitter leakage in FDD systems [3] faces a similar problem as coexistence in multi-standard devices.

To avoid desensitization in the above situations, a high dynamic range has to be implemented in the receiver, leading to high power consumption. However, because of the limited energy resources available in handheld devices, minimizing the power consumption is critical. Thus, a major challenge will be to achieve low power consumption with a high dynamic range.

This paper proposes an RF amplifier that enables a frequency-independent suppression of a 0 to +11 dBm blocker by >35 dB while consuming 7–35 mW. Thanks to this suppression, the dynamic-range requirements for the subsequent stages in the receiver are relaxed. The suppression is achieved by an adaptive nonlinear circuit: the nonlinear transfer function creates the ability to provide different gains for signals having different amplitude levels [4]. By continuously adapting the circuit's nonlinear function according to the blocker amplitude, the gain of the blocker is effectively minimized while the gain of the signal remains high. Since the method requires knowledge of the amplitude of the interferer, it is most suitable for tackling the interference due to RX/TX or FDD coupling.

14.2 Principle of Operation

Nonlinear transfer functions exhibit properties that are fundamentally different from linear transfer functions, and thereby they enable different solutions in coexistence scenarios. This is illustrated in Fig. 14.1, where the input and output signals in both frequency and time domain for various conditions are compared. When passing a strong sinusoidal signal through a conventional compressive nonlinear system, the signal gets distorted and as a result harmonics are created (Fig. 14.1b). Here only odd order harmonics result because of the point-symmetric shape of the transfer function, a situation encountered in differential circuits. Considering the special case of the third order polynomial input/output relationship as shown in Fig. 14.1c, it appears that there even exist specific situations for which only a third order harmonic is generated, and the fundamental component is completely removed. The calculations describing this effect are stated below:

14 Smart Self-interference Suppression by Exploiting a Nonlinearity 251

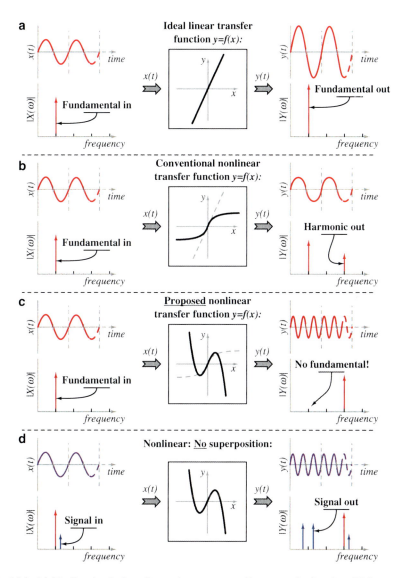

Fig. 14.1 (a) Ideally, circuits in radio receivers possess a linear transfer function. (b) In common practice however, receiver circuits generally have a nonlinear transfer function, leading to compression of the fundamental and the generation of harmonics. (c) Specifically tailored nonlinear transfers have the ability to fully suppress the fundamental for a specific input amplitude level. (d) Furthermore, weak (desired) signals, superimposed on the strong signal, are not suppressed and can even be amplified

$$y(t) = x(t) + c_3 x^3(t)$$

$$x(t) = A_{LS} \sin(\omega_{LS} t)$$

$$y(t) = \left[A_{LS} + c_3 \frac{3A_{LS}^3}{4} \right] \sin(\omega_{LS} t) + c_3 \frac{A_{LS}^3}{4} \sin(3\omega_{LS} t)$$

By choosing the third order coefficient c_3 equal to:

$$c_3 = -\frac{4}{3A_{LS}^3}$$

the output $y(t)$ becomes:

$$\rightarrow y(t) = \frac{A_{LS}}{3} \sin(3\omega_{LS} t)$$

Moreover, because nonlinear transfer functions do not obey the principle of superposition, a (much weaker) signal accompanying the strong signal undergoes a different operation. Excitation of the same nonlinear transfer function by the sum of the strong sinusoid with a weak sinusoid is shown in Fig. 14.1d. In contrast to the fundamental of the strong signal, the fundamental of the weak signal is not removed, but is still on its original location in the spectrum.

Next to the effect on the fundamental components of both large and small signals, the nonlinear operation also generates harmonics and intermodulation (IM) products. The harmonics can be removed easily by filtering at RF and because f_{LS} and f_{SS} are different, their IM products can be removed after down-conversion by filtering in the baseband. Of interest are the large- and small-signal gains, which in the rest of this paper are defined as the ratio between their fundamental output and fundamental input.

14.2.1 Strong-Signal Suppression Using a Zigzag Transfer Function

Achieving the functionality discussed in the previous section can be achieved with a wide variety of nonlinear transfer functions. Next to the example of the third order polynomial, the nonlinear transfer function shown in Fig. 14.2a (zigzag function) also achieves strong-signal suppression. The general requirements on the nonlinear transfer functions are discussed in more detail in [5]. Generally, it can be stated that the transfer must possess at least three zero-transitions, a property that is indeed seen in both the third order polynomial as well as the zigzag transfer. The zigzag transfer can be realized by combining the outputs of a linear amplifier and a clipping amplifier, and is therefore more suited considering the practicality of the concept. To demonstrate the concept using the zigzag transfer, the input spectrum consists of a strong and a weak

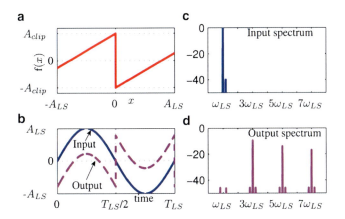

Fig. 14.2 (a) Zigzag transfer function. (b) Input (*solid blue*) and output (*dashed magenta*) signal versus time for a slope of unity for $[-A_{LS}, 0\rangle$ and $\langle 0, A_{LS}]$ in (a). (c) Input spectrum (*blue*) and (d) output spectrum (*magenta*) in dB relative to the input signal strength, illustrating the elimination of the fundamental

tone (Fig. 14.2c). This leads to an output spectrum consisting of several harmonics, but the fundamental of the strong signal is eliminated (Fig. 14.2d). To achieve this, the clipper amplitude A_{clip} must be set to:

$$A_{clip} = \frac{\pi}{4} A_{LS} G_{lin} \qquad (14.1)$$

where A_{LS} is the amplitude of the strong signal (i.e. interferer) at the input and G_{lin} is the gain of the linear amplifier. As becomes clear from Eq. 14.1, A_{clip} must be adjusted according to A_{LS} to assure zero strong-signal gain. So, successful application of this principle requires the circuitry to adapt its transfer function to the instantaneous strong-signal amplitude level (Fig. 14.2a). Furthermore, the gain of the weak signal is equal to $G_{lin}/2$ [5]. So, by assuring at least 6 dB of gain in the linear amplifier, the weak signal is amplified whereas the blocker is eliminated.

14.2.2 Application to Multi-radio Transceivers

In case two standards A and B are simultaneously active in a multi-radio transceiver, a situation is encountered where the receiver of standard A (victim) is plagued by the strong transmitted signal of standard B (aggressor). The blocking signal injected into the victim receiver is known, because the signal of the aggressor is generated in the same device. Therefore, it is possible to determine the amplitude level of the aggressor as it appears at the NIS input. This knowledge is required for proper operation, as clarified in the previous sections. In Fig. 14.3b a sub-block "Magnitude" is analyzing the baseband signal the aggressor is transmitting, resulting in the determination of the actual strong-signal amplitude A_{LS}. A sub-block "NIS control"

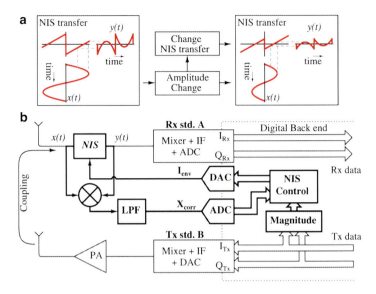

Fig. 14.3 (a) NIS input–output transfer adaptation for a change in input amplitude. (b) System level application of the NIS principle

on its turn steers the nonlinear interference suppressor (NIS) with a control current I_{env} using feed-forward, creating the desired blocker suppression.

Next to the feed-forward path, a mixer is present that multiplies the input with the output of the NIS. This operation results in the cross-correlation between these signals. The minimization of the cross-correlation means maximization of the suppression of the aggressor's signal (assuming the aggressor's signal to be dominant). The output of the mixer is being fed back into the "NIS control", and thereby it provides a measure for the residual error in the control current I_{env}. Errors in I_{env} could be caused by e.g. changes of the coupling between the aggressor and the victim. This procedure is described in more detail in [6], and will in future be extended to cases with varying envelopes. The remainder of this document will concentrate on the implementation and performance of the analog hardware that is mandatory for the NIS concept, namely the NIS circuit with the mixer followed by the low-pass filter.

14.3 Circuit Implementation in CMOS

Figure 14.4 shows the NIS circuit diagram. Firstly, transistors M_1–M_4 make up a linear amplifier resulting in a linear input–output relationship. Secondly, M_5–M_8 make up a clipper circuit with adjustable output clipping amplitude. The desired zigzag transfer function is realized by combining the outputs of these two sub-circuits with the required polarity.

14 Smart Self-interference Suppression by Exploiting a Nonlinearity

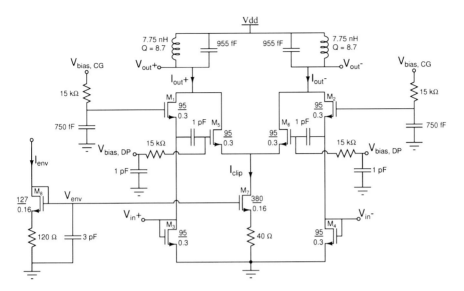

Fig. 14.4 NIS circuit diagram

$V_{bias,CG}$ is chosen such that M_1–M_4 are just conducting, resulting in a class-AB bias. In case of a large signal being present at the input, either M_1 and M_4, or M_2 and M_3 conduct causing the input resistance to be fairly constant around 60 Ω. For small and large (rail-to-rail) input signals the input return is therefore approximately -12 dB. Because the output current of the transconductor is about half the input current, the transconductance is therefore quite linear.

Next, current I_{clip} is steered through either the left or right LC tank because transistors M_5 and M_6 act as switches (clipper circuit). These transistors are configured in common-source, causing the M_5–M_6 structure to behave with opposite polarity with respect to the M_1–M_4 structure, which is configured in common-gate. By combining the output currents of both parts, the desired zigzag transfer function is created.

External control over I_{clip} is provided through current mirror M_7–M_8. By adjusting I_{env}, the adaptivity of the transfer shape is thereby provided. As Fig. 14.2 shows, the NIS concept generates several higher order harmonics. To suppress these harmonics, the circuit is loaded with an LC tank. The LC tank assures high impedance around the fundamental frequency, while it shorts the higher harmonics. In case I_{clip} is set in accordance to Eq. 14.1, the strong-signal is suppressed, and the circuit behaves in NIS mode. If there is no need for strong signal suppression, I_{clip} must set to zero. In that case the clipper circuit is not activated, resulting in a classical amplifier response (only M_1–M_4 are active).

A prototype IC, including ESD protection is implemented in 0.14um CMOS [7]. The system on the chip includes the NIS circuit and the passive mixer with a single pole low-pass filter shown in Fig. 14.5. For measurement purposes, both the RF and LPF outputs are followed by buffer circuits. The chip photo of the prototype is shown in Fig. 14.6a. The chip has been packaged in a HVQFN24 package and

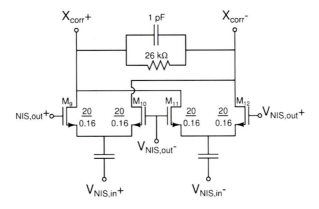

Fig. 14.5 Cross-correlation mixer circuit diagram

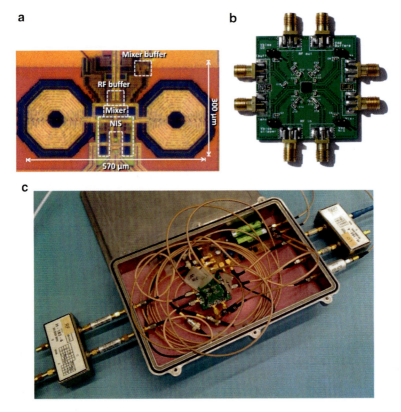

Fig. 14.6 Implemented NIS hardware. (**a**) Die photo. (**b**) PCB with packaged NIS chip (44×44 mm). (**c**) NIS PCB in Faraday cage including battery-based power supply/biasing

mounted on an FR4 PCB shown in Fig. 14.6b. Lastly, the PCB has been placed into an aluminum box making up a Faraday cage to guarantee full control over the signals applied to the system.

14.4 Experimental Results

The measured circuit transfer is maximal at 1.85GHz with a 3 dB bandwidth of 210 MHz. In both amplifier and NIS mode S_{11} is below -12 dB and S_{22} is below -13 dB. In this section first the characteristics in NIS mode, and then in amplifier mode are discussed.

14.4.1 NIS Mode

First, to demonstrate the NIS operation, a measurement is conducted by exciting the chip by the combination of a strong signal (0 dBm) and weak signal (-59 dBm). Both signals are phase modulated, and the spacing of the carrier frequency is limited to only 2 MHz. Because of the phase modulation, the envelope of the blocker is constant causing the spectral content of I_{env} to consist of only a DC signal. I_{env} is optimized such that the blocker gain is minimized. The input signal and output signal of this measurement after optimizing I_{env} are shown in Fig. 14.7.

As can be seen, the ratio between the blocker and the signal has been reduced by almost 50 dB. Next to the suppression of the blocker, also an additional tone has been created. This additional tone is an intermodulation product between the desired signal and the blocker. In general, it can be stated that the output of a memory-less nonlinear function around the fundamental when excited with a strong and a weak sinusoidal signal is equal to [8]:

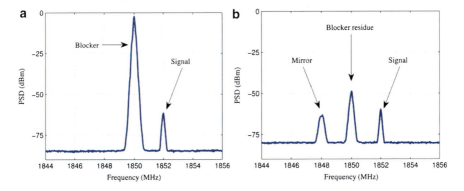

Fig. 14.7 Measured NIS response. (**a**) Input spectrum. (**b**) Output spectrum

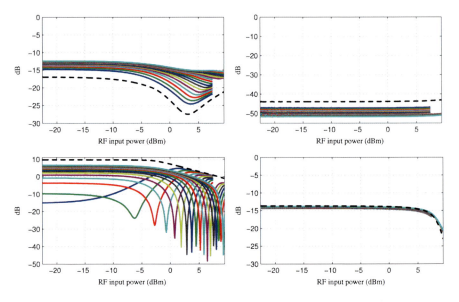

Fig. 14.8 Measured large-signal S-parameters for $I_{env} = 0.4$–3.8 mA with steps of 0.2 mA. The response for $I_{env} = 0$ mA is shown with the *black dashed lines*. (**a**) S_{11}. (**b**) S_{12}. (**c**) S_{21}. (**d**) S_{22}

$$y(t) = G_{LS} \cdot Int(t) + \frac{1}{2}\left[A_{LS} \cdot \frac{\delta G_{LS}}{\delta A_{LS}} + G_{LS}\right] \cdot s(t) \qquad (14.2)$$

$$+ \frac{1}{2}\left[A_{LS} \cdot \frac{\delta G_{LS}}{\delta A_{LS}} - G_{LS}\right] \cdot IM(t)$$

The signals in the above equations are given by:

$$Int(t) = A_{LS} \cdot \sin[\omega_{LS} t + \varphi_{LS}]$$

$$s(t) = A_{LS} \cdot \sin[\omega_{LS} t + \varphi_{LS}]$$

$$IM(t) = A_{SS} \cdot \sin[(2\omega_{LS} - \omega_{SS})t + 2\varphi_{LS} - \varphi_{SS}]$$

Here, G_{LS} is the gain of the strong signal, $\omega_{SS} = 2\pi f_{SS}$ and $\omega_{LS} = 2\pi f_{LS}$. From Eq. 14.2 it can be concluded that in case of strong-signals suppression (i.e. $G_{LS} = 0$), the magnitude of the content at ω_{SS} and $(2\omega_{LS} - \omega_{SS})$ in $y(t)$ are equal. This conclusion is in agreement with the measurement results shown in Fig. 14.7 (although the power spectral density of the mirrored component is less than that of the signal, their powers are equal).

Next, the large-signal S-parameters are measured of the NIS chip for different values of I_{env}. The results of this measurement are shown in Fig. 14.8. The transfer-controlling

Fig. 14.9 Graphical representation in the frequency domain of the strong- and weak-signal gain measurement procedure

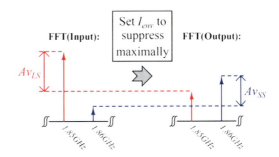

current I_{env} has been varied from 0.4 mA to 3.8 mA in steps of 0.2 mA, and the circuit is excited by a single tone at 1.85GHz, whose amplitude has been swept from -22 up to 9.5 dBm. As a reference, the response in case of $I_{env} = 0$ has been added as well, which is identified as amplifier mode. S_{21} in amplifier mode shows a P1dB of about -4 dBm as can be seen. Beyond that, the NIS operation becomes feasible. As can be concluded from the figure, S_{11}, S_{22} and S_{12} show a small variation compared to the variation observed in S_{21}, and provide sufficient performance (i.e. input and output reflection below -12 dB and reverse isolation below -40 dB). The response of S_{21} illustrates the amplitude domain filtering property that is aimed for in the NIS concept. By choosing a specific value for I_{env}, the zigzag transfer is configured such that a specific amplitude level is suppressed, according to Eq. 14.1.

Measuring both the gain for the strong blocker and the gain for the weak signal is conducted using the approach illustrated in Fig. 14.9. The input is excited by the sum of a weak signal and a strong signal. Then, the output is analyzed and I_{env} is chosen such that the strong signal is minimized. This procedure has been automated by closing the loop shown in Fig. 14.3b using an FPGA based PXI of National Instruments with AD/DA interface.

The measurement results from this procedure are shown in Fig. 14.10, for different values of $V_{bias,CG}$ in Fig. 14.4. Figure 14.10a shows the voltage gain of the strong signal, and Fig. 14.10b shows the voltage gain of the weak signal. The ratio between these two gains is identified as the strong signal suppression, which is shown in Fig. 14.10c. Although theory predicts that no cross-modulation takes place between the strong blocker and the weak signal in case of an ideal clipper/amplifier combination [5], it is seen in Fig. 14.10b that this is not fully achieved. This discrepancy is caused by the non-ideal behavior of mainly the clipper, i.e. the gain of the clipper during zero-transitions is dependent on the value of I_{env}, which on its turn depends on the amplitude of the strong signal. By lowering $V_{bias,CG}$, the dynamic range over which Av_{SS} varies, reduces.

Another observation that can be done is that although Av_{LS} is low, it never reaches zero. The cause of the limitation on the amount of suppression lies in the presence of memory effects. Memory effects in the circuit cause a phase mismatch between the linear amplifier and the clipper, causing imperfect cancellation. The ideal phase difference between the two sub-circuits of 180° is therefore not perfectly achieved. The choice of $V_{bias,CG}$ is identified here as a trade-off between

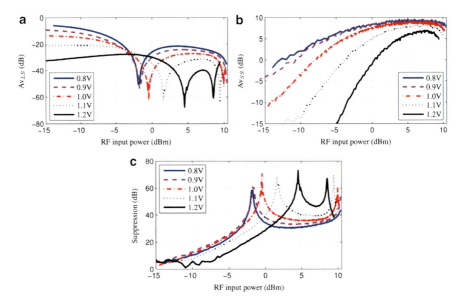

Fig. 14.10 Measured NIS behavior for different values of $V_{bias,CG}$. (**a**) Strong signal voltage gain. (**b**) Weak signal voltage gain. (**c**) Strong signal suppression

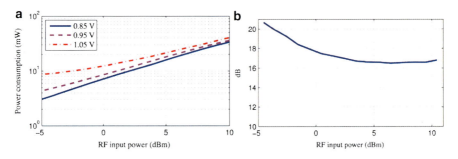

Fig. 14.11 NIS power consumption and noise figure. (**a**) NIS P_{DC} for different values of $V_{bias,CG}$. (**b**) Noise figure ($V_{bias,CG} = 1V$)

on the one hand weak signal gain (magnitude and flatness) and dynamic range, versus on the other hand the amount of suppression the circuit achieves. The power consumption and the noise figure of the NIS circuit versus the input level of the suppressed RF signal (i.e., blocker) are shown in respectively Fig. 14.11a, b.

The power consumption decreases with decreasing RF input power because of the class AB operation of the circuit, governing the absolute value of the power consumption to be low as well. The measured noise figure is just above 16 dB. As can be seen in the circuit diagram of Fig. 14.4, the input is connected to MOS

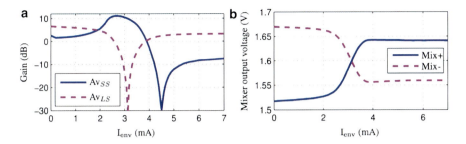

Fig. 14.12 Verification of the behavior of the cross-correlation mixer. (**a**) Weak & strong signal gain versus I_{env}. (**b**) Mixer output versus I_{env}

diodes that do not contribute to the gain of the circuit, but do dissipate the RF signal, which is not beneficial for the noise performance. Besides this effect, another aspect that complicates low noise design is the combination of an input–output path made up of M_1–M_2 configured in common gate and an input–output path made up of M_5–M_6 configured in common source. Both paths have different optimal source impedances regarding noise performance, so a trade-off occurs. Lastly, because of the spectrum mirroring effect discussed in the beginning of this section and derived using (2), the noise of $(2\omega_{LS}-\omega_{SS})$ folds into the frequency band of the desired signal, causing a minimal noise figure inherent to the NIS concept of 3 dB.

Next, the behavior of the mixer that is present on the IC is evaluated. To illustrate the behavior of the mixer, a measurement is performed by measuring its differential output voltage while increasing I_{env}, and maintaining the same input signal. The RF input of the NIS circuit is excited with a sinusoidal tone of 8 dBm during this measurement. The results are shown in Fig. 14.12, with the voltage gain of the weak and strong signal in Fig. 14.12a and the mixer output in Fig. 14.12b. As can be seen, the mixer outputs become equal in case the strong signal is minimized, indicating proper operation.

14.4.2 Amplifier Mode

By setting current I_{clip} to zero (shorting V_{env} to ground), the chip is set to amplifier mode. Characterization of the IC was done here by measuring the gain, noise figure and IIP3. In Fig. 14.13a the simulated as well as the measured voltage gain can be seen versus $V_{bias,CG}$. As the figure shows, the measured performance is worse than the simulated performance. This reduced performance is mainly explained by a lower than expected transconductance of the transistors, and a lower quality factor of the LC tank. The noise performance is shown in Fig. 14.13b. Also with respect to noise performance a reduction is seen from simulation to measurement. The IIP_3 of the chip was measured to be 6.6 dBm for a $V_{bias,CG}$ of 1.15 V.

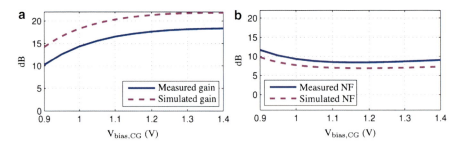

Fig. 14.13 Behavior of the chip in amplifier mode. (**a**) Voltage gain. (**b**) Noise figure

14.5 Summary and Conclusions

A 1.8 GHz RF amplifier implemented in 0.14 um CMOS with frequency-independent blocker suppression has been presented. The blocker suppression functionality is obtained by continuously adapting the nonlinear transfer function of the circuit according to the blocker amplitude. Application areas are coexistence scenarios where the envelope of the blocker to be suppressed is known, for example in multi-radio devices and standards dealing with TX leakage in FDD systems.

The circuit has two modes of operation: the NIS mode, when it provides blocker suppression, and the amplifier mode, when no blocker is present at the input. In NIS mode, a voltage gain for weak signals of respectively 7.6–9.4 dB and IIP3 >4dBm were measured in the presence of a 0 to +11 dBm RF blocker, while the blocker has been suppressed by more than 35 dB. Analysis predicted, and measurements confirmed that in case of blocker suppression using the proposed NIS method signals and noise are mirrored to the image frequency with respect to the blocker. A passive mixer has been put on the chip to derive the cross-correlation between input and output with the aim of determining the amount of blocker suppression achieved. The mixer has been used in a feedback loop, and showed the expected behavior. The noise figure in NIS mode has been measured to be just above 16 dB. The reason for this relatively high value is partly due to the specific circuit topology, and partly is inherent to the NIS concept. Future research will concentrate on optimizing the circuit topology and finding measures to counteract the spectrum mirroring taking place when using the concept.

The circuit is set to amplifier mode in case there is no blocker present at the input. In amplifier mode, the circuit provides 17 dB of voltage gain and an IIP3 of 6.6 dBm while consuming 3 mW. The performance in measurements has dropped with respect to the simulations because of a reduction in transistor transconductance and quality factor of the LC tank. The 1 dB compression point of the circuit is found to be about −4dBm, which is around the same value where the NIS concept becomes feasible. So, for interferers of up to −4dBm the circuit can be operated in amplifier mode, whereas for higher interferer levels, the NIS mode should be used.

References

1. S. Sheng, RF coexistence – Challenges and opportunities, in *Radio Frequency Integrated Circuits Symposium (RFIC) IEEE*, Baltimore, USA, 5–7 June 2011
2. J. Zhu, A. Waltho, X. Yang, X. Guo, Multi-radio coexistence: Challenges and opportunities, in *Computer Communications and Networks. ICCCN. Proceedings of 16th International Conference on*, Honolulu, USA, 13–16 Aug 2007, pp. 358–364
3. E.A. Keehr, A. Hajimiri, A rail-to-rail input receiver employing successive regeneration and adaptive cancellation of intermodulation products, in *Radio Frequency Integrated Circuits Symp. (RFIC) IEEE*, Anaheim, USA, 23–25 May 2010
4. K.J. Friederichs, A novel canceller for strong CW and angle modulated interferers in spread-spectrum-receivers, in *Military Communications Conference MILCOM. IEEE*, Los Angeles, USA, vol. 3, 21–24 Oct 1984, pp. 478–481
5. E.J.G. Janssen, H. Habibi, D. Milosevic, P.G.M. Baltus, A.H.M van Roermund, Frequency-independent smart interference suppression for multi-standard transceivers, in *IEEE European Microwave Conference (EuMC)*, Amsterdam, Oct 28th – Nov 2nd 2012
6. H. Habibi, E.J.G. Janssen, W. Yan, J.W.M. Bergmans, P.G.M. Baltus, Suppression of constant modulus interference in multimode transceivers by closed-loop tuning of a nonlinear circuit, in *Vehicular Technology Conference (VTC Spring), 2012 IEEE*, Yokohama, Japan, 6–9 May 2012
7. E.J.G. Janssen, D. Milosevic, P.G.M. Baltus, A 1.8GHz amplifier with 39dB frequency-independent smart blocker suppression, in *IEEE Radio Frequency Integrated Circuits Symposium (RFIC)*, Montreal, June 2012
8. N. Blachman, Band-pass nonlinearities. IEEE Trans Inf Theory **10**, 162–164 (1964)

Chapter 15
The Design of Ultralow-Power MEMS-Based Radio for WSN and WBAN

Aravind Heragu, David Ruffieux, and Christian Enz

Abstract Transceivers for wireless sensor networks (WSN) and wireless body area networks (WBAN) require both extreme miniaturization and ultra-low-power dissipation in order to be seamlessly integrated virtually everywhere and enable ubiquitous connectivity among persons, objects, machines and the environment. The miniaturization challenge can be addressed with a combination of system-on-chip (SoC) and system-in-package (SiP) approaches to build an ultra-compact transceiver. The confined space is also limiting the available energy, which raises several design and system issues that could severely affect the radio robustness to interferers, the link budget and the autonomy. This paper presents how innovative narrowband radio architectures devised to take advantage and circumvent the limitations of a few well-chosen MEMS devices can address the above issues and go beyond the existing solutions both in terms of miniaturization and power dissipation reduction.

15.1 Introduction

With the advent of various wireless standards like Bluetooth LE [1], MICS [2], MBAN [3] etc., wireless sensor networks (WSN) have now penetrated into the health-care aspect too and have become indispensable in today's life. The perennial integration of the wireless body area networks (WBAN) is possible only when the nodes of the network adhere to stringent power consumption and miniaturization constraints. High power consumption on these nodes either increases the size of the

---

A. Heragu (✉) • C. Enz
EPFL and CSEM, Lausanne, Switzerland
e-mail: aravind.heragu@epfl.ch; christian.enz@csem.ch

D. Ruffieux
CSEM, Lausanne, Switzerland
e-mail: david.ruffieux@csem.ch

A. Baschirotto et al. (eds.), *Frequency References, Power Management for SoC, and Smart Wireless Interfaces: Advances in Analog Circuit Design 2013*, DOI 10.1007/978-3-319-01080-9_15, © Springer International Publishing Switzerland 2014

batteries required or brings in the need for their frequent replacement. The increase in the battery size makes the system bulky and inconvenient to use especially in applications like animal tracking systems where the overall payload has to be less than a gram [4]. Although energy harvesting helps in increasing the life time of the battery, the amount of power harvested in a volume of 1 cm^3 is less than 0.1 mW [5, 6]. A realistic target power level for the wireless nodes operating at low data rate (LDR) (around 10 kbps) is at sub-100 $\mu W/cm^3$ [7]. However, commercially available radios consume few to several mW and the only way out to reduce the power consumption is by agile duty cycling.

The power consumption of a wireless sensor node can be lowered by choosing an appropriate protocol [7]. However, with the use of a standard protocol, the option ceases to exist and low power requirement gets transferred to the design and implementation of the radio. Although aggressive CMOS scaling has helped in lowering down the power dissipation of the RF circuits, and in increasing the transit frequency, it is not complemented well by the passive elements which exhibit low quality factor (Q) owing to the lossy substrate. MEMS devices like the Bulk Acoustic Wave (BAW) and Surface Acoustic Wave (SAW) resonators exhibit intrinsically high Q which can be exploited in realizing low power transceivers. The power hungry blocks like the RF front-end, PA and the frequency synthesis are the ones which get benefitted by such resonators [8–10]. BAW based oscillators exhibit superior phase noise performance compared to the LC counterparts, thanks to the intrinsic high Q of the resonators. In this work, a sub-sampling receiver architecture which exploits the high Q of the BAW resonators in the frequency synthesis and in the RF front-end to perform channel selection filtering is presented.

15.2 Receiver Architecture

A block level view of the proposed BAW based sub-sampling receiver architecture is shown in Fig. 15.1. BAW resonators are used in the frequency synthesis to provide signals with low phase noise [8] and in the RF front-end to provide channel filtering. The input RF signal from the antenna is pre-filtered and is then amplified by a low noise amplifier (LNA). The LNA is followed by a channel selection mixer which down-converts the wanted channel to the anti-resonant frequency of the BAW resonator used in the channel selection filter following the mixer. As the anti-resonant frequency of the BAW resonator is in the GHz range, low frequency LO signals are needed to down-convert the required channel. The band selection filtering profile of the pre-filter should be such that it should provide image rejection for the channel selection mixer. If $f_{if, low}$ is the channel selection LO frequency corresponding to the first channel, the image band is at an offset of $2f_{if, low}$ away from the band of interest.

The adjacent channel rejection provided by the BAW based filter can also serve as the critical anti-alias filter to support sub-sampling based down-conversion. Sub-sampling mixers require low frequency clocks to perform the down-conversion.

15 The Design of Ultralow-Power MEMS-Based Radio for WSN and WBAN 267

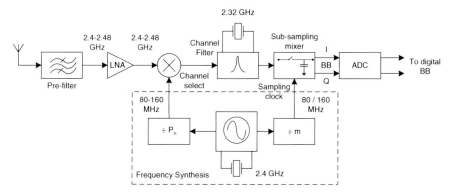

Fig. 15.1 Block level view of the proposed receiver architecture with channel filtering at RF. BAW resonators are used in the oscillator to provide low phase noise signal and in the RF front-end to provide channel filtering

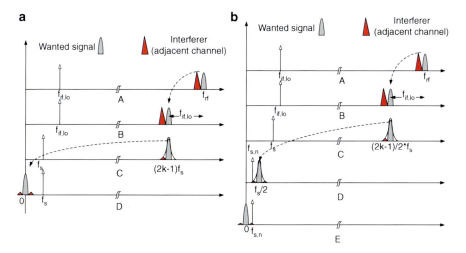

Fig. 15.2 Frequency planning in the proposed receiver architecture. The selected and filtered channel can be down-converted to baseband in quadrature by sampling in (**a**) a single step, with quadrature clocks at a rate which is an odd sub-multiple of the BAW anti-resonant frequency, or (**b**) in two steps, when differential clocks are used. The two-step down-conversion can additionally provide some discrete time filtering which improves the overall performance of the receiver

Further, some amount of discrete time filtering is also provided by the sub-sampling mixer which improves the overall performance of the receiver. The down-converted samples at baseband in quadrature are fed to a phase ADC [11] and then to a digital baseband to obtain the demodulated bit. The frequency planning of the proposed receiver architecture is shown in Fig. 15.2. The BAW digitally controlled oscillator (DCO) provides low phase noise signal around the GHz range, which can be divided to provide the clocks needed or to provide reference to PLLs which in-turn provide the required clocks in the receiver. The requirement of a PLL depends on the BAW resonators used in the DCO and in the filter and the amount of frequency tunability added to these resonators.

Although a BAW oscillator has superior phase noise performance compared with a LC oscillator, it suffers from limited tuning (typically 100 ×) [12]. With a fixed BAW oscillator frequency and a set of integer dividers it can be shown that it is impossible to address any arbitrary frequency in the band of interest. Fractional dividers are needed in such a case which in-turn demands a PLL. Sliding IF architectures have been demonstrated with a fixed DCO signal and a LC based PLL [8, 9] or a low power $G_m - C$ PLL [13]. Similarly, in the proposed architecture, when the BAW oscillator is running at a fixed frequency, there is a need for a PLL to provide the required frequencies to perform channel selection. The reference clock to the PLL can be provided by the BAW DCO as in [8]. When low power operation is required, a relaxation oscillator based PLL can be used [13]. If superior phase noise performance is required, a high frequency LC based PLL can be used [8, 9]. As shown in Fig. 15.2a, the quadrature down-conversion in a single step by sampling, requires the sampling rate to be an odd sub-multiple of the working frequency of the filter [14].

15.3   Receiver Front-End

The received RF signal from the antenna is pre-filtered and then amplified by a LNA with tuned LC load. The LNA with differential input, employs current reuse and $g_m -$ boosting to provide low power solution. The amplified signal is then fed to the channel selection filter.

15.3.1   BAW Based Channel Selection Filter

The BAW resonator intrinsically exhibits very high quality factor (Q) at GHz range which is exploited to provide the channel filtering, thus improving the in-band blocker tolerance. However, sometimes the Q might not be sufficient to achieve very narrow bandwidths. To address this issue, we propose a solution where the bandwidth of the filter can be tuned by adding a negative resistance in shunt with the resonator, the details of which are presented later.

Although the BAW resonator intrinsically offers high quality factor, its impedance is very high (capacitive) at DC (Fig. 15.3a). This poses a problem in zero IF systems as the DC offsets and flicker noise get amplified and remain at DC along with the wanted signal after it is down-converted to baseband by sampling. Moreover the attenuation obtained from a single resonator might not be sufficient and might call in for more such stages to improve the out-of-band rejection. BAW based lattice network has been previously used [10] to provide band selection.

15 The Design of Ultralow-Power MEMS-Based Radio for WSN and WBAN

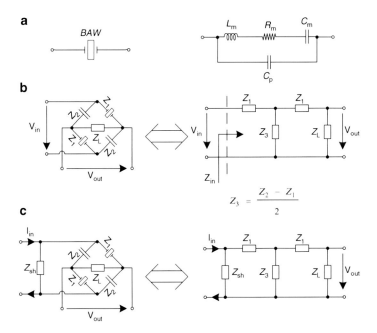

Fig. 15.3 (a) BAW equivalent model; Proposed BAW based lattice (b) with voltage input, (c) With current input

The bandwidth is however large, typically around 80 MHz for Bluetooth LE specifications. Here, we propose a lattice network built with two matched BAW resonators and two matched capacitors as shown in Fig. 15.3b to be used in the channel select filter. The voltage transfer function of the pseudo-lattice can be shown to be

$$H_{lattice} = \frac{V_{out}}{V_{in}} = \frac{Z_1 - Z_2}{Z_1 + Z_2}. \tag{15.1}$$

From the equivalent model of the BAW resonator (Fig. 15.3a it is clear that the impedance is capacitive ($\approx C_p + C_m$) from DC to the resonant frequency ω_s, then inductive from ω_s to the anti-resonant frequency ω_p and again capacitive ($\approx C_p$) from ω_p to ∞. Normally C_m is very less compared to C_p. In (15.1), if Z_1 is that of the BAW resonator and Z_2 is that of a capacitor $C_{ltc} = C_p + C_m$, then $H_{lattice} \approx 0$ at all frequencies from 0 to ∞ except at those between ω_s and ω_p. Further, it can be seen from Fig. 15.4a, the capacitance C_{ltc} crosses the inductive regime of the BAW resonator (between ω_s and ω_p) and at this point the magnitudes of Z_1 and Z_2 are equal; however Z_1 is inductive while Z_2 is capacitive. This makes H tend to infinity as $(Z_1 + Z_2)$ becomes 0 and the transfer characteristic is as shown in Fig. 15.4a.

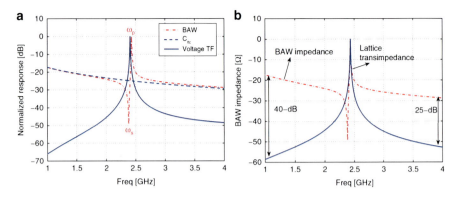

Fig. 15.4 (a) Normalized (to their respective peaks) frequency responses of the BAW impedance, the impedance of the capacitor (Z_2) used in the pseudo-lattice and the voltage transfer function of the pseudo-lattice. (b) Normalized (to their respective peaks) frequency plots of the BAW impedance, trans-impedance of the BAW pseudo-lattice

However in reality, it should be noted that the gain as given by (15.1) is not achievable by the BAW based lattice network because of impedance mismatch. The input impedance of the BAW based lattice is given by $(Z_1+Z_2)/2$. This is zero where $H_{lattice}$ is maximum. This means to achieve very high gain, the source resistance should be zero, while normally in RF systems it is 50 ohms. This poses a constraint on using voltage input to the lattice. Instead of a voltage input, if a differential current is pumped into the lattice as shown in Fig. 15.3c, then the trans-impedance of the BAW lattice using the equivalent T model [15] can be shown to be

$$Z_{tran} = \frac{V_{out}}{I_{in}} = \frac{Z_{sh}Z_L \cdot (Z_2 - Z_1)}{(Z_{sh} + Z_L) \cdot (Z_1 + Z_2) + 2 \cdot (Z_1 Z_2 + Z_{sh}Z_L)}. \qquad (15.2)$$

The trans-impedance given by (15.2) has a bandpass characteristic as shown in Fig. 15.4b with peak at the anti-resonant frequency of the BAW resonator (ω_p). This trans-impedance has better filtering characteristic compared to a single resonator especially at frequencies far away from the peak frequency. It can be shown that when the impedance Z_{sh} is a negative resistance, the bandwidth of the trans-impedance can be reduced (Q boosting) as in the case of a single resonator. Similarly, the working frequency of the lattice can be tuned by varying the load capacitance Z_L. Further from the expression (15.2), it is clear that by tuning the capacitance Z_2, a null can be placed at a frequency where Z_1 and Z_2 are equal and this property can be utilized to reject any strong blocker.

A low power solution which employs current reuse called the Amplifier-Mixer-Filter (AMF) cell is used as a trans-conductance stage to drive either a single BAW resonator or a BAW pseudo-lattice [16]. The schematic of the AMF cell with a pseudo-lattice load is shown in Fig. 15.5. The input RF from the LNA is

15 The Design of Ultralow-Power MEMS-Based Radio for WSN and WBAN

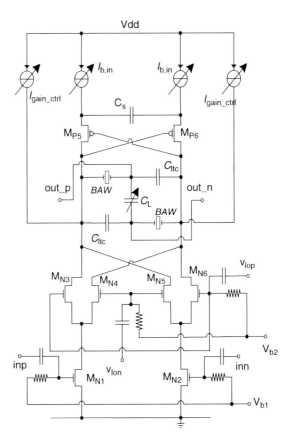

Fig. 15.5 AMF cell with BAW pseudo-lattice load: A low power solution to perform channel selection and filtering at RF

down-converted by the mixing transistors to the working frequency of the lattice. Further to boost the Q of the lattice, cross coupled PMOS pair (which share the same bias current) is added in shunt. To decouple the gain and the bandwidth settings, parallel tunable current sources are added as shown in Fig. 15.5.

15.3.2 Quadrature Sampling Mixer

The selected and filtered channel at RF (super-high IF) is down-converted to baseband in quadrature by a two stage sub-sampling mixer shown in Fig. 15.6. The first stage down-converts the wanted channel to $f_s/2$ (f_s is the sampling rate) and the anti-alias filtering required is provided by the channel filter. The second stage of the sampling mixer re-samples at a rate of $f_s/4$ which down-converts the wanted signal to baseband in quadrature [17]. The anti-alias filtering required for the second stage is built into the sampling mixer itself. The second stage in the

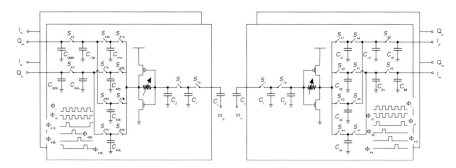

Fig. 15.6 Quadrature two-stage sub-sampling down-conversion mixer. The second stage also provides discrete time filtering and decimates the sampling rate

process of down-converting, also provides discrete time filtering (FIR/IIR) and decimation of the sampling rate. The FIR response of the mixer is given by,

$$H_{FIR}(z) = (1 + z^{-2}) - (z^{-1} + z^{-3}), \quad \text{with} \quad z = e^{j2\pi f/f_s}. \tag{15.3}$$

The FIR response has nulls at all multiples of $f_s/4$ except at odd multiples of $f_s/2$, where the signal is present and this acts as an inherent anti-alias filter for the second stage of sampling. The total discrete time response (FIR + IIR) is shown in Fig. 15.10. To improve the overall conversion gain of the receiver, self biased inverter based IF amplifiers are placed in between the two stages of the sampling mixer as shown in Fig. 15.6.

15.4 PLL-Free Frequency Synthesis: The Integer-N Approach

Instead of running the BAW DCO at a fixed frequency as in [8], if some frequency tuning is enabled by adding a switchable capacitor array across the resonator, then let us say each channel center frequency $f_{rf,ch,n}$ in the required band can be expressed as,

$$f_{rf,ch,n} = f'_{baw,f} + \frac{f'_{baw,o}}{P_n}, \tag{15.4}$$

where P_n is an integer with values ranging from N_l to N_h. The frequency $f'_{baw,o}$ is the tuned BAW DCO frequency and $f'_{baw,f}$ is the tuned working frequency of the BAW channel filter. Mathematically these are given by,

$$f'_{baw,o} = f_{baw,o} - \Delta f_o \quad \text{and} \quad f'_{baw,f} = f_{baw,f} - \Delta f_f. \tag{15.5}$$

15 The Design of Ultralow-Power MEMS-Based Radio for WSN and WBAN 273

In order to sample the filtered channel with a clock derived integer division of the BAW DCO signal and down-convert it in quadrature, the initial sampling rate f_s is given by [17],

$$f_s = \frac{f'_{baw,o}}{m} = \frac{2f'_{baw,f}}{2k-1}, \quad \text{with} \quad m, k = 1, 2, 3 \ldots . \tag{15.6}$$

It is clear from (15.6) that, the intrinsic anti-resonant frequencies of the BAW resonators used in the DCO ($f_{baw,o}$) and in the filter ($f_{baw,f}$) have to be different and the offset between these is at a minimum when $m = k$ in (15.6). Small offsets between the anti-resonant frequencies can be easily generated by depositing an oxide layer on the electrodes [18]. When the offset is minimum, the frequencies $f'_{baw,o}$ and $f'_{baw,f}$ are related as,

$$f'_{baw,f} = \left(1 - \frac{1}{2m}\right) \cdot f'_{baw,o}, \quad m = 1, 2, 3 \ldots . \tag{15.7}$$

Using (15.7) in (15.4), we get,

$$f_{rf,ch,n} = f'_{baw,o} \cdot \left(1 + \frac{1}{P_n} - \frac{1}{2m}\right). \tag{15.8}$$

The minimum tuning required on the BAW DCO ($\Delta f_{o,\,min}$) to address any arbitrary frequency in the band of interest can be shown to be given by,

$$\Delta f_{o,min} = \frac{f_{baw,o}}{\left((N_l + 1)^2 - \frac{N_l \cdot (N_l+1)}{2m}\right)}. \tag{15.9}$$

For nominal values of $f_{baw,\,o}$, N_l and m to be 2.4-GHz, 15 and 15 respectively, the tuning required is 9.68-MHz (0.4 %) which is a feasible requirement [12]. With minimum difference between the intrinsic anti-resonant frequencies of the resonators used in the DCO and in the filter, the sampling rate can be expressed as,

$$f_s = 2 \cdot (f'_{baw,o} - f'_{baw,f}). \tag{15.10}$$

Taking nominal values for the intrinsic anti-resonant frequencies $f_{baw,\,o}$ and $f_{baw,\,f}$ at 2.4- and 2.32-GHz respectively, the sampling rate is 160-MHz with $m = 15$. The range of frequencies that can be addressed with this approach is,

$$f'_{baw,o} \cdot \left(1 + \frac{1}{N_h} - \frac{1}{2m}\right) \leq f_{ch} \leq f_{baw,o} \cdot \left(1 + \frac{1}{N_l} - \frac{1}{2m}\right). \tag{15.11}$$

Taking the minimum and maximum integer division ratios for the channel selection LO to be 15 and 30 respectively, the range of frequencies that can be addressed with this approach is 89.68-MHz which exceeds the bandwidth of the 2.4-GHz ISM band. This way, the proposed frequency synthesis without a PLL combines both the positive aspects of low phase noise and low power which is absent in other BAW based receivers [8, 13]. The BAW DCO and the integer dividers used are similar to those presented in [8] and for the purpose of brevity, the implementation details are not provided here.

15.5 Experimental Results

The proposed receiver is designed and integrated in 0.18-μm CMOS process. The BAW (FBAR) resonators are wire bonded to the chip as shown in Fig. 15.7. The complete receiver including the frequency synthesis consumes 5.94-mA current from a 1.8-V supply. The power breakdown is given in Table 15.1. The integer divider providing the clock for channel selection can divide from 12 to 36. To address specific frequencies in the 2.4-GHz ISM band, the anti-resonant frequency of the BAW resonators used in the DCO and in the filter have to be chosen accordingly.

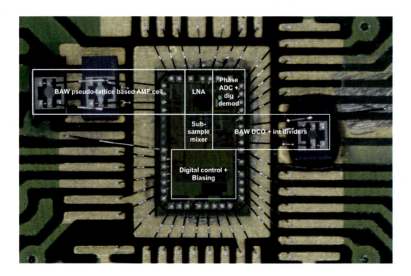

Fig. 15.7 Chip Photomicrograph

Table 15.1 Current consumption, $V_{DD} = 1.8$-V

Block	Current (mA)
LNA	1.5
BAW DCO	1.21
AMF cell	1.08
IF amplifier	4 ×0. 1
Dividers, buffers, ADC, demodulator	1.75
Total	5.94

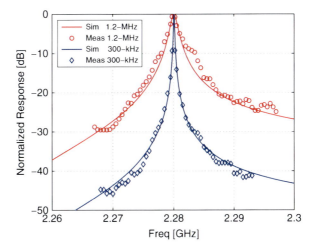

Fig. 15.8 BAW pseudo-lattice based AMF cell filtering response with different tuning settings. A maximum tuning of 8.1-MHz is observed for the AMF cell. The plot is normalized to 30.4-dB (LNA + AMF cell)

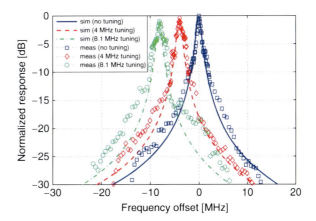

Fig. 15.9 Measured filtering response of the AMF cell with a single BAW resonator load

15.5.1 Filtering in the Receiver

The intrinsic Q of the BAW resonators used is around 400. With a single BAW resonator load (working at 2.28-GHz), the bandwidth of the AMF cell can be brought down to 300-kHz ($Q = 7,600$) as shown in Fig. 15.8. The frequency response of the BAW pseudo-lattice filter for different tuning settings is shown in Fig. 15.9. The AMF cell with the BAW pseudo-lattice load peaks at 2.433-GHz with no additional tuning capacitance. This Q is boosted by the AMF cell to provide bandwidths down to 1-MHz ($Q = 2,430$). The AMF cell consumes 1.08-mA with no additional capacitance and for the maximum tuning it increases by 0.35-mA. With the BAW pseudo-lattice load, the initial sampling rate is 68.5-MHz which is in accordance with (15.10), while the decimated sampling rate is at 17.12-MHz. The BAW DCO without any additional capacitance (for tuning), works at

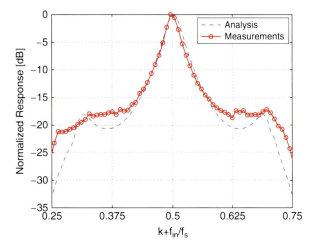

Fig. 15.10 Discrete time filtering by the sub-sampling mixer. The maximum measured conversion gain of the sub-sampling mixer is 17-dB

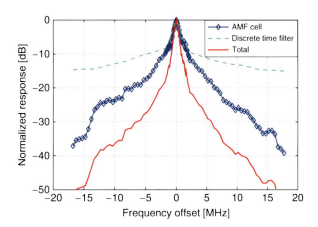

Fig. 15.11 Overall measured response (normalized to 46.2-dB) of the receiver

2.467-GHz and consumes 1.21-mA. A maximum tuning of 8.95-MHz is observed on the DCO and the current consumption for this maximum tuning is 1.74-mA.

The maximum conversion gain of the RF front-end including the LNA and the AMF cell (with the pseudo-lattice load) is 30.4-dB. The measured discrete time filtering by the sampling mixer matches well with the response obtained by the analytical model as shown in Fig. 15.10. The overall filtering response of the receiver including the AMF cell and the discrete time filtering is shown in Fig. 15.11. A rejection of around 50-dB is measured at an offset of 15-MHz from the center frequency. The measured conversion gain of the receiver from RF to baseband is 46.2-dB.

Owing to the mismatches between the mixing transistors in the AMF cell, the large-amplitude, low-frequency channel selection clock feeds through to the output and appears close to the wanted signal after down-conversion by sub-sampling. It should be noted here that the DC offset or flicker noise from the LNA and from the input

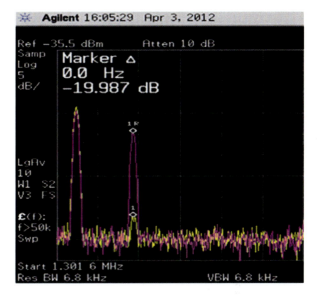

Fig. 15.12 LO (channel selection) feed-through owing to mismatch between the mixing transistors in the AMF cell (in *magenta*). By tuning the capacitors C_{ltc} this LO feed-through is attenuated by 20-dB (in *yellow*)

transistors of the AMF cell also get up-converted to the same frequency of the LO used for channel selection. The measured spectrum with this LO feed-through at the output of the sub-sampling mixer is shown in Fig. 15.12. It is detailed before that, the capacitances C_{ltc} can be tuned to place the null at the required frequency and this feature of the proposed pseudo-lattice is exploited to reject this LO signal as shown in Fig. 15.12. It can be seen that the wanted signal remains at the same level as before while the unwanted signal arising from LO feed-through is attenuated by 20-dB.

Linearity tests with a two tone input of ±500-kHz offset from the center frequency were conducted. The receiver exhibits an IIP3 of −38.75-dBm and IIP5 of −37-dBm at maximum gain setting. The measured quadrature between the I and Q paths is 93.4°. The receiver exhibits a global noise figure of 8.6-dB including the external balun.

15.5.2 Receiver Bit-Error Rate

The measured BER performance of the receiver for BFSK modulated signal with a modulation index of 0.7 is shown in Fig. 15.13. The data rate is at 268-kbps which corresponds to division of the initial sampling rate by 256. The phase ADC is clocked at 2.14-MHz, corresponding to an over-sampling ratio of 8. A sensitivity of −78-dBm is measured for a BER of 10^{-3}. The measured performance of the receiver is summarized in Table 15.2.

The receiver sensitivity for the measured noise figure has to be more than what is measured. The reason for the degrading of the sensitivity might be owing to the loading effect of the phase ADC which follows the sampling mixer. The mixer load

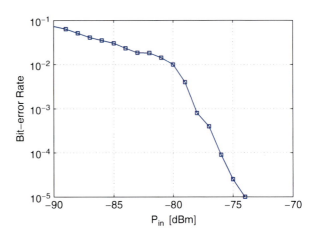

Fig. 15.13 BER measurements. A sensitivity of -78-dBm is observed for a BER of 10^{-3} and a data rate of 268-kbps

Table 15.2 Measured performance summary

Parameter	Value	Comment
BAW DCO tuning	8.96-MHz	@ 2.467-GHz (0.37 %)
AMF cell tuning	8.1-MHz	@ 2.433-GHz (0.34 %)
Filter bandwidth	1-MHz	with intrinsic Q of 400
Max conversion gain	46.2-dB	RF to baseband
I/Q phase	93.4 °	–
Noise figure	8.6-dB	@ Max gain
Linearity	-38.75- and -33-dBm	IIP3 and IIP5
Sensitivity	-78-dBm	BER of 10^{-3} and data rate of 268-kbps

capacitance might be small to drive the phase ADC, which results in attenuation of the down-converted signal. This can be readily improved by placing a buffer stage between the sampling mixer and the phase ADC. Further, by employing pulse shaping like in GFSK modulation, the sensitivity might get slightly improved as the bit transitions are smoother and any undesirable short duration spikes get filtered and the BER decreases at the given signal strength. The sensitivity of the receiver can be further improved by choosing higher modulation indices [10].

15.6 Conclusion

The BAW based receiver proposed in this work has been the first of its kind, with channel selection and filtering at RF followed by quadrature sub-sampling down-conversion mixer. The receiver exhibits low power consumption when compared with the other state-of-the-art multi-channel receivers. Further, the integer-N PLL-free approach for the frequency synthesis and the sub-sampling approach makes the receiver to be easily scalable from one technology node to another.

The use of a DCO and a series of integer dividers makes the receiver to be suitable for duty cycling [8, 10]. The digital words used for the DCO and the dividers can be stored in a memory and when the radio is woken up, the stored digital control can be read out and the radio is instantaneously ready to start receiving the data. The receiver has been validated to be a low power approach suitable for the WSN and WBAN applications.

References

1. M. Honkanen, A. Lappetelainen, K. Kivekas, Low end extension for bluetooth, in *Radio and Wireless Conference, 2004 IEEE*, Atlanta, Sept 2004, pp. 199–202
2. Medical Device Radiocommunications Service (MedRadio), FCC. Available: http://www.fcc.gov/encyclopedia/medical-device-radiocommunications-service-medradio
3. R. Krasinski, Medical body area networking (MBAN), IEEE 802.15.4j standardization, in *1st Invitational Workshop on Body Area Network Technology and Applications*, Worcester, June 2011
4. D. Yeager, F. Zhang, A. Zarrasvand, B. Otis, A 9.2 ua gen 2 compatible UHF RFID sensing tag with − 12 dbm sensitivity and 1.25 uVrms input-referred noise floor, in *IEEE International Solid-State Circuits Conference, ISSCC 2010, Digest of Technical Papers*, San Francisco, Feb 2010, pp. 52–53
5. R. Vullers, R. Schaijk, H. Visser, J. Penders, C. Hoof, Energy harvesting for autonomous wireless sensor networks. Solid State Circuits Mag. IEEE **2**(2), 29–38, Spring
6. S. Arms, C. Townsend, D. Churchill, J. Galbreath, B. Corneau, R. Ketcham, N. Phan, Energy harvesting, wireless, structural health monitoring and reporting system, in *Proceedings of the Asia-Pacific Workshop on SHM*, Melbourne, Dec 2008
7. C. Enz, J. Baborowski, J. Chabloz, M. Kucera, C. Muller, D. Ruffieux, N. Scolari, Ultra low-power MEMS-based radio for wireless sensor networks, in *18th European Conference on Circuit Theory and Design, ECCTD 2007*, Sevilla, Aug 2007, pp. 320–331
8. D. Ruffieux, M. Contaldo, J. Chabloz, C. Enz, Ultra low power and miniaturized MEMS-based radio for BAN and WSN applications, in *2010 Proceedings of the ESSCIRC*, Sevilla, Sept 2010, pp. 71–80
9. M. Contaldo, B. Banerjee, D. Ruffieux, J. Chabloz, E. Le Roux, C. Enz, A 2.4-GHz BAW-based transceiver for wireless body area networks. IEEE Trans. Biomed. Circuits Syst. **4**(6), 391–399 (2010)
10. J. Chabloz, A low-power 2.4 GHz CMOS receiver using BAW resonators, Ph.D. Dissertation, STI, Lausanne, 2008. Available: http://library.epfl.ch/theses/?nr=4141
11. B. Banerjee, C. Enz, E. Le Roux, A 290 ua, 3.2 mhz 4-bit phase adc for constant envelope, ultra-low power radio, in *NORCHIP, 2010*, Tampere, Nov 2010, pp. 1–4
12. J. Hu, R. Parkery, R. Ruby, B. Otis, A wide-tuning digitally controlled FBAR-based oscillator for frequency synthesis, in *2010 I.E. International Conference on Frequency Control Symposium (FCS)*, Shillong, June 2010, pp. 608–612
13. D. Ruffieux, J. Chabloz, M. Contaldo, C. Muller, F.-X. Pengg, P. Tortori, A. Vouilloz, P. Volet, C. Enz, A narrowband multi-channel 2.4 GHz MEMS-based transceiver. IEEE J. Solid State Circuits **44**(1), 228–239 (2009)
14. A. Heragu, V. Balasubramanian, C. Enz, A concurrent quadrature sub-sampling mixer for multiband receivers, in *European Conference on Circuit Theory and Design, ECCTD 2009*, Antalya, Aug 2009, pp. 271–274
15. G. Temes, J. LaPatra, Introduction to circuit synthesis and design (McGraw-Hill, 1977). Available: http://books.google.ch/books?id=fPIiAAAAMAAJ

16. A. Heragu, D. Ruffieux, C. Enz, A 2.4-GHz MEMS-based PLL-free multi-channel receiver with channel filtering at RF, in *Proceedings of the ESSCIRC (ESSCIRC), 2012*, Bordeaux, Sept 2012, pp. 137–140
17. A. Heragu, D. Ruffieux, C. Enz, A 2.4 GHz MEMS based sub-sampling receiver front-end with low power channel selection filtering at RF, in *Radio Frequency Integrated Circuits Symposium (RFIC), 2012 IEEE*, Montreal, June 2012, pp. 257–260
18. M.-A. Dubois, J.-F. Carpentier, P. Vincent, C. Billard, G. Parat, C. Muller, P. Ancey, P. Conti, Monolithic above-IC resonator technology for integrated architectures in mobile and wireless communication. IEEE J. Solid State Circuits **41**(1), 7–16 (2006)

Chapter 16
mm-Wave Silicon: Smarter, Faster, and Cheaper Communication and Imaging

Ali M. Niknejad, Amin Arbabian, Steven Callender, JiaShu Chen, Jun-Chau Chien, Shinwon Kang, Jungdong Park, and Siva Thyagarajan

Abstract This paper will highlight three mm-wave integrated circuit systems appropriate for communication and imaging. The first chip is an efficient transmitter for the realization of a mm-wave system using digital modulation and spatial quadrature power combining. The second system is a prototype 260 GHz short range chip-to-chip communication system in CMOS using on-chip antennas. The final system is a 3D imager with a 90 GHz carrier and 25-parsec pulse width, potentially applicable for HCI (gesture recognition) and medical imaging. These systems represent new application domains for mm-wave electronics where the high volume/low cost of silicon technology can be exploited to realize new functionalities and data rates, addressing important hurdles in the expansion of our communication networks and opening up the ability to see objects in a completely new way using 3D mm-wave imaging.

16.1 Introduction

The past decade has witnessed much growth in interest and research in the mm-wave frequency bands, in particular the 60 GHz bands. Several standards, including WiGig, 802.11ad, WiHD, and even few products [1], enable high data rate communication in the 60 GHz band with much higher energy efficiency compared to today's radios. To put this into perspective, imagine a relatively

---

A.M. Niknejad (✉)
Berkeley Wireless Research Center, University of California, Berkeley CA, USA
e-mail: niknejad@eecs.berkeley.edu

A. Arbabian
Stanford University, Stanford Berkeley Wireless Research Center, Berkeley CA, USA

S. Callender • J. Chen •
J.-C. Chien • S. Kang • J. Park • S. Thyagarajan
Berkeley Wireless Research Center, Berkeley, CA, USA

A. Baschirotto et al. (eds.), *Frequency References, Power Management for SoC, and Smart Wireless Interfaces: Advances in Analog Circuit Design 2013*, DOI 10.1007/978-3-319-01080-9_16, © Springer International Publishing Switzerland 2014

short link, 1–10 m in range, and a 802.11 g radio operating at 54 Mbps, compared to a 60 GHz radio operating at 7 Gbps. From a power consumption perspective, both would require $\sim$ 200 mW of power for the RF PHY/MAC circuitry, which means that the 60 GHz radio is more than 100 ×more efficient. While an 802.11 ac radio has a throughput of 1 Gbps, it comes at the cost of channel bonding (160 MHz) and spatial multiplexing (MIMO), which drives the power to > 1 W. In contrast, a 60 GHz radio can also do channel bonding and spatial multiplexing to achieve in excess of 28 Gbps, which makes it at least 30 ×more energy efficient.

Given the successful demonstrations of 60 GHz circuitry and systems by academic and industrial organizations, one can see a clear path for commercialization in years to come. As the need for bandwidth quickly exceeds the supply below 6 GHz carriers, consumer demand will drive the market towards 60 GHz. What other applications can one envision for mm-wave communication?

Another bandwidth bottleneck is the 4G mobile spectrum, not only for mobile users but for the backhaul network. To meet the demand for mobile traffic, cell sizes are shrinking, which means the cost, power consumption, and footprint of base-stations has to follow. If a wireless mm-wave back-haul link is used, then a base-station only needs power and physical space, which makes it easier to deploy on rooftops and other areas without a wired connection to the back-bone network.

At the other spectrum of range we can envision other exciting applications for mm-wave communication, namely in the domain of chip-to-chip or centimeter range very high data rates (30–100 Gbps). If such communication circuits can be realized in CMOS with small footprints, then it is possible that for many applications one can build wireless links between chips. It may seem more efficient to simply use a wire or trace on a PCB for such a short distance, but many applications are space and form factor constrained, and a wireless link is preferred. In fact, if such a link were possible, one can dream of building complete systems as "lego" building blocks, with power and ground connections snapping IC's onto a simple 1 or 2 layer PCB. The chips would them form a nano-network and communicate with each other.

Imaging at mm-wave frequencies is a well known application, especially for observation of interstellar phenomena in space sciences. Since the wavelength approaches 3 mm at 100 GHz, images at this frequency are not sharp as optical images, but have enough resolution to show the bulk features of an object. For example, mm-wave imaging is used for security applications to detect concealed objects at airports. This imaging is favored over X-ray systems since the radiation is non-ionizing. In fact, the black body emission from a person can be used directly without an illumination source to directly image a scene in a passive manner. Other emerging applications for this imaging include human-computer interaction (HCI) using gestures, intelligent surfaces that can detect the presence of human and non-human occupants in a room, gaming, and medical imaging. Here we will describe a 90 GHz time-domain pulsed imaging system that is appropriate for 3D imaging.

16.2 mm-Wave Communication

From a research perspective, most of the challenges related to the 60 GHz RF and baseband circuitry have been resolved, even using inexpensive CMOS technology. Transceivers in started to appear in 130 nm CMOS [2], and today 65 nm seems like an optimal operating frequency to realize good performance [3]. To overcome the high path loss of 60 GHz, phased-arrays (a larger aperture) are used, which comes with advantages and disadvantages. Due to the directionality of the antenna, tracking algorithms are required to track the angle of an incoming signal (beam steering) so that the maximum SNR can be realized. A side benefit, though, is that unwanted signals from other incoming angles are attenuated by the antenna pattern, and in theory a given direction can be nulled out (null steering), although in practice the attenuation is limited by the array phased resolution ($\sim$30 dB of nulls for $\sim$10°). Another side benefit is that the unwanted multi-path propagation paths from source to destination are also attenuated, reducing the delay spread of the signal. This is particularly important for high data rate communication since even short delays can lead to inter-symbol interference (ISI). This simplifies the design of the baseband equalizer, which can easily be as power hungry as the RF circuitry for high data rates ($>$10 Gbps).

While much attention has focused on short and medium range communication in the 60 GHz band, relatively little research has been done on realizing a kilometer range link using CMOS technology. As alluded to earlier, one of the main challenges is to raise the average efficiency of transmitter.

16.2.1 Power Generation via Digital Quadrature Spatial Combining

The demands on the backhaul are very stringent, in excess of 2 Gbps communication at a range of a kilometer, and the signal must contend with outdoor conditions, namely heavy rain which can result in an additional 31 dB/km of attenuation in the E-band ($\sim$80 GHz) under heavy rain.

To realize such a long range link, a very high gain antenna is needed. Given the sharp pencil beam that would emerge, antenna steering accuracy and mechanical stability make it desirable to use a phased array instead. An array with hundreds or even thousands of elements can be utilized to reach these distances. While possible, the efficiency of the array is likely to be small, mostly due to the power amplifier. To realize high data rates, complex modulation schemes are needed, which require back-off from peak power in the power amplifier, and the overall efficiency suffers greatly as a result. Here we propose techniques to improve the efficiency of the transmitter chain.

For example, recent demonstration of phased arrays at 60 GHz in CMOS and SiGe all realize transmitter efficiencies of less than 5 % [3, 4]. The main bottleneck

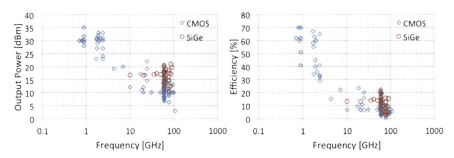

Fig. 16.1 Output power and efficiency of published CMOS/SiGe PAs in the literature

is the low efficiency of the CMOS PA. As shown in Fig. 16.1, published mm-wave PAs have lower efficiencies, typically below 20 %. Compared to a 5 GHz PA, the efficiency is lower due to the lower available gain, which means that the pre-PA stage is also high power. The lower gain ($G_p \approx f_{max}/f$) is a result of operating close to the activity limits of the technology. While scaling technology helps, the f_{max} of the process is not improving in step with the scaling due to increased resistive parasitics in the device (gate resistance, source/drain resistance). Most published PAs are linear Class A stages, since biasing a transistor closer to cut-off results in even less gain. The capacitive parasitics of a transistor also limit the maximum device size, which requires power combining, which is lossy on a silicon substrate, about 1-dB of loss for a two way combiner. In switching amplifiers this is a more severe limit, since a large device is needed to minimize the on-resistance of the switch.

The AM-AM and efficiency characteristics of a 60 GHz CMOS PA are shown in Fig. 16.2 [5]. Given the low peak efficiency, one is tempted to use the simplest modulation schemes that do not require much back-off, such as BPSK, to maintain peak efficiency. But due to the low spectral efficiency of BPSK, to realize high data rates with limited bandwidth (say in the licensed E-bands), higher order modulation will be used (e.g. 64-QAM). For example, with a 6-dB back-off from peak, the resulting efficiency of the PA is only 2 %, which is the upper bound for the efficiency of the entire system. This low efficiency is partly a result of the quadratic efficiency drop typical of a Class-A PA. Class B PA's have a more gradual linear drop-off in efficiency.

In our research we have focused on improving both the peak and average efficiency of a mm-wave PA. Polar [6], Cartesian, and Outphasing [7] transmitter architectures are popular ways to realize high efficiency and linearity at lower frequencies (Fig. 16.3). One of the disadvantages of the polar architecture is the phase and amplitude path should be matched. Moreover, the phase path experience bandwidth expansion due to the arctangent non-linearity, requiring approximately 7 × more bandwidth in the phase modulation path. For applications requiring several gigahertz of bandwidth, this is prohibitive and so a Cartesian architecture is preferred.

16 mm-Wave Silicon: Smarter, Faster, and Cheaper Communication and Imaging

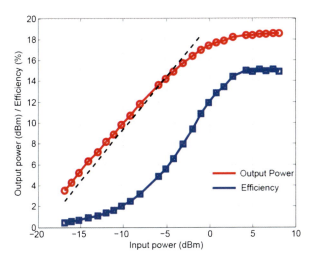

Fig. 16.2 AM-AM compression characteristics and efficiency of a mm-wave CMOS PA

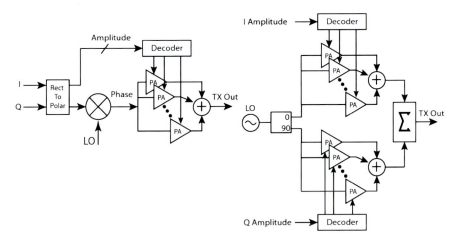

Fig. 16.3 Cartesian and polar digital PA architectures

In a Cartesian architecture the in-phase (I) and quadtrature-phase (Q) signals are amplified separately and combined. Each PA can be realized as an "RF-DAC", where each unit PA element is realized as a constant envelope switching amplifier. Combing a large number of unit elements, though, is problematic. The capacitive parasitics cause excessive loading. Moreover, any interaction between the I and Q paths leads to undesired distortion. Unlike a normal PA where the AM-AM distortion can be corrected with a simple lookup table (assuming no memory in the non-linearity), an I/Q PA with interaction requires a 2D table since the Q (I) PA

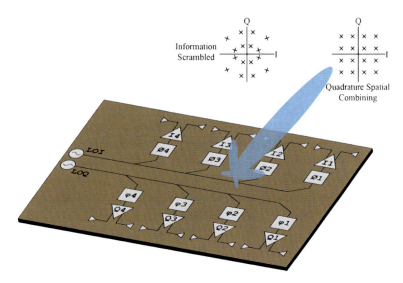

Fig. 16.4 Proposed power combining and modulation using quadrature spatial digital arrays

can drive the I (Q) PA into compression earlier than if it were operating in isolation. An isolated combiner solves this problem but incurs a loss penalty.

The proposed solution, show conceptually in Fig. 16.4, employs quadrature spatial power combining and digital modulation. The spatial combining of the signals alleviates both the number of units connected together and the unwanted interaction between the I and Q signals. Each element can also be phase shifted to produce a phased array, so the beam of the array can point to different directions. Unlike a normal phased array, the useable antenna beamwidth is determined not by only the antenna, but also by the requirement of orthogonality between the and Q patterns, which occurs over a narrow "Information Beamwidth", as shown in Fig. 16.5. The beamwidth is in turn a function of the spacing between the I and Q arrays, which are spaced apart to minimize I/Q coupling.

A prototype has been designed and fabricated to demonstrate the idea at 60 GHz [8]. The block diagram in Fig. 16.6 shows 4 I and 4 Q channels comprised of a Class E/F$_2$ switching amplifier, driven by a baseband phase rotator. Each amplifier is further divided into an array of devices sized in a non-uniform fashion to pre-distort for the compression characteristics. Elements are switched into and out of the circuit using a source side switch, which results in linear back-off characteristics.

The digital modulation is performed by the DSP unit, which takes an oversampled version of the I/Q bits (4 ×over-sampling), filters them using digital FIR filters to attenuate the spectral clock image. The outputs are fed into the digital PA using an early and late branch (generated from a synchronous delay), so the data at the output ramps similar to a first-order interpolation. Together with the FIR filters, more than 40 dB of attenuation is observed for the first clock image while consuming 20 mW of power.

16 mm-Wave Silicon: Smarter, Faster, and Cheaper Communication and Imaging

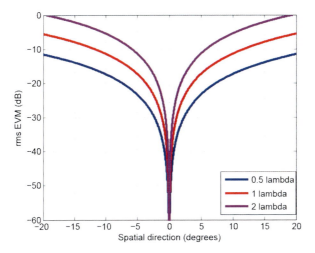

Fig. 16.5 Information beam width of a quadrature spatial array as a function of spacing between I/Q array elements

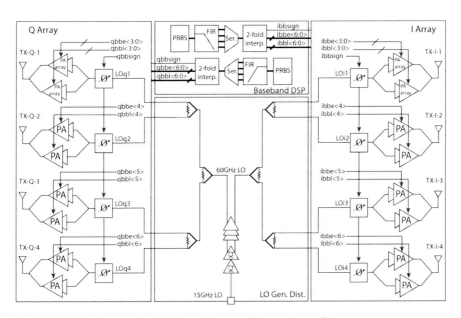

Fig. 16.6 Block diagram of a digital quadrature spatial array transmitter

Each PA stage uses Class E/F_2 tuning to maximize the efficiency and optimize the waveforms. Second harmonic tuning improves the tolerance to parasitic capacitance at the drain of the differential amplifier. To avoid using a second LC tuning network, the second harmonic tuning is realized using the common-mode

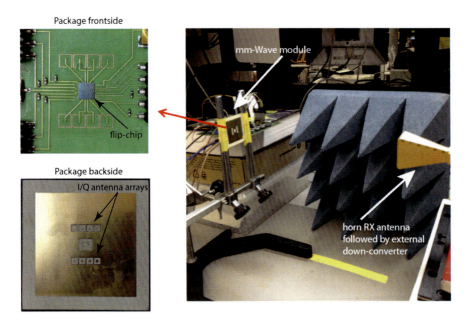

Fig. 16.7 Testing setup for the measurement of the prototype digital quadrature spatial array

impedance of a 2:1 transformer. The transformer has the triple functionality of impedance matching, differential-to-single ended conversion, and also the ability to resonate the second harmonic. Since the second harmonic signal is a common mode excitation to the transformer, the effective inductance is much lower (currents flow into center tap), and thus the transformer size is optimized to realize this dual resonance. The transformer has a simulated loss of 1.2 dB, and the PA has the ability to drive 11 dBm into the antenna with an efficiency of 46 % while realizing 10-dB of power gain. Taking the transformer loss into account, the peak efficiency is 35 %.

A test setup is shown in Fig. 16.7 where the die is packaged on a Rogers board with 4 I + 4 Q antennas. Measurements of a stand-alone (probed) unit element confirm the performance, realizing a peak efficiency of 30 % (simulated 35 %) and a peak power of 10 dBm, very close to simulations. The AM-AM characteristics are also very linear, confirming the effectiveness of the pre-distortion circuitry. Measurements of the 4 + 4 packaged transmitter constellation (off the air measurements) and the transmitter beam width are shown in Fig. 16.8. The overall efficiency of the entire transmitter, including baseband and the LO paths, is better than 17 %. Even in a 6-dB back-off mode, supporting a peak data rate of 6 Gb/s with 16-QAM modulation, the measured average transmitter efficiency is 7 %. To put this number into perspective, recall that for a single PA the 6-dB back-off efficiency was 2 %, so an entire transmitter built with the same PA would be likely to have a much lower efficiency.

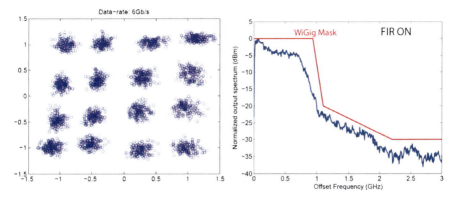

Fig. 16.8 Measured 16-QAM constellation and transmitter mask for prototype transmitter

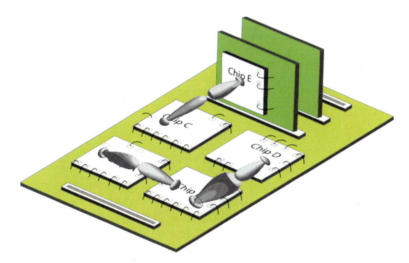

Fig. 16.9 Conceptual chip-to-chip link using a directional mm-wave carrier

16.2.2 Chip-to-Chip Communication

Almost every integrated circuit requires a digital interface for transporting data into and out of a chip sub-system. The data rates for these interfaces are increasing, driven by the need for moving high quality video and large amounts of date to off-chip memory storage. At the same time, many of these chips are in area constrained environments (such as inside of a handheld mobile phone), and it is desirable to eliminate as much as the physical wiring as possible.

A wireless chip-to-chip link, as shown conceptually in Fig. 16.9, would not only solve this problem, but enable new applications and new form factors. For example,

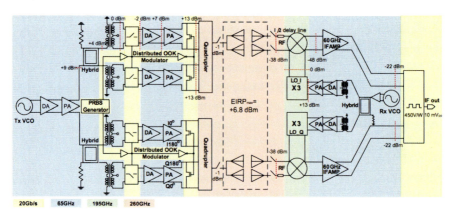

Fig. 16.10 Block diagram of a prototype 260 GHz transceiver with on-chip antenna

one can imagine building an entire system out of chips that only need power and ground connections, with all information transfer occurring through the air. This helps to lower the cost of PCB and minimize the need for materials on the PCB. A paper thin PCB made from flexible substrates could house thin integrated circuit packages into a low cost "disposable" tablet computer.

To realize such short range links, we need an energy efficient radio (wires consume < 5 pJ/bit for short range links) that also has a small physical footprint. No matter how small we make the radio itself, the footprint will be dominated by the antenna, which is on the order of the size of the wavelength at the carrier frequency. The only way to achieve a reasonable footprint and to avoid high cost packaging is to scale the frequency > 100 GHz and to eliminate all costly (and lossy) off-chip connections. On chip antennas are thus an ideal candidate for such a radio [9].

Our team has demonstrated such a radio operating at 260 GHz using 65 nm CMOS technology [10]. The antennas are realized using leaky wave antennas, which are broadband and can support very high data rates. One of the key limitations to operating at such high frequencies is that it's beyond the activity limits (f_{max}) of the technology. Even with technology scaling down to 28 nm, we are still very close to activity limits and so power gain and power generation are key challenges. These issues are solved by using frequency multiplication in the transmitter and a mixer-first architecture in the receiver, as shown in the block diagram, Fig. 16.10. By operating the core transmitter at ~ 60 GHz, very high power can be generated to overcome the losses of frequency multiplication. Four carriers at 0°, 90°, 180°, and 270° are generated and combined in a quadrature-push architecture [11, 12], generating 4 ×the carrier frequency before the antenna while rejecting the fundamental. The antenna is fed from one end while the signal is received at the other end, thus utilizing the antenna as a T-R switch without incurring the penalty of a physical switch. At the receiver, a 67 GHz VCO output is tripled to 201 GHz to downconvert the signal from 260 GHz to a baseband of 59 GHz. The 59 GHz IF signal is amplified and the amplitude is detected (on-off keying, OOK) directly.

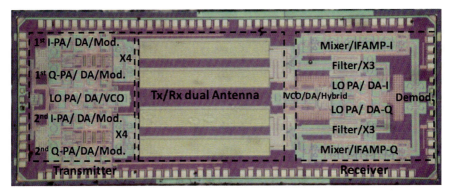

Fig. 16.11 Die photo of prototype 260 GHz transceiver

The die photo of the prototype is shown in Fig. 16.11, with a footprint of 4 × 1.5 mm. The T ×consumes 688 mW and the R ×consumes 488 mW of power. The measured transmitter EIRP is + 5 dBm, close to the expected power. Non-coherent OOK modulation up to 14 Gb/s is verified from a toggling signal captured over the air. A complete link at 4 cm using two identical chips using a toggling signal is verified up to 10 Gbps. This chip was the first demonstration of a complete system realized in CMOS with a carrier frequency over 200 GHz. Only an off-chip reference signal clock is needed in the measurements. While only a first step, this clearly shows that there is a great opportunity to exploit frequencies above 100 GHz for new and exciting applications. We believe that both the power consumption and data rate can be improved dramatically, approaching 10 pJ/bit, which is competitive with wired links. This will only be realistic if the power generation efficiency and the sensitivity can be improved, goals that can drive research in the coming years.

16.3 mm-Wave Imaging

Using mm-waves for imaging has many new and relevant applications, including gaming, gesture recognition (HCI), and even medical imaging and tumor detection [13]. Our research has focused on building a 3D imaging system, shown schematically in Fig. 16.12. By exploiting both the high spatial resolution of a 90 GHz carrier and the timing resolution of silicon circuitry, we can realize a 3D imaging system. By transmitting a short pulse (25 parsec), one can build a radar to image the depth (z-direction) by observing the time of arrival of the reflected pulse. A timed-array, similar to a phased array, can also focus the energy in a given direction on the xy-plane by timing the pulses to arrive coherently on a given point. Thus a voxel in 3-dimensional space can be illuminated and imaged using such a system.

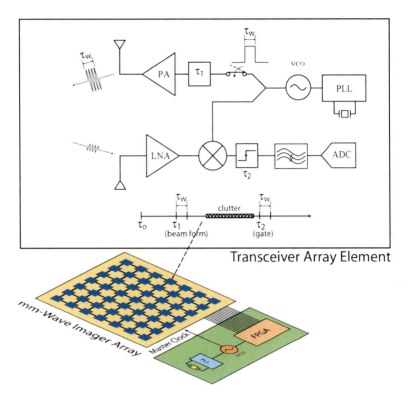

Fig. 16.12 Block diagram of a 3D imager

The key elements in this camera are a coherent and high power 90 GHz source, the ability to gate the source into a short pulse, to control the coherence between the pulse envelope and the carrier, and the ability to control the pulse time of flight and phase with picosecond resolution. The need for both time and phase coherency can be understood as follows. To image a given point on the x y-plane, adjacent array elements need to be delayed by the difference in time of flight from source to target, on the order of picoseconds for imaging an object at a few centimeters with millimeter resolution. Likewise, to steer the beam over a volume, the absolute delay has to vary about 30 parsec for every centimeter of dimension. The need to vary the delay both in fine time steps and large time steps is solved by using a DLL with both variable size and interpolation [14]. By interpolating between elements within the DLL, fine delay accuracy of 2.85 parsec (worst case) was measured. By changing the size of the DLL or by interpolating from more distant neighbors, the delay can be increased, in our prototype up to 365 parsec. Coherency between the pulse envelope and carrier is achieved by locking the DLL and PLL to a common reference. To coherently combine signals at the target, a phase shifter would also be needed to adjust the phase of the carrier appropriately.

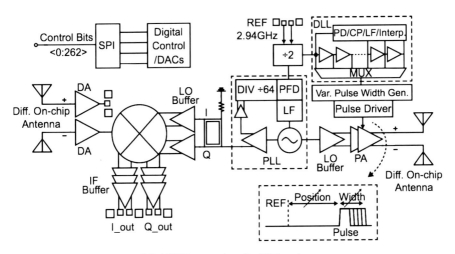

Fig. 16.13 Block diagram of the TUSI transceiver for 3D imaging

To generate a high power switched carrier, a continuous carrier is switched using differential steering at the output of the PA, which obviates the need for high bandwidth circuitry. At the receiver, a broadband (> 110 GHz) traveling wave amplifier is used in conjunction with broadband I/Q mixers (BW >25 GHz). The complete Timed-array Ultra-wideband Silicon Imager (TUSI) chip, shown in Fig. 16.13, implements most of the transceiver, including on-chip Rx and Tx antennas, PLL, DLL, PA, pulser, LNA, mixer, LO and IF stages [15]. The chip is realized in a 130 nm SiGe process and measures 1.4 × 4.4 mm. Complete loop-back testing using a reflector confirm the functionality as a radar imager, with the capability to distinguish two objects less than 6 mm apart. Using an external down-converter, the transmitted pulse waveform is captured and displayed on an oscilloscope and spectrum analyzer (Fig. 16.14). Coherency is verified by time averaging pulses and observing an increase in pulse SNR. PLL locking is observed on the spectrum analyzer by analyzing the sinc comb shape with sharp spectral peaks at the pulse repetition frequency (PRF). The chip is also used as an interferometer to detect distances less than 300 μm (limited by mechanical resolution and not electronics) by observing the phase of the received signal.

The next step is to realize a broadband baseband data acquisition system to build a complete RF-to-digital pixel, appropriate for building a timed array camera. Each pixel only needs to be programmed with the appropriate delay and PRF reference signal, thus simplifying the design of a large array. Each pixel element can perform time averaging to improve the SNR of the signal and in unison the scene can be imaged efficiently using multiple receivers.

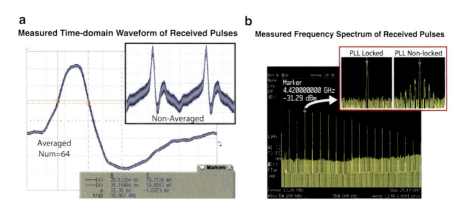

Fig. 16.14 (a) Measured down-converted pulse after a round trip through the transceiver. (b) Measured spectrum of the pulses shows the characteristic comb spectral lines at the PRF

16.4 Conclusion

This paper has highlighted exciting new application for mm-wave silicon electronics, including large range high data rate backhaul for 4G/5G base-stations, chip-to-chip communication using sub-THz frequencies, and 3D imaging for medical, gaming, and smart surfaces applications. These applications together help us to build a smarter and more flexible world of objects, capable of directional high data rate communication and imaging. When these devices are realized in CMOS in large volumes, they are inexpensive and can be deployed to make smart surfaces (walls, tables, automobiles), capable of not only streaming multi-gigabit per second data streams to the users, but even detecting the presence and perhaps even the health and identity of the occupants.

References

1. Emami et al., A 60 GHz CMOS phased-array transceiver pair for multi-Gb/s wireless communications, in *ISSCC Digest of Technical Papers*, San Francisco, Feb 2011, pp. 164–165
2. S. Emami, C. Doan, A.M. Niknejad, R. Brodersen, A highly integrated 60 GHz CMOS front-end receiver, in *ISSCC Digest of Technical Papers*, San Francisco, Feb 2007, pp. 190–191
3. M. Tabesh, J. Chen, C. Marcu, L.-K. Kong, E. Alon, and A.M. Niknejad, A 65 nm CMOS 4-channel sub-34mW/channel 60 GHz phased array transceiver, in *ISSCC Digest of Technical Papers*, San Francisco, Feb 2007, pp. 166–167
4. Valdes-Garcia et al., A fully integrated 16-element phased-array transmitter in SiGe BiCMOS for 60-GHz communications. IEEE J. Solid State Circuits **45**, 2757–2773 (2010)
5. J. Chen, A.M. Niknejad, A compact 1V 18.6 dBm 60GHz power amplifier in 65 nm CMOS, in *ISSCC Digest of Technical Papers*, San Francisco, Feb 2011, pp. 432–433

6. D. Chowdhury, Y. Lu, E. Alon, A.M. Niknejad, A 2.4 GHz mixed-signal polar power amplifier with low-power integrated filtering in 65 nm CMOS, in *IEEE Custom Integrated Circuits Conference (CICC)*, San Jose, 2010
7. Xu et al., A flip-chip-packaged 25.3 dBm class-D outphasing power amplifier in 32 nm CMOS for WLAN application. IEEE J. Solid State Circuits **46**, 1596–1605 (2011)
8. J. Chen et al., A digitally modulated mm-wave cartesian beamforming transmitter with quadrature spatial combining, *ISSCC Digest of Technical Papers*, San Francisco, Feb 2013, pp. 232–233
9. Floyd et al., Intra-chip wireless interconnect for clock distribution implemented with integrated antennas, receivers, and transmitters. IEEE J. Solid State Circuits **37**, 543–552 (2002)
10. J.-D. Park, S. Kang, S.V. Thyagarajan, E. Alon, A.M. Niknejad, A 260 GHz fully integrated CMOS transceiver for wireless chip-to-chip communication, in *Symposium on VLSI Circuits (VLSIC)*, Honolulu, 2012, pp. 48–49.
11. D. Huang, T.R. LaRocca, M.C.F. Chang, L. Samoska, A. Fung, R.L. Campbell, M. Andrews, Terahertz CMOS frequency generator using linear superposition technique. IEEE J. Solid State Circuits **43**, 2730–2738 (2008)
12. J.-D. Park, S. Kang, A.M. Niknejad, A 0.38 THz fully integrated transceiver utilizing a quadrature push-push harmonic circuitry in SiGe BiCMOS. IEEE J. Solid State Circuits **47**, 2344–2354 (2012)
13. X. Li, S. Hagness, A confocal microwave imaging algorithm for breast cancer detection. IEEE Microw. Wirel. Compon. Lett. **11**(3), pp. 130–132 (2001)
14. S. Callender, A. Niknejad, A phase-adjustable delay-locked loop utilizing embedded phase interpolation, in *2011 RFIC Symposium*, Baltimore, June 2011
15. A. Arbabian, S. Kang, S. Callender, J.-C. Chien, B. Afshar, A. Niknejad, A 94 GHz mm-wave to baseband pulsed-radar for imaging and gesture recognition, *Symposium on VLSI Circuits (VLSIC)*, Honolulu, 2012, pp. 56–57

Chapter 17
An IEEE 802.15.4A Ultra-Wideband Transceiver for Real Time Localisation and Wireless Sensor Networks

Dries Neirynck

Abstract The combination of wireless sensor networks and real-time localisation systems can enable many new exciting applications. Ultra-wideband is the natural choice for such location aware wireless sensor networks. DecaWave's ScenSor IC implements the IEEE 802.15.4a standard to meet the needs of WSN and RTLS manufacturers. This chapter discusses the design of the ScenSor chip.

17.1 Introduction

Starting with Marconi's original spark gap transmission, the earliest wireless communications could be seen as impulse radio ultra-wideband (UWB). Later, this was abandoned in favour of band-limited, carrier-based modulation which allowed a more spectral efficient sharing of the ether. Much of the following was a quest for more spectral efficiency, squeezing ever higher data rates in narrow bandwidths.

Since the turn of the century though, spurred by a change in the FCC regulations [1], UWB has made a comeback as an underlay technology. Initially, most of the renewed interest was sparked by the high bandwidth, allowing the pursuit for ever higher data rates to continue. While very high data rate UWB seems to be struggling in the market, UWB is still alive for lower data rate applications. Here, commercial interest is particularly driven by its inherent robustness to multipath fading and its potential for high precision localisation.

In this chapter, the design of an IEEE 802.15.4a [2] compliant impulse radio UWB transceiver for real-time localisation systems (RTLS) and wireless sensor networks (WSN) is described. In the next section, WSN and RTLS are described and the corresponding system requirements highlighted. Section 17.3 explains the

---

D. Neirynck (✉)
DecaWave Ltd, Adelaide Chambers, Peter Street, Dublin 8, Ireland
e-mail: dries.neirynck@decawave.com

A. Baschirotto et al. (eds.), *Frequency References, Power Management for SoC,
and Smart Wireless Interfaces: Advances in Analog Circuit Design 2013*,
DOI 10.1007/978-3-319-01080-9_17, © Springer International Publishing Switzerland 2014

motivation for choosing UWB for such systems. Section 17.4 describes the main, relevant features of the IEEE 802.15.4a standard. The design of the ScenSor chip based on this standard is discussed in Sect. 17.5.

17.2 Context and Background

Wireless sensor networks (WSN) are a relatively well-known concept, where a collection of sensors wirelessly connect to form a network that can pass the information from one node to another. Personal and body area networks, which wirelessly connect a number of devices on or around a person, can be seen as examples of WSN. Other applications, for example in industrial monitoring or agriculture, can have a much larger number of sensors.

One example of a wireless sensor network is electronic shelf labelling. When the management of a large supermarket chain decides to change the price of their products, this used to involve informing every branch, where an employee had to go to replace the shelf labels. With electronic shelf labelling, the paper labels are replaced by small displays wirelessly connected to the supermarket's intranet. Now, price information can be set in the head office and will be automatically updated at all the branches.

Real time localisation systems (RTLS) augment sensor networks with an awareness of the location of the sensor nodes. At its most basic, systems can flag when items are near a particular point. For example, when an animal with an RFID tag passes through a gate, the RFID reader at the gate could register the identity of the animal. The Global positioning system (GPS) is a much more advanced example, where each node is able to calculate its own position. While GPS works well outside, indoors its accuracy becomes very poor because the direct link to the satellites is often attenuated or even lost.

A number of techniques can be used to estimate a transceiver's location. A simple technique is to look at the strength of the incoming signal. However, obstructions and multipath propagation can have a much greater influence on the signal strength than the actual distance between the devices, so the accuracy of signal strength based ranging is poor. Much better distance estimates can be achieved by systems that are based on the time of flight.

As an example, consider the group of fire fighters who have been called to a fire in a block of flats and need to search through the building for people trapped inside. Suddenly, one of the fire fighters is knocked unconscious by a falling beam. Luckily, the radios the fire fighters use for communications are able to form a network and nodes are able to establish their location relative to other nodes in the network. This way, the other fire fighters are aware of their injured colleague's position and can evacuate him from the building.

In a security application, the potential to measure the time of flight can be used to prevent relay attacks in wireless authentication systems [3]. Even though the relay enables the attacker to answer all higher layer security challenges correctly, the extra delay in the relay channel allows the system to detect that the authenticated user is not near the access point. However, in order to be able to do this, the physical layer must be able to determine the communications range within decimetre level.

As the example of electronic shelf labelling demonstrates, low cost, small form factor and long battery life are crucial for the success of WSN. From the perspective of the wireless transceiver, the need to support a potentially very large number of nodes and the requirement to ensure reliable communications even in rich multipath environment are two additional challenges.

In the example of the fire fighters, the ability to communicate and determine the range within a non-line-of-sight, multipath environment is also crucial. Note that the capability of the radios to form their own network and to determine their relative locations is crucial in this scenario since buildings can't be counted upon to have RTLS infrastructure in place.

17.3 The Case for UWB

The recent interest in UWB can be traced back to an FCC 2002 Report and Order [1] that authorised the unlicensed use of the frequency range between 3.1 and 10.6 GHz for UWB transmissions. In the document, UWB is defined as a transmission with a fractional bandwidth above 20 % or an absolute bandwidth of at least 500 MHz. However, since most of the spectrum between 3.1 and 10.6 GHz was already licensed to specific users, the transmit power was restricted to -41.3 dBm/MHz for most scenarios.

Many of the advantages and disadvantages of UWB can be related back to three characteristics that are contained in the FCC Report and Order:

- Large bandwidth;
- Limited transmit power;
- Unlicensed operation

By 2002, the power of unlicensed operation had been amply demonstrated, most notably by the commercial success of both Bluetooth and Wi-Fi. However, whereas Bluetooth and Wi-Fi operate in separate unlicensed bands, the spectrum made available to UWB was already occupied. The limit on transmit power was required to ensure the acceptance by licensed spectrum users. While the transmit levels allowed are well below the noise floor, many remained hostile towards intentional interference in their bands. The transmit level of -41.3 dBm/MHz corresponds to the level of unintentional impulsive noise those licensed user already had to tolerate.

Since the UWB transmission has to be low power, it is attractive to applications where lower power consumption and long battery life are essential. Many wireless sensor networks consist of an asymmetric topology where large numbers of transmit-only end-nodes pass information to a few central receiving nodes. Even if the end-nodes are not transmit only, most WSN communication will originate from the sensor node, which sends a packet to a central hub and then listens for a short time to get an acknowledgement.

By design, UWB minimises interference to other systems, avoiding coexistence issues with other technologies. Since the transmission takes place below the noise level, they are very hard to detect, which is an advantage in applications where secrecy and security are valued.

Shannon's theorem [4], which relates the maximum error-free communication rate possible in a channel to its bandwidth and the signal-to-noise ratio, explains part of the excitement about the large bandwidth. The amount of information, noted as the channel capacity C, that can be reliably transmitted over a channel with bandwidth B, is given by Shannon's theorem as

$$C = B log_2 \left(1 + \frac{S}{N} \right)$$

where S/N is the signal-to-noise ratio. The relationship between bandwidth and throughput potential is linear, whereas increases in transmit power, and therefore the signal-to-noise ratio, only logarithmically correspond to increased data rates. Hence, UWB provides a power efficient way to serve high throughput applications.

Because the bandwidth of UWB communications is so large, the constructive and destructive phase additions that lead to multipath fading in narrowband systems average out and the received power is far more constant [5]. It can therefore provide reliable communications in highly reflective environments, e.g. industrial complexes with lots of metal constructions and machinery.

17.3.1 UWB Technologies

In its Report and Order [1], the FCC only defined UWB in terms of bandwidth and emission limits. This has led to a wide variety of systems being investigated.

Historically, UWB was often synonymous with pulse position modulation. Particularly the idea of generating very short pulses that could be fed directly to the antenna seemed attractive since they didn't require modulation on a carrier. However, the difficulty to find pulse shapes that comply with the FCC's strict spectral emission mask seems to have put an end to this carrier-less variety of UWB.

Therefore, practically all UWB systems currently considered are carrier based. The required modulation with a sinusoid is well understood from narrowband communications. By allowing a number of different carrier frequencies, the capacity of the resulting systems can be increased. The potential to operate at different frequencies also allows systems to avoid potential interference from licensed spectrum users.

Two different technologies were proposed to IEEE 802.15.3a task group [6], which was looking to specify a high data rate, UWB based standard. The first, multiband OFDM, extended orthogonal frequency division multiplexing (OFDM)

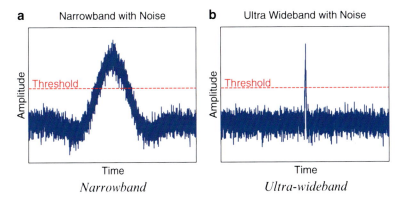

Fig. 17.1 Pulses in the presence of noise

known from digital video broadcasting, ADSL and Wi-Fi by increasing the bandwidth to 500 MHz. A second proposal used more traditional pulse based UWB in a system called direct sequence UWB.

With agreement within the IEEE out of reach, multiband OFDM was commercialised by an industry consortium under the name WiMedia. However, the high expectations for commercial success haven't been met yet.

In parallel, the IEEE802.15.4a [2] task group started developing work on a PHY that supported localisation. One of the technologies selected there was an impulse radio UWB PHY that was later merged with 802.15.4 [7] and will be discussed further in Sect. 17.4.

17.3.2 UWB Ranging

Consider a scenario where both transmitter and receiver share a common time base. If the transmitter encodes the time of transmission in the payload of a frame, the receiver can derive the time of flight by measuring the time of arrival of the packet.

However, since the speed of light is about 300 million metres per second, an accuracy down to decimetre level requires the receiver to estimate the arrival time of the signal with $1/3$ ns resolution.

In a multipath environment, we want to detect the first signal reaching the receiver, since that gives the best possible estimate of the distance. This algorithm is known as leading edge detection. To do that, a high sampling rate and, ideally, large bandwidths enabling steep rising edges are required, such that the time the signal rises above the noise floor can be measured precisely. Mathematically, this is expressed by the Cramer-Rao bound [8], but an easy intuitive understanding can be gained from Figs. 17.1 and 17.2.

Figure 17.1a, b represent a narrowband and an UWB pulse respectively. The time axis is scaled such that if the UWB pulse corresponds to 500 MHz bandwidth,

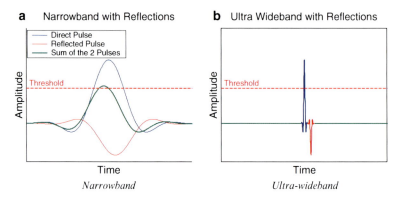

Fig. 17.2 Pulses in the presence of multipath

the narrowband pulse has a 20 MHz bandwidth. The amplitude axis is scaled such that both have the same peak amplitude. The noise power is equal in both figures.

The receiver's task is to estimate the arrival time of the signal. The most basic algorithm simply calculates a threshold, represented by the red line in the figure, used to distinguish signal from noise. The leading edge detection algorithm should then simply estimate when the signal crosses the threshold.

Figure 17.1a shows how the noise combined with the slow rising time introduces considerable uncertainty about the precise crossing time. In Fig. 17.1b on the other hand, a steep rising edge drastically reduces the uncertainty about the exact crossing time.

Figure 17.2 shows the same narrowband and UWB pulses, this time in a simple multipath environment, consisting of one direct path, represented in blue, and a reflected path, represented in solid red. The receiver perceives the sum of the two multipath components, shown in green. For clarity, a noise-free case is considered.

In the narrowband case, represented in Fig. 17.2a, phases differences between the multipath components lead to fading of the received signal. Crucially, for a leading edge detection algorithm, it also changes the slope of the signal and will corrupt the time-of-arrival estimate.

Figure 17.2b shows the UWB signal passing through the same multipath environment. Because of the narrow pulse width, the receiver perceives two separate pulses. Leading edge detection algorithms are not affected by multipath propagation.

Also, note that whereas multipath propagation degrades the performance of narrowband communication systems, UWB systems are able to exploit the multiple independent copies of the signal to improve the reliability of the communications.

17.4 The IEEE 802.15.4a Standard

The 802.15.4a standard [2] was originally an amendment of the IEEE 802.15.4 standard, which is used in the technology commercialised as ZigBee [9]. In 2011, it was merged into the main 802.15.4 standard as the UWB PHY [7].

Fig. 17.3 IEEE 802.15.4a PHY frame format

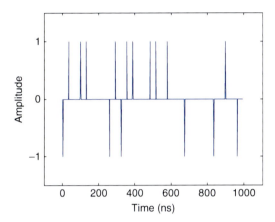

Fig. 17.4 Example preamble code

The goal of the extension was to provide wireless sensor networks with increased range, better mobility, enhanced coexistence and precision ranging capability. Two additional physical layers were defined, one based on chirp spread spectrum and one based on UWB. It is the latter that is discussed here.

The standard prescribes the format of the waveforms to be transmitted and leaves a lot of flexibility to the implementers, especially when it comes to the receiver architecture. It was carefully written such that both coherent and non-coherent receivers are supported.

17.4.1 Frame Format

A standard compliant frame consists of three main parts: a synchronisation header (SHR), followed by a PHY header (PHR) and a data field (Fig. 17.3).

The synchronisation header is transmitted first. Its first part, SYNC, consists of the repetition of a known preamble sequence such that the receiver can detect the presence of a transmission. For each frequency band, a number of codes was chosen such that the cross correlation between the codes is minimal. This allows for multiple networks operating on the same frequency.

In order to support both coherent and non-coherent receivers, a set of ternary preamble codes are defined in the standard. Ternary refers to the fact that the codes consist of pulses, often represented as ' + ', silences, represented as '0', and phase inverted pulses, represented as ' − '. Each preamble code element is transmitted as a single pulse, spaced a fixed distance apart. An example is shown in Fig. 17.4.

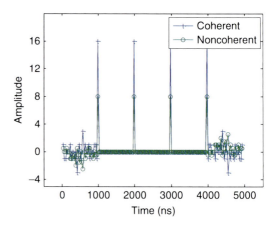

Fig. 17.5 Preamble autocorrelation

A non-coherent receiver can use the patterns of signal – silence to detect transmissions. A coherent receiver can also exploit the phase of the signal to obtain a further 6 dB performance gain from the correlations.

The codes are members of a set of codes known as Ipatov Sequences discovered by Valery Ipatov in 1979. Both the magnitude of the code and the code itself have perfect periodic autocorrelation, as shown in Fig. 17.5. Once the presence of a transmission has been detected, the receiver can use the remainder of the preamble to reconstruct the channel impulse response for ranging.

The second part of the SHR preamble is a start of frame delimiter (SFD), indicating to the receiver that the preamble is coming to an end and the transmission of the data part of the frame will follow. Like the SYNC preamble, the standard defines ternary SFD sequences. The pattern of absence and presence of the preamble symbol is intended to be used by non-coherent receivers, while coherent receivers get extra help from the phase inversion of some of the preamble symbols.

After the SFD, the modulation format changes. Pulses are no longer transmitted separately, but grouped in continuous bursts. In order to flatten the spectrum of the transmission and improve coexistence of networks, the burst sequences and positions are determined by a pseudo-random spreading code. Coherent receivers can use the same pseudo-random code to improve their performance.

The information to be transmitted is encoded by a combination of burst position modulation (BPM) and binary phase shift keying (BPSK). This means that each symbol contains two bits of information, one encoded in the position of the burst, another in its phase. Before transmission, the PHR and data field pass through a systematic, rate $1/2$ convolutional encoder. The systematic output is mapped to the position of the burst. Since both coherent and non-coherent receivers can detect the position, both are able to receive the packet. The parity bit is used to determine the phase of the burst. Coherent receivers achieve superior performance by exploiting this extra error correction information.

The part immediately after the SFD is called the PHY header (HDR). It informs the receiver about the length of the data field and the rate used to transmit it.

Since this information is crucial for successful decoding of the data, it is protected by a single error correcting, double error detecting (SECDED) Hamming code.

Finally, the data field is transmitted at the rate specified in the PHR. To help the receiver with error corrections, a systematic (63,55) Reed Solomon code over Galois field 6 is applied to the data field.

17.4.2 Front-End Specifications

The digital baseband of the transmitter produces the standard frame as described above. From an analogue baseband perspective, the frame can be seen as a long train of ternary pulses, where either a silence (0), a pulse ($+1$) or a phase inverted pulse (-1) should be transmitted. The standard specifies that this pulse train should be transmitted with a peak pulse repetition frequency of 499.2 MHz, corresponding to a pulse spacing of just over 2 ns.

The actual baseband pulse shape used is largely left up to the implementer. The standard only specifies certain constraints in terms of the cross-correlation between the pulse shape implemented and a root-raised cosine pulse. A transmit spectral mask is also included to avoid interference between neighbouring channels.

Besides the default 2 ns pulse duration, three optional shorter pulse shapes are defined. These result in a wider bandwidth, and hence allow for more power to be transmitted. Wider bandwidth can also be used to achieve higher ranging resolution.

The standard defines three bands of channels that can be used: a sub GHz band, with one channel centred on 499.2 MHz, a low band, comprising four channels with centre frequencies between 3.5 and 4.5 GHz, and a high band, which includes ten channels with centre frequencies from 6.5 to 9.5 GHz. These different channels can be used to increase the number of coexisting networks and/or to avoid interference from licensed users.

Since impulse UWB is a time-based system, with occasional short pulses of energy separated by long silences, it is important that the transmitter places the pulses at the right positions. To ensure that is the case, the standard specifies a ± 20 ppm tolerance on the 499.2 MHz clock and RF carrier.

17.4.3 Ranging Provisions

A ranging capable device must be equipped with a ranging counter and be able to report a figure of merit associated with the ranging estimate.

A ranging exchange is just a standard sequence of a data transmission and an acknowledgement sent in return. The transmitter records the counter value when the first pulse leaves its antenna and when the first pulse of the acknowledgement arrives back at the antenna. Similarly, the receiver needs to record the counter value when the first pulse of the data packet arrives at its antenna and the value at the moment the first pulse of the acknowledgement is sent from its antenna.

Optionally, the device may characterise the crystal offset. To avoid having to perform division at PHY level, the offset is reported in terms of a time interval and the amount of timing correction the device had to apply to track the timing of the incoming signal.

The figure of merit is a way for the PHY to signal to the higher layers how confident it is about the reported ranging counters. Its value can vary according to the signal to noise ratio of the communication link, the length of the preamble over which the channel impulse response was estimated and the accuracy with which the internal delays in the chip are measured.

After the ranging exchange, the ranging start and stop counter values, the figure of merit and the timing tracking interval and offset value are grouped in a standardised timestamp report. This is meant to be communicated to the higher layers, which perform the actual range calculation.

17.5 DecaWave ScenSor

DecaWave's ScenSor [10] is an IEEE 802.15.4a compliant transceiver aimed at wireless sensor networks (WSN) and real-time localisation systems (RTLS). The chip enables customers to replace proprietary solutions based on discrete components with a standard based integrated circuit (IC). It builds on the advantages of UWB, such as unlicensed operation, robustness in multipath environments, high precision ranging and low power transmission. It benefits from the low cost and small form factors of IC technology. The fully coherent receiver architecture ensures maximum communications range and positioning precision.

ScenSor is an acronym, for Seek, Control, Execute, Network and Sense, Obey, Respond. Sense, Obey, Respond corresponds to an IEEE reduced functionality device, which can be part of a personal area network but lacks the capacity to control it. Seek, Control, Execute, Network provides the network coordinator capability required from a full functionality device.

Practically, the chip itself implements a coherent 15.4 compliant UWB PHY, providing all the modulation, demodulation and error correction required. The chip also contains the necessary timers and a leading edge detection algorithm to accurately timestamp transmit and receive messages. Most of the MAC functionality has to be implemented by the host processor, which communicates with the chip via a SPI connection. Address filtering and cyclic redundancy check (CRC) are implemented on the chip itself in order to support automatic acknowledgement of frames.

The integrated circuit is manufactured using TSMC CMOS 90 nm technology. A functional block diagram is given in Fig. 17.6. The following subsection gives more detail about the transmitter. Subsection 17.5.2 discusses the receiver implementation. Since impulse UWB is essentially a time-based system, accurate

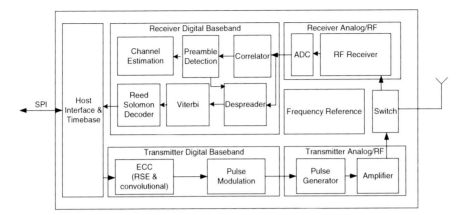

Fig. 17.6 ScenSor block diagram

frequency references are very important. These are discussed separately in subsection 17.5.3. Finally, subsection 17.5.4 discusses the performance of the chip with regards to the criteria from Sect. 17.2.

17.5.1 Transmitter

The transmitter digital baseband is mainly a straightforward implementation of the prescriptions in the standard. Three data rates, 110 kbps, 850 kbps and 6.8 Mbps, are supported.

Some custom additions to the standard have been included. One simply allows the payload to be extended up to 1,023 bytes, rather than the 127 allowed by the standard, to reduce protocol overhead in certain applications.

The front-end converts the digital ternary sequence to RF modulated pulses. The pulse generator makes the actual baseband pulse shape. Some optimisation is applied to ensure that the pulse shape fulfils the spectral regulations.

17.5.2 Receiver

The receiver uses a highly linear, highly selective strip. The main concern driving this choice was high selectivity. Interference from the 2.4 GHz ISM and 5 GHz U-NII bands or licensed users is much stronger than the UWB signal power at the receiver, so out-of-band suppression has to be as high as possible.

The low noise amplifier achieves a sub-3 dB noise figure and has an output voltage compression point, $0V_{1dB}$ of 3 V_{pk} differential. The architecture includes a high degree of programmability such that performance can be optimised for the selected band.

A programmable gain amplifier ensures that the signal and potential interferers are within the dynamic range of the adaptive, high sampling rate ADC.

Once digitised, the incoming signal is correlated with the expected preamble sequence. Individual pulses can be buried deep below the noise floor. Therefore, several correlation results have to be accumulated over a long period in order to average out the noise. Thanks to the perfect autocorrelation properties of the preamble codes, the result of this accumulation will be an approximation of the channel impulse response. This is used in combination with the system timers to extract the ranging information from the packet.

After the synchronisation header, the receiver switches to the demodulation of the BPM-BPSK symbols. A despreader gathers all the energy in both possible burst positions. In combination with the Viterbi decoder, the most likely transmitted symbol is chosen. A Reed Solomon decoder detects and corrects errors before passing the data up to the higher layers.

17.5.3 Frequency References

The combination of pulse position modulation and weak signal levels make accurate timing references essential to high performance UWB systems.

Signal detection relies on recognising the pattern of signal and silences. At the limits of the communication range, the actual signal will be buried well below the noise floor. In order to make it stand out, accumulation over a long time period is required.

Any offset in the 499.2 MHz clock that determines the pulse spacing at the transmitter and the sampling clock at the receiver will cause the sampling point to drift from the peak of the pulse shape. Once timing has drifted too much, more noise than signal will be added to the accumulator, hindering signal detection instead of helping it.

For example, the IEEE 802.15.4a standard specifies a ± 20 ppm accuracy for the clock reference [2], meaning the worst case mismatch between transmitter and receiver can correspond to ± 40 ppm. If the main lobe of the pulse is assumed to be 2 ns wide, this implies that the maximum integration interval is about 50 μs [11].

This limit to the integration period affects both coherent and non-coherent receivers. Because the latter can't suppress the noise as effectively, the maximum achievable range for non-coherent receivers will be more limited.

The offset between the RF frequency in the transmitter and the receiver causes coherent receivers to experience an increasing phase drift between successive pulses.

The effect of this phase drift is twofold. Firstly, it will degrade the signal power perceived after correlation. Once the phase shift exceeds 180°, accumulation becomes destructive and the receiver may have been better off using non-coherent preamble detection.

Secondly, ranging relies on the perfect periodic autocorrelation properties of the preamble code sequences. However, if for some reason the pulses don't demodulate properly as ± 1, but with slight differences, the correlation peak is reduced and a small residue remains when the sequences are shifted. When the preamble is long, amplitude differences should average out. However, the differences due to the phase offset are persistent and will accumulate. This can lead to spurious peaks in the estimated channel impulse response that may confuse leading edge detection.

The phase noise on the RF carrier frequencies causes random phase variations in the demodulated pulses. From a system perspective, this can be seen as an extra noise source, which particularly affects the higher data rates where a symbol consists of only one or two pulses.

The ScenSor chip uses a low jitter frequency synthesiser driven from an external frequency reference. It operates with crystal offsets up to ± 20 ppm and allows use of temperature controlled crystal oscillators (TCXO) for superior performance.

17.5.4 Performance in WSN and RTLS

Section 17.2 highlighted a number of requirements for WSN and RTLS technologies. ScenSor has been carefully designed to meet those.

The integrated circuit implementation ensures a low unit cost and a small form factor. Because the device operates at higher frequencies, up to 6.5 GHz, the antenna size can also be small when compared with sub-GHz and 2.4 GHz applications.

Whereas ZigBee supports data rates up to 250 kbps [9], ScenSor's maximum data rate is 6.81 Mbps. This higher data rate ensures that the time each device occupies the ether is much shorter. From the device's perspective, power consumption can be reduced, while at a network level, ScenSor can support many more devices, up to 1,500 packets per second.

Since ScenSor is based on UWB technology, it is able to benefit from multipath propagation, rather than to suffer from frequency selective fading. The sensitivity of the receiver goes down to -105 dBm, about 20 dB below the noise floor. The device is capable of reliable communication in long-range (several hundred metres) line-of-sight scenarios, as well as handling severe non-line-of-sight signal attenuation.

Thanks to the time-of-flight based ranging, ScenSor's location performance is accurate down to centimetre levels. This performance solely relies on the presence of the line-of-sight component of the signal, not on its strength. ScenSor ranges effortlessly in multipath environments, even in obstructed line-of-sight scenarios.

The chip is very flexible and can support a wide range of localisation topologies. If anchors with known locations are present, software can determine the absolute position of the tags. If a time-difference-of-arrival scheme is used, the tags can be transmit-only, which leads to an even longer battery life. If no fixed infrastructure is present, two-way ranging between tags is still able to locate the relative position of the devices.

17.6 Conclusion

Wireless sensor networks (WSN) are a well-known concept, with many applications in industrial and environmental monitoring as well as in the personal sphere. When combined with real-time localisation systems (RTLS), a whole new set of exciting applications can be supported.

Ultra-wideband is the natural choice for such location-aware WSN. The ultra-wide bandwidth allows for higher data rates and shorter transmission time which increases network capacity and battery life. The large bandwidth provides robustness against multipath fading, which boosts the reliability of the wireless link, and allows for accurate time-of-flight ranging. By design, UWB avoids interference to other systems by limiting the transmit power. This also reduces the power consumption in the sensor nodes.

DecaWave's implementation of the IEEE 802.15.4a standard meets the needs of WSN and RTLS manufacturers. The coherent transceiver combines the benefits of an UWB based system with those of standard based solutions. The integrated circuit implementation fulfils the requirements for low cost and small form factor.

This chapter motivates the choice of UWB for WSN and localization and highlights the technology issues and decisions made when designing the ScenSor IC to comply with the IEEE 802.15.4a standard.

References

1. Federal Communications Commission, Revision of part 15 regarding ultra-wideband transmission systems: First report and order, ET docket FCC 0248 (2002)
2. Standard IEEE 802.15.4a-2007, *in Part 15.4: wireless medium access control (MAC) and physical layer (PHY) specifications for low-rate wireless personal area networks (WPANs): Amendment to add alternate PHY*, IEEE Computer Society, New York (2007)
3. F. Lishoy, G. Hancke, K. Mayes, K. Markantonakis, Practical relay attack on contactless transactions by using NFC mobile phones, cryptology ePrint Archive, Report 2011/618, (November 2011)
4. C.E. Shannon, Communication in the presence of noise. Proc. Inst. Radio Eng. **37**(1), 10–21 (1949)
5. B. Gaffney, *Considerations and Challenges in Real Time Locating Systems Design*, DecaWave white paper, Dublin (2008)
6. IEEE 802.15 WPAN high rate alternative PHY task group 3a, http://grouper.ieee.org/groups/802/15/pub/TG3a.html
7. Standard IEEE 802.15.4-2011. Part 15.4: low-rate wireless personal area networks (LR-WPANs), New York (September 2011)
8. S. Gezici, Z. Tian, G. Giannakis, H. Kobaysahi, A. Molisch, H. Poor, Z. Sahinoglu, Localization via ultra-wideband radios: a look at positioning aspects for future sensor networks. IEEE Signal Process. Mag. **22**(4), 70–84 (2005)
9. ZigBee Alliance, ZigBee specification, ZigBee document 053474r06, version 1, (2006)
10. DecaWave, DecaWave Sensor Product Brief. www.DecaWave.com/downloads.html
11. V. Brethour, Crystal Offsets and UWB, IEEE 802.15-05-0335-01-004a. Huntsville, Alabama (June 2005)

Chapter 18
Architectures for Digital Intensive Transmitters in Nanoscale CMOS

Mark Ingels

Abstract Thanks to nanoscale CMOS, the computational power of digital integrated circuits has increased tremendously. For wireless communication systems, this resulted in increased transmission speeds using complex modulation schemes. The speed of nanoscale CMOS allowed to integrate the analog RF transmitter together with the digital baseband and brought high bitrate wireless communication to the consumer. Complex modulation schemes have to be supported by performant RF transceivers though. The design complexity of the analog transceivers has increased while their scalability is poor. Furthermore, many transistor parameters are degrading for traditional analog techniques. Calibration is therefore essential to achieve the required performance in traditional transmitters, but this increased tunability also offers new opportunities. Concurrently, the speed of nanoscale CMOS brought the digital closer to the antenna and enabled a new transmitter architecture: the direct digital modulator, which comes with its own set of challenges and solutions.

18.1 Introduction

The evolution of CMOS technology is mainly driven by the need to decrease the cost of computing power. This results in ever decreasing device sizes combined with increasing operating speed. Powerful baseband processors handle complex modulation schemes and bring ever increasing communication bitrates to the consumer market. This created a new family of handheld wireless devices, the smartphone, that combines a multitude of communication standards. As these phones are targeting the consumer market, cost and form factor are important

---

M. Ingels (✉)
Imec, Kapeldreef 75, 3001 Heverlee, Belgium
e-mail: Mark.ingels@imec.be

A. Baschirotto et al. (eds.), *Frequency References, Power Management for SoC,*
and Smart Wireless Interfaces: Advances in Analog Circuit Design 2013,
DOI 10.1007/978-3-319-01080-9_18, © Springer International Publishing Switzerland 2014

parameters. The analog functions are therefore best integrated together with the processor, in the digital technology.

Unfortunately, while the CMOS technology is continuously optimized for digital circuits, many of its analog performances are degrading. As an example the supply voltage is reducing much faster than the transistors' threshold voltage. The output impedance and the intrinsic gain of the transistors are also decreasing with their lengths. Analog functions need new circuits to provide the required performances, as they can no longer rely on stacking of transistors with their relatively large threshold in a limited supply voltage. New architectures have to leverage on the strength of the new technologies, whose main assets are their small feature size and high speed. A multitude of small devices can be added to support and improve the main analog function, while the available computing power can be used to *program* the complex analog circuit in its optimal operating point. The intrinsic speed of nanoscale CMOS also enabled completely new architectures, such as direct digital RF modulation.

This paper focuses on how nanoscale CMOS can be used to the designer's advantage in wireless transmitters. First, it will be demonstrated how transmitters based on traditional analog architectures can take advantage of nanoscale CMOS. In a Software Defined Radio transmitter, the tunability of the transmitter is increased tremendously compared with more traditional realizations, without concession on performance. The second part of this paper presents a new family of transmitters: the Direct Digital RF Modulators. These leverage on the high speed and low size of nanoscale CMOS to bring the digital domain closer to the antenna. This architecture comes with specific challenges. Several realizations will be presented to address some of these challenges.

18.2 Digitally Assisted Analog Transmitter

Today's dominant wireless communication device is a smartphone that gives its user ubiquitous access to a multitude of services provided trough an even wider range of wireless standards, covering short and long range communication combined with small or wide data rates. Depending on the use case, a link is set up on the most appropriate channel, but several links may also be active in parallel. To accommodate the various standards, flexibility of the transmitter is a must. Indeed, adding every new wireless feature by hardware multiplication is no longer possible. Instead, a transmitter that is capable of transforming itself satisfying the requirements of any desired communication protocol while still providing competitive power consumption, is required. This is even enhanced with the recent introduction of LTE that combines variable baseband bandwidths with variable RF frequencies. A Software Defined Radio (SDR) Transmitter, provides this functionality. Such a transmitter can be designed despite but mainly thanks to nanoscale CMOS.

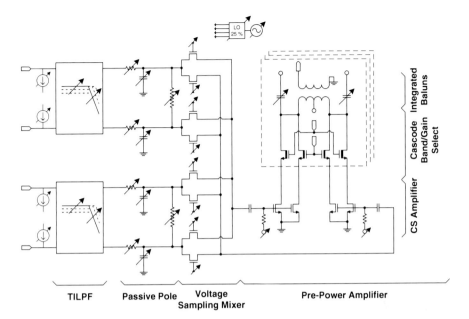

Fig. 18.1 40 nm CMOS SDR transmitter circuit diagram

18.2.1 A 40 nm CMOS Software Defined Radio Transmitter

Figure 18.1 shows the schematic diagram of a SDR transmitter realized in 40 nm CMOS. The transmitter is based on a direct up-conversion architecture which is the most suitable to build a SDR radio [1]. Besides its powerful performance, it has the potential to allow flexible operation by including reconfigurability into the circuit blocks. The transmitter consists of a low-pass filter followed by a passive mixer and a pre-power amplifier. While these are traditional building blocks, they benefit from nanoscale integration which provides programmability to improve the overall performance by calibration and tuning.

The baseband section consists of an active transimpedance low-pass filter (TILPF) based on a flexible Tow-Thomas architecture (Fig. 18.2) that offers independent programming of the transimpedance gain, the bandwidth and the quality factor [1]. The active filter is followed by a programmable passive filter to further reduce the out-of-band noise. All resistors and capacitors are split in small units that can be added or removed from the circuit as needed. This results in a programmable bandwidth from 400 kHz up to 20 MHz. The transimpedance gain of the filter can be programmed from 1 to 8 KΩ. This accommodates the various standards targeted. But the tunability reaches further. Both amplifiers in the active filter consist of several small opamps connected in parallel. They can be turned on or off individually to adjust the amplifier's gain-bandwidth product in eight steps from 60 to 480 MHz. The current consumption of the amplifier is now linearly

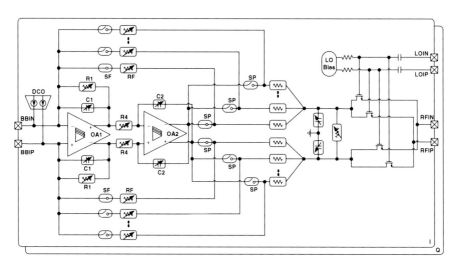

Fig. 18.2 SDR baseband followed by the passive voltage sampling mixer

proportional to its gain-bandwidth. This allows to trade power for performance and vice versa. Maximal performance is achieved at maximal power, but when the requirements are less stringent, power can be saved. The programmability is achieved thanks to nanoscale CMOS, that offers good, small switches and little parasitic overhead.

The DC offset of the TILPF can be compensated by injecting a small DC current at its input through integrated calibration current DACs. DC offset has thus not to be guaranteed over PVT at design time, which would result in the need for larger, better matching transistors and would require more design effort and consume more power. Instead, nanoscale CMOS allows the compact integration of simple DACs, and offers the possibility for automatic integrated tuning at run-time with little overhead.

The output of the active low pass filter is further filtered by a programmable passive RC filter before being up-converted in the subsequent quadrature voltage sampling mixer. Passive mixers require good switches and are widely used in nanoscale CMOS. In the presented implementation, a passive voltage sampling mixer is chosen for its good out-of-band noise performance as required for FDD operation.

The final block in the transmitter is the pre-power amplifier. Its circuit schematic is included in Fig. 18.1. It consists of a cascoded differential Common-Source amplifier loaded with two on-chip baluns with programmable center frequency. A wideband differential output is also provided. The amplifier transistor is split in multiple units that can be turned on or off by one of the three the thick-oxide cascode transistors topping each amplifier branch. These cascode transistors also determine to which of the three outputs the amplifier's output is derived. Here again, nanometer CMOS enables to split the main amplifier into small parts with little overhead, to obtain the extra functionality.

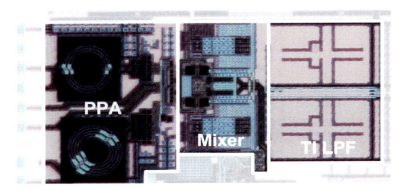

Fig. 18.3 40 nm LP CMOS SDR transmitter chip photograph

The 25 % duty cycle LO generator benefits from the speed of the technology and is based on conventional CMOS logic. This eases the implementation and lowers the power consumption. It is proportional to the LO frequency as the generator does not consume static power. The phase noise is low thanks to the large LO voltage swing, which is also beneficial to drive the passive mixer's switches. Note that the sizing of the devices has to be performed in an analog way, to achieve the extremely low phase noise required for SAW-less FDD.

The presented Software Defined Radio transmitter was implemented in 40 nm LP CMOS. Its core area is 1.4 × 0.7 mm2 (Fig. 18.3) and is mainly dominated by the baseband capacitors and the on-chip baluns. The transmitter consumes 13–44 mA from the 1.1 V supply (TILPF + LO generation) depending on the selected bandwidth and the LO frequency. The PPA consumes less than 43 mA from the 2.5 V. This is proportional to the required output power and linearity. The SDR concept is demonstrated in the modulator's performance summary of Fig. 18.4. A single transmitter achieves SOA performance comparable to dedicated solutions for a multitude of standards, including the toughest WCDMA/LTE bands as well as GSM, WLAN and WiMAX. The output P1dB is better than 10 dBm in all modes except GSM, where lower PPA linearity is traded for current consumption. An EVM better than 2.5 % is measured in WCDMA, LTE GSM and WiMAX while the CNR in the RX band is better than -160 dBc/Hz, which is sufficient for SAW-less operation in all WCDMA/LTE FDD bands.

The SDR concept may also be used to further reduce the transmitter's average power consumption in two ways. First, the user's communication is not limited to a single protocol as in a dedicated terminal. Instead, the most optimal link can be chosen in any given situation. What's more, when the optimal link has been selected, the power budget can be further optimized by programming the hardware for the best trade-off for noise, filter order, linearity, etc. for the actual channel conditions. Traditionally, these trade-offs were fixed at design time but are now performed at run-time, allowing for the best compromise possible between user experience and battery life.

Band # Fcarrier Δ TX-RX	Mode	BW [MHz]	OP1dB [dBm]	Pout [dBm]	EVM [%]	ACLR1/2 [dBc]	CNR [dBc/Hz]	Imax [mA] 1.1V/2.5V	DG.09 [mW]
Band I 1.92GHz 190MHz	UMTS	4	10.4	3.79	2.0	−40.2/67	−162	21/41	30.9
	LTE	20	10.4	2.1	2.5	−39/−58	−160	25/40	
Band II 1.85GHz 80MHz	UMTS	4	10.4	4.39	2.3	−40.2/−63	−164	20/40	30.2
	LTE	20	10.4	2.6	2.4	−40.1/−59	−162.5	24/40	
Band V 0.82GHz 45MHz	UMTS	4	10.9	4.45	1.7	−41/−68	−161.7	14/37	24.8
	LTE	10	10.9	2.45	1.7	−41.4/−63	−160.5	17/37	
Band VII 2.5GHz 120MHz	UMTS	4	13.5	7.1	1.5	−46/−72	−159	24/40	38.5
	LTE	20	13.5	5.1	2	−42/−58	−158	28/40	
Band XI 1.42GHz 48MHz	UMTS	4	11.7	5.2	1.5	−41/−67	−162.6	18/37	27.8
	LTE	20	11.7	3.4	2.0	−42/−58	−160.4	21/36	
Band XII 0.7GHz 30MHz	UMTS	4	10.5	2.41	2.1	−39.5/−65	−160	13/34	24.8
	LTE	10	10.5	0.4	1.8	−41/−67	−159	17/33	
0.9GHz 20MHz	GSM	0.2	8.1	4.5	1.7	−54/−67	−160	18/25	
2.4GHz 100MHz	WLAN	40	13.6	4.6	3.3	−41/−54	−159	27/40	
3.5GHz 100MHz	WiMAX	20	12.6	3	1.8	−41/−54	−155	38/43	
4.8GHz 100MHz	WLAN	20	10.47	1	8.1	−42/−48	−156	44/39	

Fig. 18.4 SDR TX performance summary (Measurements with on-chip balun, except Band VII, WiFi and WiMAX where the wideband output is used)

18.3 Direct Digital RF Modulator

Switching amplifiers have the potential to realize high efficiency. They eliminate internal device dissipation by avoiding the simultaneous occurrence of current through and voltage over the switch. Switching amplifiers typically feature a constant output amplitude and rely on a polar architecture in non-constant envelope modulation schemes. Figure 18.5 presents a basic polar transmitter. The Cartesian I and Q components are first converted into phase and amplitude. The phase modulated LO is applied to the amplifier, while the amplitude is modulated through the supply voltage. While this architecture aims for the high efficiency of the switching amplifier, it suffers from several implementation issues. As the Cartesian to polar conversion is non-linear, the bandwidth of both the phase and the amplitude are increased by at least a factor 4. This impacts the implementation of both the phase and amplitude modulator.

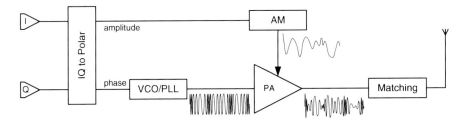

Fig. 18.5 Polar modulator architecture

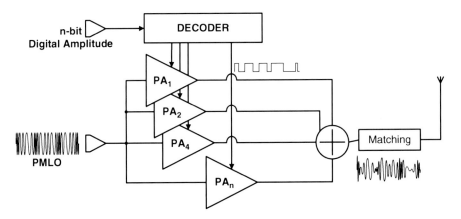

Fig. 18.6 Polar modulator with digital amplitude modulation

Phase modulators based on a digital PLL have been presented [2], but their bandwidths are still not sufficient to deal with modern, high bandwidth communication schemes. Alternatively, a phase modulated LO can be generated with a constant amplitude Cartesian up-converter, but this obviously creates quite some overhead. The AM modulator has to deliver modulated power to the amplifier, but high efficiency is difficult to combine with high bandwidth. A modulated DC-DC converter can be used for high efficiency, but its bandwidth is very limited. It may be combined with a linear regulator, which offers wide bandwidth, but bad efficiency. As a result, the combination of the switching PA with its amplitude modulator results in average efficiency and RF bandwidth.

Nanometer CMOS enables a new, digital architecture for the amplitude modulator. The switching amplifier is split up in a multitude of smaller parallel amplifiers whose outputs are summed. The amplitude of the output signal is modulated by turning on or off more or less unit amplifiers as needed (Fig. 18.6) [3, 4]. The transmitter's DAC has thus effectively been moved to the antenna. Amplitude modulation speed is no longer a bottleneck. It is now a transfer of information rather than power. The digital modulator is the base for Direct Digital RF Modulators (DDRM), a new family of RF transmitters which are likely to gain in importance, as they are extremely suited for integration in scaled CMOS.

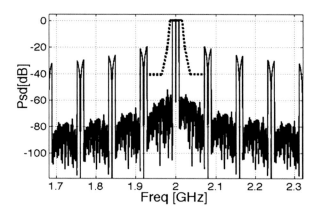

Fig. 18.7 DDRM spectral response

Obviously, as the DDRM is in essence a DAC at the antenna, it lacks an anti-aliasing filter. As a result, both aliases and quantization noise reach the RF output un-attenuated (Fig. 18.7). They can be reduced by increasing the resolution and the baseband sampling rate though, which is more likely in the newest technologies. Furthermore, new architectures can specifically address the DDRM drawbacks. As technology scaling improves the DDRM's performance and as porting the DDRM to any new technological node is relatively easy thanks to its digital nature, its importance will continue to grow.

18.3.1 A Compact 90 nm CMOS Polar Amplitude Modulator

A compact 90 nm CMOS polar amplitude modulator is presented in this paragraph [5]. It is based on a cascoded common source amplifier that is divided in multiple unit amplifiers. All units are driven by a common RF phase-modulated signal. Each unit amplifier has a dedicated cascode transistor which is also used as a switch that turns the unit amplifier on and off based on a digital control bit (Fig. 18.8). The output currents of all units are summed to shape the RF envelope. As the bias current is switched off in the unused branches, a good efficiency is obtained. Furthermore, the open loop structure and the absence of high impedance nodes give this amplitude modulator a wide bandwidth.

An 8 bit prototype has been implemented. This gives 2 bit margin to transmit a WLAN-like 64 QAM modulated OFDM signal [4]. Monotonicity of the DAC may be obtained by using thermometric coding. However this would result in a complex decoding scheme, and a large area overhead. Binary and thermometric coding are therefore combined to get the advantages of both schemes, while limiting the disadvantages [6] (Fig. 18.8). In the prototype, 4 LSBs are addressed binary while four MSBs are thermometric coded. The size of the unit CS transistor is a compromise between loading of the phase modulated LO and mismatch. It is scaled to obtain monotonicity of the amplitude response in presence of device mismatch.

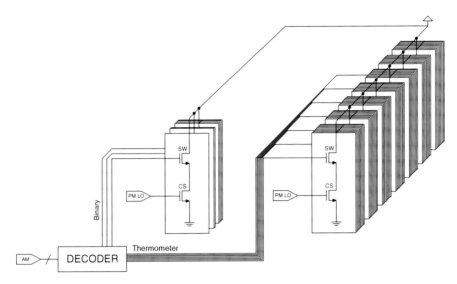

Fig. 18.8 Polar DDRM based on CS amplifier units and segmented digital control

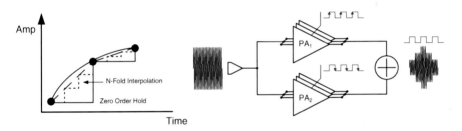

Fig. 18.9 N-fold interpolation principle & twofold DDRM schematic

As discussed before, a DDRM behaves as a DAC without anti-alias filter and the aliases of the transmitted signal are present at multiples of the clock frequency around the LO frequency. They may thus violate the out-of-band emission requirements. The aliases can be spread and attenuated by increasing the baseband sampling frequency. However, the required oversampling may be too high. The aliases can also be reduced by applying linear interpolation between two consecutive samples. This adds a supplemental sinc filter to the DAC output spectrum [7]. In practice, linear interpolation can be approached in discrete steps, resulting in an intermediate filtering (Fig. 18.9). The presented prototype features twofold interpolation by clocking two parallel DDRM arrays at opposite baseband clock edges.

The layout of the DDRM matrix is primordial for its performance. As the DDRM is a DAC, matching is important. Both the phase modulated LO and the output RF are distributed to the various cells. Coupling between both should be minimized to limit LO feed-through. Digital baseband data is distributed throughout the matrix as well and coupling to the RF lines may result in spurs at the output. Finally, long RF

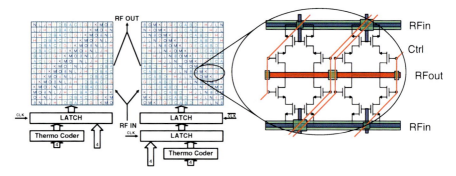

Fig. 18.10 (*Double*)DDRM matrix & Compact DDRM basic cell

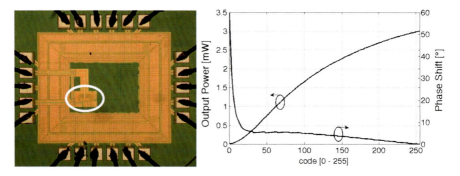

Fig. 18.11 (**a**): Compact DDRM chip photo with 110 × 65um circuit highlighted. (**b**): DDRM AM/AM and AM/PM response

lines increase the losses and reduce the global efficiency of the transmitter. All these points have to be considered carefully to layout the DDRM matrix.

The 8 bits amplitude modulator is segmented into four binary coded LSBs and four thermometric coded MSBs. The thermometric unit cell consists of 16 LSB units. To reduce errors due to process gradients, these are distributed along the central diagonal resulting in a matrix of 16 × 16 LSBs. An extra border of dummy cells is added. The binary units are placed on the central diagonal and distributed with their center of gravity in the middle. The thermometric unit cells are flipped both vertically and horizontally to share the horizontally routed RF and supply lines. RF input and output are routed on alternating horizontal lines to avoid their crossing. The digital control lines are routed along the diagonals. This strategy results in a compact structure as presented in Fig. 18.10. The complete modulator consists of two identical flipped DDRM arrays. The digital amplitude words are applied through latches clocked on opposite clock phases to obtain the twofold interpolation.

The chip has been realized in 90 nm CMOS [5]. The microphotograph of the chip is presented in Fig. 18.11a. The effective area of the polar amplitude modulator is an extremely compact 110 × 65 μm^2.

To measure the prototype, it was bonded directly on a printed circuit board (PCB). Figure 18.11b shows the measured static AM/AM and AM/PM responses of the modulator. The non-linear AM/AM curve is mainly due to the varying output impedance of the amplifier with the digital code. This is a common problem with DDRM, as the number of active cells varies with the output amplitude. The AM/PM curve is flat at the highest codes, but changes rapidly at the lowest codes due to a higher contribution of LO feed-trough. Fortunately, the digital nature of the transmitter makes it easy to compensate both curves through pre-distortion, beit that this reduces the effective number of bits.

The modulator is tested with a WLAN-like 64 QAM OFDM signal. The amplitude data is pre-distorted based on a static look-up table and applied at a rate of 40 MHz to the modulator. Note that this baseband clock speed was limited by the available equipment, rather than by the modulator. With a phase modulated LO at 2.45 GHz, an EVM better than -26 dB is obtained for -2.5 dBm output power. The maximal drain efficiency is 23 % at 5 dBm Pmax.

18.3.2 A Class E Impedance Modulated DDRM

Previous paragraph described a digital amplitude modulator which is the base for many polar DDRMs. In this paragraph an amplitude modulator that acts on the amplifier's load through a digitally programmable impedance matching network is presented. This architecture is demonstrated with a class-E based polar amplifier.

The class-E amplifier is a switching amplifier that achieves high efficiency by avoiding any overlap of voltage over and current through the switch [8]. As a result, no power is dissipated and a theoretical efficiency of 100 % is achievable. In a class-E amplifier, the output power is inversionally proportional to the load impedance. Depending on the supply voltage and the required output power an impedance transformation network is placed between the amplifier and the antenna. Its parameters are traditionally fixed at design time, but in this work, the matching network is used for the dynamic amplitude modulation of the polar amplifier. As the network is controlled directly from the digital data stream, the achievable modulation bandwidth is large. Furthermore, the efficiency of the presented scheme is only limited by second order effects, mainly the deviation from the optimal class-E operating point during operation and in less extend the parasitics in the matching network. The choice of the transformation network is a compromise between tunability, efficiency and practical realization. The ratio between minimal and maximal transformed load impedance determines the dynamic range of the resulting amplifier. It is limited, as the losses in the matching network increase with its complexity and reduce the efficiency. For 10 dB dynamic range, the ratio between the maximal and the minimal transformed load impedance should be a factor 10. To keep the matching network realizable, tuning of inductors is not considered. Figure 18.12 shows the simulated real impedance and phase of the chosen π-type matching network at 2.4 GHz when sweeping the digitally controlled

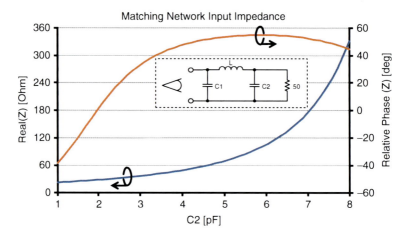

Fig. 18.12 Pi type impedance transformation network response

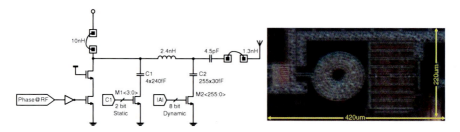

Fig. 18.13 Impedance modulated class-E amplifier circuit diagram and photograph

capacitor C2. The real impedance seen by the amplifier varies from approximately 20Ω up to 300Ω. The non-linear shape implies that pre-distortion is needed, both for amplitude and phase.

The impedance modulation concept is demonstrated with a 90 nm CMOS polar amplifier prototype [9]. Its circuit diagram and microphotograph are presented in Fig. 18.13. The main amplifier is protected from the high voltage swing which may reach above the supply voltage due to the class-E operation, by a thick-oxide cascode transistor. A 1 cm bondwire to the PCB is used as a 10 nH load inductor. The total active area of 420 μm × 220 μm is dominated by the tunable impedance transformation network.

For the dynamic measurements, amplitude and phase are generated with Matlab and pre-distorted based on static measurements. The eight bits wide amplitude information is sent to the digital impedance modulator at a rate of 100MS/s, while the LO is phase-modulated with the pre-distorted phase information.

A dynamic range of 11dBm is measured. Figure 18.14 shows a measured vector diagram after the receive filter for a π/4 DQPSK modulated Bluetooth-like signal

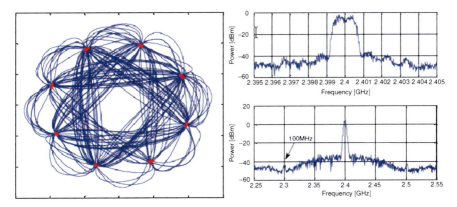

Fig. 18.14 Pi/4 DQPSK diagram & spectral plots

with a symbol rate of 1MS/s (bitrate of 2 Mb/s) at 2.4 GHz. The transmitter features an EVM of 2.6 % at 6 dBm RMS power. Both the narrow and the wide band spectral plots of the modulated output are shown in the same figure. Some alias power is visible at 100 MHz offset due to the direct digital modulation.

18.3.3 An IQ DDRM with RF FIR-Based Quantization Noise Filter

The previous paragraphs introduced direct digital modulation for a polar modulator. However, due to the bandwidth increase introduced by the Cartesian to polar conversion and the complexity of a digital LO phase modulator, the polar architecture is less suited for modern wide bandwidth communication standards. Two digital amplitude modulators can be combined into a Cartesian digital modulator to solve this problem. One of both is modulated with a fixed LO to up-convert the I baseband data, while the other has a fixed 90° LO for the Q data. Differential modulators accommodate the sign change of I and Q. A compact IQ DDRM cell that combines all phases into a single unit is presented in Fig. 18.15. It consists of 4 RF switches (RF0, RF90, RF180, RF270) modulated with fixed 25 % duty cycle LO phases in series with the digitally controlled baseband switches (EN0, EN90, EN180, EN270). The latter determine whether a certain LO phase is active in a given cell. The eight thin oxide switches are cascoded with a thick-oxide current source that determines the gain of the cell and protects the low voltage switches from the large output swing. The output currents of all DDRM units are summed at their outputs and dumped into the load. To transmit the quantized code $a + jb$ for example, an equivalent of a EN0 and b EN90 switches are closed. The resulting drain current is depicted in Fig. 18.15.

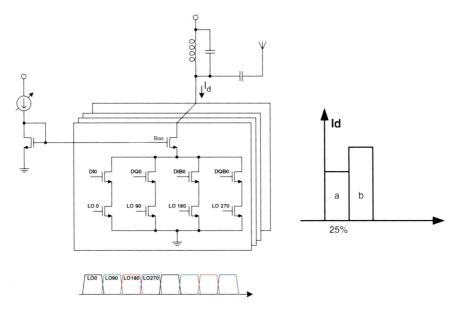

Fig. 18.15 Compact 25 % duty cycle IQ DDRM

The presented cell is compact as it combines all thin oxide switches under a single thick oxide current source, the latter consuming the largest area in the unit. Its bias voltage can be used to control the average output power.

As the architecture is based on Cartesian modulation, it has potentially a wide bandwidth. However, besides the benefits, this architecture obviously shares the main disadvantages of Direct Digital RF Modulation, being quantization noise and aliases. Both can be reduced by increasing the baseband oversampling [10] and increasing the number of bits [11]. For saw-less FDD operation the quantization noise requirements are very stringent though. The noise floor should be reduced below −160 dBc/Hz [12] in the RX band. In FDD, the most stringent noise requirements are localized in the RX band associated with the TX band. Instead of aiming to reduce the global out-of-band quantization noise, it is sufficient to filter the quantization noise specifically at the RX frequency. This can be achieved by combining a number of correctly sized DDRMs to implement a FIR filter that acts directly at RF(Fig. 18.16). The bias of the DDRMs' current sources determine the FIR filter coefficients and can be tuned to adjust the filter's shape and notch position. Note that the depth of the notch will be limited by the thermal noise of the current sources and the LO phase noise.

A digital transmitter prototype was realized in 130 nm CMOS and contains four 8 bit DDRM matrices [13]. They are fed from a 2.7 V supply. The digital data between each matrix is delayed by flip-flops (FFs) to form a fourth-order RF FIR filter. The 4 LO phases with 25 % duty cycle are generated on chip by a digital frequency divider. The digital circuits, the LO generator and the switches in the

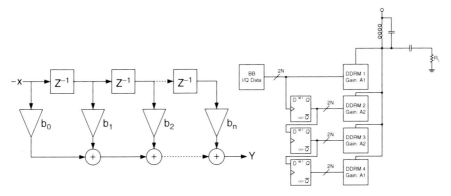

Fig. 18.16 FIR and its implementation in an IQ DDRM

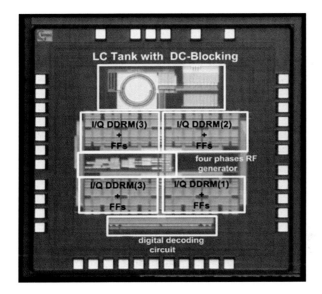

Fig. 18.17 Compact 25 % duty cycle IQ DDRM

DDRM are all powered from 1.2 V The resulting chip occupies an area of 1.5 × 1.5 mm^2 (Fig. 18.17).

The transmitter achieves a peak power of 15.4 dBm with a drain efficiency of 13 %. The power consumption of the drivers and the digital part (the flip-flops and the digital decoding) is 55 mW. Pre-distortion is applied to compensate the non-linear AM/AM behavior of the IQ DDRM. This pre-distortion is more complicated than for the polar DDRM where AM/AM and AM/PM can be considered independently. In the IQ DDRM, I and Q are linked, and a pre-distortion matrix is used rather than a vector.

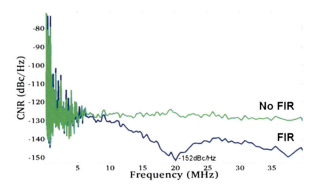

Fig. 18.18 Quantization noise notch with FIR DDRM

The effect or the FIR notch is demonstrated in Fig. 18.18. For a 200 kHz baseband tone transmitted at 900 MHz, the noise floor reaches −152 dBc/Hz @ 20 MHz offset, an improvement of 22 dB compared with a FIR-less modulator. When applying a 64 QAM signal with 10 MHz RF BW to the transmitter, an EVM of −27.2 dB is measured at 4.1 dBm RMS output power. The transmitter then consumes 48 mW from the 2.7 V supply

18.3.4 A CMOS IQ Doherty DDRM with Modulated Tuning Capacitors

Many challenges that exist for traditional transmitters still hold for DDRM based transmitters and some of the solutions for these challenges can take advantage of the DDRM architecture. Modern communication systems use complex modulation schemes with a high peak to average power ratio (PAPR). For a good average transmitter efficiency, its drain efficiency should be preserved at lower power levels such as at 6 dB back-off or lower. The regular DDRM typically has a Class-B like efficiency curve [13]. In traditional power amplifiers, the Doherty architecture can be used to increase the efficiency at back-off [14]. It combines a main amplifier with a peaking amplifier which is turned on when the main amplifier is at maximal output swing and efficiency (Fig. 18.19a). The main amplifier still works at maximal efficiency at higher power and the power for the peaks is provided by the auxiliary amplifier.

A major design challenge of this architecture is the accurate biasing and modulation of the auxiliary amplifier. In a DDRM, the transmitted amplitude is exactly controlled at any moment in time, so it makes sense to apply the Doherty scheme on the DDRM transmitter. Indeed, the digital modulation of both the main and the peaking DDRM allow a perfect control of the cooperation between them.

A Doherty transmitter requires a power combining impedance transformation network. which is traditionally realized using an off-chip λ/4 Transmission Line.

18 Architectures for Digital Intensive Transmitters in Nanoscale CMOS 327

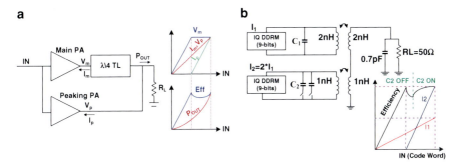

Fig. 18.19 (**a**): Doherty principle. (**b**): Doherty DDRM with transformer based power combiner

However, an on-chip impedance transformation network using integrated transformers can be used for power combining and impedance transformation [15, 16]. Two transformers are combined to create a fully integrated Doherty amplifier [17]. The analog amplifiers can be replaced by DDRMs though [18] (Fig. 18.19b). For output powers below 6 dB back-off, the peaking amplifier is turned off and ideally the peaking transformer is shorted. The transformed output impedance is seen as a load by the main amplifier. At 6 dB back-off the main amplifier has maximal swing, and the transformed load is optimal. Beyond this power, the peaking amplifier gradually increases its contribution, and feeds part of the output load. As a result, the load seen by the main amplifier decreases gradually. The latter delivers more output power at maximal voltage swing and stays optimally loaded. The power delivered by the peaking amplifier continues to increase until both amplifiers deliver their power at full swing on the output impedance, which is then divided between both.

In practice, it is impossible to short the second transformer completely when it is not used. A low impedance would require very large switches that would introduce too large parasitics. An alternative is to open the primary of the second transformer. When no current flows through one transformer winding, the other acts as a single inductor. However, the tuning capacitor on the primary of the second transformer then still creates a high impedance in the signal path. This can be solved by disabling this capacitor with a series switch when the auxiliary amplifier is not in use. Again, this is easily achievable in a DDRM based amplifier, as it is exactly known when the auxiliary amplifier is in use.

A 9 bit prototype of the IQ digital Doherty transmitter is realized in 90 nm CMOS [18]. Its microphotograph is shown in Fig. 18.20. The chip measures 1.9×1.9 mm^2. The main and auxiliary DDRMs are fed from 2.4 V while the digital circuits, including the LO are powered from 1.2 V. At 2.4 GHz, the digital transmitter achieves a peak power of 24.8 dBm with a drain efficiency of 26 %. This efficiency is also achieved at 6 dB back-off. From the measurements in Fig. 18.20 the benefit of the switched transistor in the second transformer is clearly visible.

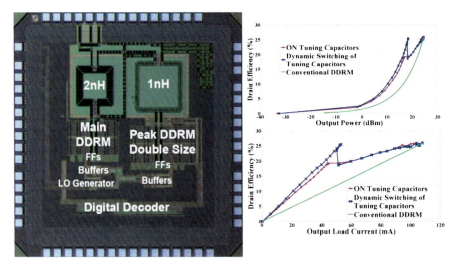

Fig. 18.20 Doherty IQ DDRM chip photo with measured efficiency

18.4 Conclusions

Nanoscale CMOS increased computing power tremendously and introduced complex modulation schemes in wireless communication. The limited analog performances of the technology forced the development of new transceiver architectures to deal with the new parameters and take advantage of the new potentials. The traditional analog transmitters are now supported by more calibration while increased configurability improves both the functionality and the performance. In parallel, the digital domain is moving towards the antenna in the wireless transmitter. This is expected to be a trend as the technology scales further and innovative architectures improve the DDRM even beyond the technological scaling.

References

1. V. Giannini et al., A 2mm2 0.1-5GHz software-defined radio receiver in 45-nm digital CMOS. IEEE J. Solid State Circuit **44**(12), 3486–3498 (2009)
2. P.-E. Su, S. Pamarti, A 2.4 GHz wideband open-loop GFSK transmitter with phase quantization noise cancellation. IEEE J. Solid State Circuit **46**(3), 615–624 (2011). 615
3. P.T.M. van Zeijl et al., A digital envelope modulator for a WLAN OFDM polar transmitter in 90 nm CMOS. IEEE J. Solid State Circuit **42**(10), 2204–2211 (2007)
4. A. Kavousian et al., A digitally modulated polar CMOS power amplifier with a 20-MHz channel bandwidth. IEEE J. Solid-State Circuit, **43**(10), 2251–2258 (2008)

5. V. Chironi et al., A compact digital amplitude modulator in 90nm CMOS, in *Design, Automation & Test in Europe Conference & Exhibition (DATE)*, Dresden – Germany, pp. 702–705, 8–12 Mar 2010
6. C.-H. Lin, K. Bult, A 10-b 500MSamples/s CMOS DAC in 0.6mm2. IEEE J. Solid State Circuit **33**, 1948–1958 (1998)
7. Y. Zhou, J. Yuan, A 10-Bit wide-band CMOS direct digital RF Amplitude modulator. IEEE J. Solid State Circuit **38**(7), 1182–1188 (2003)
8. N. Sokal, A. Sokal, Class E-A new class of high-efficiency tuned single-ended switching power amplifiers. IEEE J. Solid State Circuit **10**, 168–176 (1975)
9. M. Ingels et al., An impedance modulated class-E polar amplifier in 90nm CMOS, in *IEEE Asian Solid State Circuits Conference (A-SSCC)*, Jeju – Korea, 2011, pp. 285–288
10. Antoine Frappé, An all-digital RF signal generator using high-speed $\Delta\Sigma$ modulators. IEEE J. Solid State Circuit **44**(10), 2722–2732 (2009)
11. Z. Boos et al., A fully digital multimode polar transmitter employing 17b RF DAC in 3G Mode, in *IEEE International Solid-State Circuits Conference (ISSCC)*, San Francisco – California, 2011, pp. 376–377
12. C. Jones et al., Direct-conversion WCDMA transmitter with 163dBc/Hz noise at 190MHz Offset. in IEEE International Solid-State Circuits Conference (ISSCC) 2007. pp. 336–607
13. W. Gaber et al., A CMOS IQ direct digital RF modulator with embedded RF FIR-based quantization noise filter, in *IEEE European Solid State Circuits Conference (ESSCIRC)*, Helsinki – Finland, 2011, pp. 139–142
14. W.H. Doherty, A new high efficiency power Amplifier for modulated waves. Proc. IRE **24**, 1163–1182 (1936)
15. P. Reynaert, A.M. Niknejad, Power combining techniques for RF and mm-wave CMOS power Amplifiers, in *IEEE ESSCIRC*, Munich – Germany, 2007, pp. 272–275
16. Peter Haldi et al., A 5.8 GHz linear power Amplifier in a standard 90nm CMOS process using a 1V power supply, in *IEEE Radio Frequency Circuits Symposium (RFIC)*, Honolulu Hawaii, 2007, pp. 431–434
17. Ercan Kaymaksut, Patrick Reynaert, CMOS transformer-based uneven Doherty power amplifier for WLAN applications, in *IEEE ESSCIRC*, Helsinki – Finland, 2011, pp. 135–138
18. W. Gaber et al., A CMOS IQ digital Doherty transmitter using modulated tuning capacitors, in *IEEE European Solid State Circuits Conference (ESSCIRC)*, Bordeaux – France, 2012, pp. 341–344

Printed by Publishers' Graphics LLC